Data Storage Architectures and Technologies

Jiwu Shu

Data Storage Architectures and Technologies

Jiwu Shu
Department of Computer Science and Technology
Tsinghua University
Beijing, China

ISBN 978-981-97-3533-4   ISBN 978-981-97-3534-1 (eBook)
https://doi.org/10.1007/978-981-97-3534-1

© Posts and Telecom Press Co., Ltd 2024
Jointly published with Posts & Telecom Press, Beijing, P.R.China.
The print edition is not for sale in Mainland China. Customers from Mainland China please order the print book from: Posts & Telecom Press.
Posts & Telecom Press, Beijing, China

This work is subject to copyright. All rights are solely and exclusively licensed by the Publisher, whether the whole or part of the material is concerned, specifically the rights of reprinting, reuse of illustrations, recitation, broadcasting, reproduction on microfilms or in any other physical way, and transmission or information storage and retrieval, electronic adaptation, computer software, or by similar or dissimilar methodology now known or hereafter developed.

The use of general descriptive names, registered names, trademarks, service marks, etc. in this publication does not imply, even in the absence of a specific statement, that such names are exempt from the relevant protective laws and regulations and therefore free for general use.

The publishers, the authors, and the editors are safe to assume that the advice and information in this book are believed to be true and accurate at the date of publication. Neither the publishers nor the authors or the editors give a warranty, express or implied, with respect to the material contained herein or for any errors or omissions that may have been made. The publishers remain neutral with regard to jurisdictional claims in published maps and institutional affiliations.

This Springer imprint is published by the registered company Springer Nature Singapore Pte Ltd.
The registered company address is: 152 Beach Road, #21-01/04 Gateway East, Singapore 189721, Singapore

If disposing of this product, please recycle the paper.

Foreword 1

Storage technologies have gained greater significance with the exponential explosion in data. Developing high-performance storage systems that efficiently and securely store massive amounts of data over the long haul is of utmost importance.

Since the 1990s, we had played a pioneering role in researching network storage system technologies in China. However, our storage technologies were far behind those of other markets during that period. Thanks to the hard work of so many dedicated professionals, we were able to overcome a wide range of challenges, by boosting the scalability, reliability, and performance of storage systems. We diligently built prototype systems and successfully applied and enhanced key technologies.

After over two decades of tireless efforts, it has been gratifying to witness the remarkable growth of China's storage industry, which now becomes one of the global leaders. In terms of storage hardware, Chinese enterprises like Huawei have delivered industry-leading high-performance storage solutions tailored to diverse application scenarios. On the storage software side, the Tsinghua storage team has developed massive and scalable distributed storage software systems based on large-scale data centers and exascale supercomputing centers. With regard to scientific research, Chinese universities have showcased impressive academic achievements in the storage field at international conferences. These contributions have had a significant impact on advancing scientific research in the storage field.

However, up until now, a comprehensive book summarizing the foundational principles and cutting-edge storage research in China had yet to be written. Such a book would be of enormous benefit to researchers, students, and professionals by serving as a valuable resource. Therefore, when Mr. Shu Jiwu set about to write this book on data storage architectures and technologies, I was extremely enthusiastic about the project.

This book impresses me with two highlights: First, it provides both a comprehensive overview of storage hardware and software, as well as detailed insights into the intricacies of each field. With the emergence of storage hardware like NVMe SSDs and persistent memory, understanding each technology's own unique set of features and benefits is now crucial to build high-performance storage software systems. Second, this book not only intricately explores the complementary relationship between storage software and hardware but also incorporates perspectives from both academia and industry. This book's compiler and author Shu Jiwu has more than 20 years of research experience in the storage field and is one of the first professors to teach network storage technologies in China. He is a keen observer of the latest trends in the storage field and an excellent forecaster of where the industry is headed.

Furthermore, this book incorporates the invaluable expertise and insights of industry leaders like Huawei, delving into hot topics such as storage maintenance and the efficient implementation of storage solutions tailored to diverse application scenarios.

I hope you find this book as insightful and engaging as I do, and that it inspires you to pursue a career in an industry that promises to reshape our digital world.

Academician of Chinese Academy of Engineering Zheng Weimin
Tsinghua University, Beijing, China

Foreword 2

Building Excellent Data Infrastructure by Mobilizing Storage Industry Professionals

Human society has evolved through the Agricultural Revolution and the Industrial Revolution and is already in the digital age. Digitalization has touched every aspect of our lives and reshaped finance, telecommunications, manufacturing, transportation, energy, education, and healthcare. We are embracing a fully connected, intelligent world.

Data is the driver of digitalization, powering the information age, and providing the insights and intelligence we need to stay ahead of the curve. Huawei predicts that by 2030, human beings will enter an era of unprecedented data growth—a yottabyte (YB) era. Powerful data infrastructure is essential for collecting, transmitting, storing, processing, and using the mountain of data at our disposal. As one of the core elements of data infrastructure, advanced storage systems are the key to store and manage high-value data related to research and development (R&D), production, and operations across different industries.

Data storage capacity is the foundation for enterprise digital transformation, as it allows organizations to accommodate vast amounts of data and pursue groundbreaking innovation. The amount of data an enterprise can store and manage directly impacts its ability to leverage digital technologies and accelerate digital transformation. Storage performance determines the ease of data access and efficiency of storage. Extracting the maximum value from data requires seamlessly aligned storage, network, and computing resources. Storage reliability determines the persistence and availability of data. The loss or inaccessibility of critical data can cause irreparable damage to individuals and society at large. Storage security is the last line of defense for data. It identifies and protects critical data, detects and responds to unpredictable attacks and threats, and restores key data. Together, these functions form the basis for a comprehensive data security architecture.

The storage industry has long been active at the forefront of innovation, evolving from database storage to file storage, from storage arrays to distributed storage, and from mechanical hard disk storage to all-flash storage. A broad range of storage applications and system architectures of storage have emerged as well and helped bring a new digital landscape into being. Emerging data applications like big data, AI, and distributed databases are driving the development of new computing, network, and media technologies. The storage industry is riding a new wave of innovation as data centers are deployed with a growing focus on green and low-carbon solutions.

In recent decades, storage industry professionals have taken on a variety of different roles, working closely with experts in fields such as computer science and engineering. The close cooperation between scholars and experts, with expertise in theories, architectures, engineering technologies, and industry scenarios, is driving the continuous development of the storage industry. As the amount of data generated continues to skyrocket, the need for skilled storage professionals has never been greater.

This book will systematically walk you through the different storage architectures and technologies, delving into detailed technical issues like components, system architectures, key technologies, solutions, implementation, and application scenarios. It is ideal for students, researchers, enterprise R&D personnel, enterprise IT teams, and other practitioners who are interested in storage and would like to contribute to the storage industry.

Huawei is one of the leading global ICT infrastructure providers, with a stated mission to bring digital to every person, home, and organization for a fully connected, intelligent world. Huawei Storage has promoted the company's vision by developing cutting-edge products that have drawn global acclaim. It is a great honor that our efforts in improving our products, strategies, and customer experience have been recognized by Gartner, an international technological research and consulting organization. Since 2016, Huawei has been named a leader in Gartner® Magic Quadrant™ for primary storage. Furthermore, Huawei is committed to providing students with the opportunity to hone their skills by working with storage research professionals and assisting in building robust data infrastructure for all.

Let us work together to unleash the power of data and forge a brighter digital future.

ICT Infrastructure Managing Board
Huawei, China

David Wang

Preface

Data is the core asset in information technology (IT). Efficient and reliable storage of data lays the foundation for various applications, such as the Internet, big data, AI, and high-performance computing.

In recognition of the significance of data storage, our team has offered a course on Network Storage Technologies for postgraduate students at Tsinghua University since 2003. This course provides foundational knowledge and information on the latest advancements in the field of data storage. Over these years, I have found that there is no suitable, comprehensive book or textbook that my students could reference or learn during the course. Data storage is often relegated to a brief and insufficient chapter in most computer system books, and this fails to provide the comprehensive knowledge that they need. The data storage books that are currently available are largely outdated, having failed to keep up with the latest breakthroughs in the storage field. I believed that a well-written, up-to-date book could greatly benefit my students and readers across the globe.

As I explored the idea, I came across Huawei Storage experts who were just as passionate about sharing their expertise, and the perfect collaboration was born. This book has been compiled with the help of Huawei Storage experts and contains our team's years of accumulated teaching and research expertise in the data storage field, as well as insights on the advanced solutions developed by the Huawei Storage team.

The book consists of 14 chapters:

Chapter 1 Data Storage Background

This chapter provides background on data storage, detailing the importance of data storage in today's Information Age and describing the major storage performance indicators.

Chapter 2 Storage Disks and Media

In a computer, data is stored in different storage media and devices. This chapter introduces storage disks and media, including hard disk drives (HDDs), solid-state drives (SSDs), and main memory, detailing their historical development, structure, and performance characteristics. By leveraging firmware techniques such as address mapping, caching, and scheduling, storage disks are able to fully exploit hardware capabilities and achieve optimal performance. This chapter also explores other storage media, such as optical storage and tapes.

Chapter 3 Storage Arrays
Independent storage devices cannot fulfill application demands for capacity and reliability. To address this, storage devices are organized into centralized storage pools to form storage arrays, and ensure efficient resource allocation and high availability. This chapter describes the hardware architecture of storage arrays, including controller and interface modules. It also delves into software architecture, with a deep dive on redundant arrays of independent disks (RAID) algorithms. We will also discuss how to design a storage array that provides optimal performance and reliability.

Chapter 4 Storage Protocols
Storage protocols are responsible for communication and data exchange between hosts and storage devices. This chapter introduces typical storage protocols, including Small Computer System Interface (SCSI) and Non-Volatile Memory Express (NVMe), as well as a new cache-coherent interconnect Compute Express Link (CXL).

Chapter 5 Key-Value Stores
Massive amounts of data can be expressed in key-value pairs, and thus key-value stores are widely used. This chapter introduces common index structures of key-value stores, such as hash tables, B+ trees, and LSM trees, explores the data layout in key-value stores, and discusses crash consistency mechanisms that allow key-value stores to recover to a consistent state after crashes.

Chapter 6 File Systems
This chapter introduces the basic operations of file systems, including file and directory operations. It also uses practical case studies to analyze the key design modules of file systems, including namespace management, cache, and consistency, to provide a fuller view of the operating principles behind a file system. We will also examine the key design modules of file systems through practical case studies.

Chapter 7 Network Storage Architectures
Compared with storage arrays, network storage systems have obvious advantages in terms of scalability, stability, and shared access. This chapter explores the evolution of network storage architectures, spanning from direct-attached storage (DAS) to centralized network storage, such as network-attached storage (NAS) and storage area network (SAN), parallel storage, P2P storage, and cloud storage. We will also cover cutting-edge technologies like storage virtualization, software-defined storage, and hyper-converged architectures.

Chapter 8 Distributed Storage Systems
A distributed storage system stores data on multiple servers to cope with increasing data volumes. This chapter introduces the typical architecture and key measurement indicators of distributed storage systems and looks into distributed key-value stores,

distributed object storage systems, distributed block storage systems, and distributed file systems. We will also provide practical examples of different types of distributed storage systems to help readers get a better sense of how each type of system works.

Chapter 9 Storage Reliability
High reliability is a fundamental requirement of data storage. This chapter covers the basics of storage reliability, compares the reliability of HDDs and SSDs, examines the working principles and trends of erasure coding technologies, and analyzes the reliability of distributed storage systems.

Chapter 10 Storage Security
Storage security incidents can lead to severe privacy breaches and catastrophic data loss. This chapter delves into the basic concepts and cutting-edge technologies of storage security, encompassing system security, data security, and advanced security management. We will also cover the latest hardware security technologies, such as trusted execution environment (TEE).

Chapter 11 Data Protection
This chapter analyzes best practices for protecting data, with an in-depth look at technologies like mirroring, snapshots, and cloning. It covers three major data protection scenarios—backup, archiving, and disaster recovery—with illuminating case studies and descriptions for each.

Chapter 12 Storage Maintenance
As storage systems become increasingly complex, maintenance becomes increasingly difficult as well. This chapter describes preventive and corrective maintenance, the former of which is used to prevent storage system faults and the latter of which is used to rectify faults that have occurred.

Chapter 13 Storage Solutions
This chapter introduces storage solutions for different industries, including carriers, government, finance, healthcare, and education. You will learn how to design efficient data storage solutions based on differing requirements through practical examples.

Chapter 14 Storage Technology Trends and Development
This chapter introduces cutting-edge storage technologies, including near-data computing, persistent memory, online storage, intelligent storage, edge storage, blockchain storage, disaggregated data center architecture, and high-density new storage.

Beijing, China   Jiwu Shu

Acknowledgments

I would like to express my sincere gratitude to all of those who helped contribute to this book:

Associate professor Lu Youyou, postdoctoral scholars Wang Qing, Chen Youmin, and Zhang Yuhao, and PhD candidates and postgraduates Cheng Zhuo, Lv Wenhao, Yan Bin, Lin Jiazhen, Feng Yangyang, Li Junru, Gao Jian, Li Zeqi, Fan Ruwen, and others, from the storage research team at Tsinghua University;

Associate professors Shen Zhirong, Li Qiao, Gao Congming, and others, from Xiamen University;

Zhou Yuefeng, Pang Xin, Zhang Fupeng, Zhang Guobin, Wang Zhen, Ding Zhibin, Liang Jiani, Yang Tianwen, Dong Wei, Gu Xuehu, Zhang Nandong, Wang Sheng, Yan Peng, Li Zhaonan, Li Guojie, Qin Guo, Yang Juntao, Liao Zhijian, Yan Dahong, Pan Hao, Wang Peng, Li Chu, Zhang Peng, Rao Chengli, Cao Changbin, Qiu Youcheng, Wang Wei, Li Chao, Li Qiang, Zhang Ying, Xu Xudong, Chen Weiping, Zeng Yangyang, Liu Jian, Xia Bingxin, Du Xiang, Zeng Hongli, Chen Keyun, Zhou Xifeng, Zhou Xiaofeng, Yuan Qizhao, Tang Guohui, Lu Jiangang, He Jie, Dai Wei, Huang Rong, Sun Lin, Qi Chenrui, and others, from Huawei Technologies Co., Ltd.

Summary

This book introduces data storage architectures and technologies, including storage disks, storage media, storage arrays, storage protocols, key-value stores, file systems, network storage architectures, distributed storage systems, storage reliability, storage security, and data protection. In addition to these basic storage technologies and concepts, this book also discusses topics such as storage maintenance, storage solutions, and storage technology trends, as well as the latest research case studies on related topics.

This book is intended to undergraduates or postgraduates in computer science and related fields and can also be used as a reference for industry professionals.

Contents

1. Data Storage Background ... 1
2. Storage Disks and Media ... 9
3. Storage Arrays ... 43
4. Storage Protocols ... 75
5. Key-Value Stores ... 113
6. File Systems ... 131
7. Network Storage Architectures ... 149
8. Distributed Storage Systems ... 185
9. Storage Reliability ... 225
10. Storage Security ... 271
11. Data Protection ... 311
12. Storage Maintenance ... 349
13. Storage Solutions ... 365
14. Storage Technology Trends and Development ... 379

Chapter 1
Data Storage Background

Data storage is the process of recording and storing information on various media for subsequent access. In the distant past, knotted strings were used to store information before writing even existed, as described in an ancient Chinese manual called the *Book of Changes* (Zhouyi). With the progress of human and technological evolution, words began to be stored on oracle bones, bamboo tubes, and later paper. More recent advancements have led to the explosion of information technology (IT), which has, in turn, fueled the development of various storage hardware devices, such as magnetic tapes and disks. Computers then became a major tool for storing data, allowing us to manage and access data on the storage hardware through storage software.

Today's world is producing data at an unprecedented rate. It is projected that by 2025 the total data volumes generated around the world will reach 181 zettabytes (ZB), requiring roughly 200 billion 1 TB disks to store. Storage will play a critical role in handling such massive amounts of data. In this chapter, we will introduce the importance of data storage in the information age and its main objectives.

1.1 The Importance of Data Storage

It goes without saying that data storage is crucial for any individual. We store precious, important photos and videos on portable devices like mobile phones and laptops and upload some of them to reliable storage systems like cloud disks for long-term data retention. Similarly, on a macroscale, data storage plays a critical role in economic and social development in today's information age.

Data storage contributes to economic growth. In terms of direct economic benefits, the consulting firm Gartner predicts that the global storage market will reach US$150 billion by 2025. The indirect economic benefits generated by data storage are even more incalculable. Industries across the world are undergoing digital transformation, a process which is inseparable from data storage. Take the e-commerce sector as an example. During the world's busiest shopping days, like Black Friday in the USA or Double Eleven in China, storage systems must be fast enough, reliable enough, and stable enough to support sky-high numbers of transactions per second (TPS). In the field of oil exploration, high-performance storage systems are needed to process the huge data volumes generated by reservoir simulations. Access to such systems has slashed the time needed to explore potential energy reserves and drastically improved production efficiency across the industry.

Data storage improves public governance. The State Council of China stated in their 2015 Action Outline for Promoting the Development of Big Data that, "Data has become a national basic strategic resource, and big data is increasingly having a significant impact on global production, circulation, distribution, and consumption activities as well as the economic operation mechanism, social lifestyle, and the state's capacity in governance." Big data can effectively help governments deal with complex social problems. For example, deeper analysis of public health data can potentially be used to predict the outbreak and spread of infectious diseases. In this process, the big data centers need advanced storage systems to collect, store, and analyze data while supporting transparent and efficient computing and analysis.

Data storage fuels breakthroughs in other fields. High-performance storage has been the key to recent breakthroughs in many disciplines. As pointed out by Li Guojie, an academician from the Chinese Academy of Engineering, "The emergence of massive amounts of data has given rise to a new scientific research model. Now, researchers only need to find or mine the information and knowledge directly from the data, rather than directly contact the objects themselves." In 2019, for example, the astronomy field released its first-ever photo of a black hole, an important feat to verify the theories related to black holes. This photo would not have been possible without high-capacity data storage, since the radio telescopes that captured the image needed to store 5 PB of data in real time on high-performance helium-filled hard disks.

1.2 The Objectives of Data Storage

Humans have constantly sought to innovate new tools and technologies for data storage, always striving for higher performance, usability, and reliability. This section describes the hardware and software design used to achieve these objectives. Further exploration will be provided in later sections.

1.2.1 High Performance

Data storage performance is typically measured in terms of throughput and latency. Throughput refers to the number of operations that a data storage system can process per unit of time, whereas latency is the time required to complete a single

1.2 The Objectives of Data Storage

operation. High-performance storage aims to offer best-in-class throughput at ultra-low latency to meet the growing demands of upper-layer applications. For example, China's Double Eleven shopping festival produces ever-increasing transaction volumes, requiring enterprise storage systems with the throughput to match this demand. In terms of latency, users are also increasingly concerned about the quality of the services supported by these storage systems. Storage systems are therefore being designed to process every data access request as fast as possible to minimize perceived user latency.

To achieve higher performance, storage hardware and software must be designed to operate collaboratively. Figure 1.1 shows the diverse types of storage hardware currently in use and their related advantages and disadvantages in terms of performance, capacity, and price. In general, a solid-state drive (SSD) runs with latency of about 10–100 microseconds (μs), which far outperforms the 10–millisecond (ms) latency of a hard disk drive (HDD). Switching to faster storage hardware is an easy way to upgrade storage system performance, but it can result in two main problems: first, high-performance hardware is expensive to procure and pushes up a system's total cost of ownership (TCO), hindering broad adoption; second, increased throughput of storage hardware is often not matched by an increase in capacity, which limits the amount of data that the entire system can accommodate. The design of any paired storage software must also be adapted to match hardware upgrades.

Mainstream storage software often supports a wide range of hardware types to improve throughput. One of the earliest storage virtualization technologies that uses the performance of multiple disks while appearing logically as a single storage unit is the redundant arrays of independent disks (RAID) [1] technology, which was developed by a group of scientists at the University of California, Berkeley. RAID 0 distributes data across multiple disks for better performance. Another common technology is the distributed storage system, which interconnects multiple storage servers over a network to provide extremely high aggregate throughput. Figure 1.2 shows the distributed file system architecture of China's supercomputer Sunway

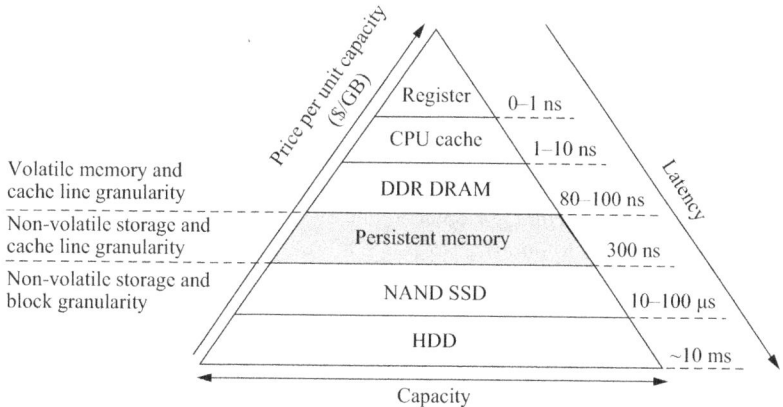

Fig. 1.1 Storage hierarchy pyramid

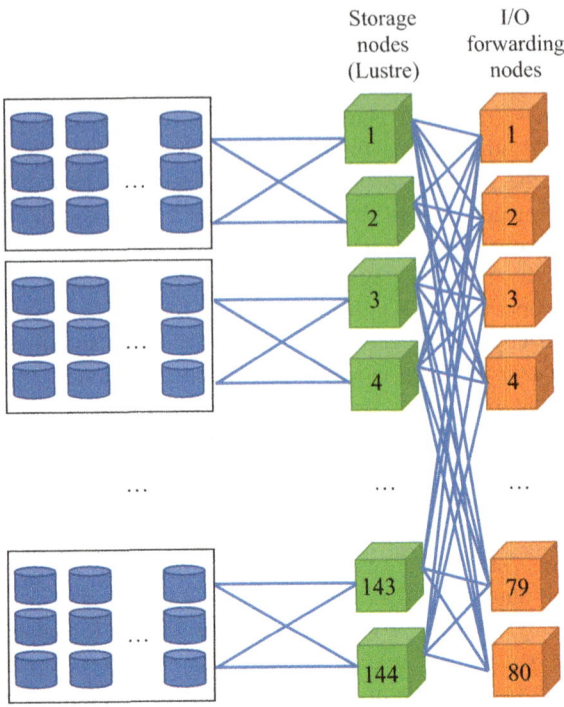

Fig. 1.2 Distributed file system architecture of Sunway TaihuLight [2]

TaihuLight. It consists of 80 I/O forwarding nodes, 144 storage nodes, and 72 disk arrays (each of which contains 60 disks).

Performance can also be improved by leveraging data locality principles to design schemes like caching and prefetching. Data locality includes temporal and spatial locality. Temporal locality refers to the tendency for the same data to be accessed repeatedly over a short period of time. An example of this is that data of some popular products on an online shopping platform is frequently read and written. Spatial locality means that nearby data is likely to be accessed when a piece of data is accessed. For example, if one file in a folder is accessed, other files in the same folder are likely to be accessed as well.

Storage software uses temporal locality-based caching to store frequently accessed data on fast storage components, boosting access throughput while slashing latency. In a single storage server, a Linux file system can use the page cache to cache file data in the memory. Replacement algorithms like least recently used (LRU) can be employed to improve the cache hit ratio. In a data center, a distributed cache cluster built with software like Memcached can provide a read cache for back-end HDD-based storage systems.

Storage software uses spatial locality-based prefetching at multiple layers to move data from low-speed to high-speed storage components before the data is needed. One common example is the readahead prefetching algorithm used by Linux file systems. In such systems, if a set of files is being accessed sequentially, the remaining files in the sequence will be moved to the page cache before they are accessed.

1.2.2 High Usability

Another goal of data storage is high usability, which improves interaction between upper-layer applications and storage systems. This includes faster data writes and more efficient reads. Figure 1.3 shows common data storage interfaces that can improve usability. The simplest of these semantics is the block interface, in which the entire storage space is abstracted into equal-sized data blocks that are each assigned a unique digital identifier. Those identifiers are then used by an application to read and write the corresponding data blocks. Due to its inability to express the data semantics of most applications, the block interface usually serves as a base for other storage interfaces and is not directly exposed to upper-layer applications.

In comparison, the key-value interface is easier to use as it maintains a set of key-value pairs. Each key-value pair consists of a key and a value, which are data blocks of variable length. Applications locate corresponding key-value pairs by using keys to perform insertion, update, query, and deletion. Thanks to its advantages in clear semantics and flexibility, the key-value interface is widely used by companies like Amazon in their key-value system Dynamo [3] to store online shopping data.

The file interface, on the other hand, organizes data into a directory tree structure, which consists of directory and ordinary files. Applications can create, delete, read, and write these files. Ordinary files store file data, while directory files record the names of ordinary files under their directory as well as other directory files. Each file has attributes like permissions that determine access control. File systems are hierarchical, making them suitable for common users. This is why most desktop operating systems (OSs) use file systems to help users easily manage their own data. In addition, extension functions for key-value and file interfaces are supported to improve usability. Some key-value stores support multiple keys to query the same data, which is called the secondary index function. Some file systems support transactional semantics that enable atomicity of multiple file operations.

In addition to storage interfaces, data sharing also contributes to better usability. Despite recent innovations in storage array technology that partially address capacity and performance constraints, data sharing remains a common bottleneck that prevents users in different locations from accessing the same data. Recent

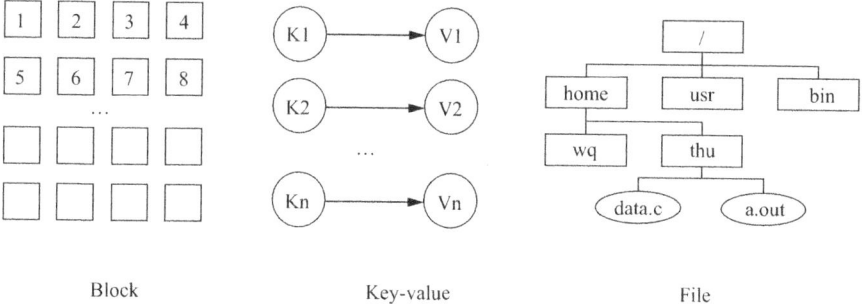

Fig. 1.3 Three common data storage interfaces

developments in network technologies and storage protocols have made data sharing possible over networks for storage systems. One example of this is network-attached storage (NAS), which supports file sharing over a local area network (LAN). These sharing capabilities are further improved in large-scale distributed storage systems, such as Facebook's Tectonic [4], a data center-wide distributed file system that provides excellent scalability and data sharing to maximize resource utilization.

1.2.3 High Reliability

High-reliability storage prevents data loss and service interruption in the event of system exceptions, including disk, server, and network failures, as well as human error. One report from Facebook reveals that up to 110 servers can crash a day in a 3000-server production cluster. In 2017, Amazon's S3 storage system went offline for 5 h, cutting access to thousands of websites and apps and causing huge losses. The consequences of such reputational and financial losses make storage reliability a high priority for many enterprises.

One of the most basic approaches to achieve high reliability is through data redundancy. Common data redundancy schemes include multi-copy and erasure coding (EC) schemes. Figure 1.4 shows a multi-copy scheme where multiple copies of single pieces of data are stored in multiple locations. This means if a copy in one location cannot be accessed, the copies in other locations will still be available. In contrast, an EC scheme uses some coding methods to calculate additional parity blocks from multiple data blocks and stores the data and parity blocks in different locations. Any data that cannot be accessed can be rebuilt using the parity blocks and other data blocks. Compared to a multi-copy scheme, EC has lower storage space overheads but higher computing and recovery overheads.

Well-developed storage systems offer reliability assurance at multiple levels. At the storage device level, error correction methods, such as low-density parity-check

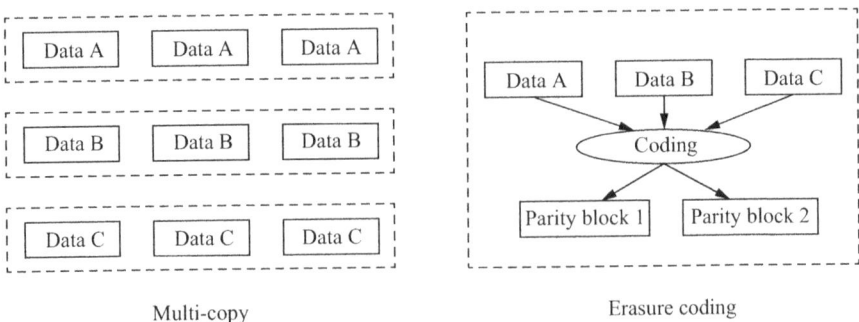

Fig. 1.4 Multi-copy and erasure coding schemes

(LDPC) code, detect and attempt to repair data errors. At the server level, RAID-based disk arrays can be used to tolerate storage component failures. At the cluster level, distributed multi-copy or EC schemes handle abnormal server events. In addition to all of this, data is often periodically backed up and archived.

Theoretically, the preceding schemes are sufficient to ensure data storage reliability, but in practice, data content errors may still occur due to storage software code vulnerabilities caused by programmer errors. For example, enabled storage software that cannot provide concurrent data access may be suffering from corrupted data. To achieve high reliability, storage software must undergo frequent correctness tests and even adopt some formal methods. When Amazon built its next-gen cloud object storage system, it used lightweight formal methods to evaluate whether the system is working correctly in terms of interface semantics, data consistency, and concurrent access [5].

1.2.4 Other Objectives

In addition to high performance, usability, and reliability, data storage often needs to fulfill the following objectives:

Cost-effectiveness. Data storage aims to provide higher performance at a lower price per unit. Storage tiering is a common method to increase efficiency and decrease costs based on data value and access requirements. For example, one common method for achieving cost-effectiveness is storage tiering, which involves storing hot data on high-speed storage devices (e.g., SSDs) and cold data on low-speed storage devices (e.g., magnetic tapes).

High security. Data storage must protect core data from cyber events, be it data loss or sensitive information leakage. Common security practices include data encryption, access control, and privacy-preserving computation.

High scalability. Distributed storage systems combine multiple servers to achieve high-performance data access. This allows scaling by adding more servers while increasing the overall system performance linearly. To achieve high scalability, the storage software needs to prevent typical server-based bottlenecks, and this can be achieved by schemes like load balancing and dynamic data migration.

References

1. Patterson D A, Gibson G, Katz R H. A case for redundant arrays of inexpensive disks (RAID) [R/OL]. (1987–12) [2023-04-05].
2. Lin H, Zhu X, Yu B, et al. Shentu: processing multi-trillion edge graphs on millions of cores in seconds. In: SC18: international conference for high performance computing, networking, storage and analysis. New York: IEEE; 2018. p. 706–16.
3. Decandia G, Hastorun D, Jampani M, et al. Dynamo: Amazon's highly available key-value store. ACM SIGOPS Oper Syst Rev. 2007;41(6):205–20.

4. Pan S, Stavrinos T, Zhang Y, et al. Facebook's tectonic filesystem: efficiency from exascale. In: 19th USENIX Conference on File and Storage Technologies (FAST 21); 2021. p. 217–31.
5. Bornholt J, Joshi R, Astrauskas V, et al. Using lightweight formal methods to validate a key-value storage node in Amazon S3. In: Proceedings of the ACM SIGOPS 28th Symposium on Operating Systems Principles; 2021. p. 836–50.

Chapter 2
Storage Disks and Media

In computers, storage media record the status of binary data. Programs and data in computers are recorded in binary mode. Different storage media use the different physical states of components to represent "0" and "1" and support switching between them.

There are magnetic, electrical, and optical storage media. Magnetic storage media use the magnetic poles of magnetic particles to record data. The two magnetization directions represent data "0" and "1." Common storage disks that use magnetic storage media include magnetic disks and tapes. Electrical storage media use the number of electrons stored in a storage cell to record data. The number of electrons affects the level on the bit line, representing data "0" or "1." Common storage disks that use electrical storage media include flash memory and dynamic random-access memory (DRAM). For optical storage media, a laser is used to irradiate the media. This causes physical or chemical changes to the properties of the media to represent "0" and "1." Common storage disks that use optical storage media include compact discs (CDs), digital versatile discs (DVDs), Blu-ray discs (BDs), and archival discs (ADs).

In a computer system, a storage hierarchy combines different storage devices to virtualize high-speed and high-capacity storage resources for processors. Figure 2.1 shows several common storage media and their latencies. Broadly speaking, storage media comprise memory (also called main memory) and storage (also called secondary storage). Currently, memory mainly refers to DRAM, which has a latency of dozens of nanoseconds (ns). Storage is usually a combination of flash memory and hard disk drives (HDDs), which have latencies of dozens of microseconds (μs) and milliseconds (ms), respectively.

This chapter describes several major storage disks and media, including HDDs, solid-state drives (SSDs), and main memory.

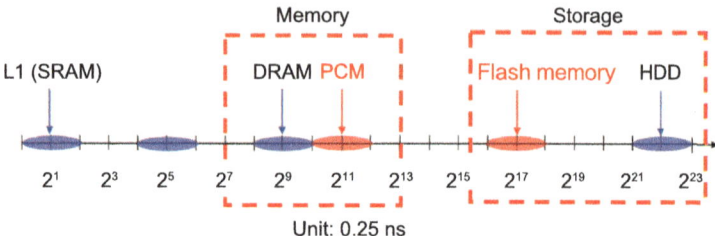

Fig. 2.1 Common storage media and their latencies

2.1 HDDs

An HDD uses magnetic storage media to store data. It is also known as a magnetic disk or mechanical hard disk. In 1956, IBM released the world's first disk drive product, the IBM 350. It had a storage capacity of only about 5 MB contained on 50 24-in. (1 in. ≈ 2.54 cm) diameter disks and an access latency close to 1 s. This product was not integrated into a system as a component but ran independently as a standalone device.

In 1973, IBM unveiled the IBM 3340 "Winchester" disk drive. Its design incorporated heads, platters, and spindles into a sealed cartridge that was filled with inert gas. It used magnetic heads that floated and moved over the spinning platters to read and write data. This design made it possible to mass-produce these disk drives. The design of the "Winchester" disk drive is still used today, and it has inspired the designs of many modern HDDs.

In 1997, the giant magnetoresistance (GMR) technology was introduced and used in HDDs. By using materials with a strong magnetoresistance effect and a multilayer thin-film structure, this technology was able to increase the HDD storage density from 3–5 GB/in^2 to 10–40 GB/in^2. This technology has played a significant role in the design of high-capacity HDDs. The 2007 Nobel Prize in Physics was awarded to Albert Fert and Peter Grünberg for the discovery of GMR.

Over the past 70 years, there has been vast improvement in the capacity, performance, and reliability of HDDs. HDDs continue to play a major role in large-scale storage systems, and their capacity advantages make them suitable for capacity-oriented storage systems.

2.1.1 Components and Structure

An HDD consists of mechanical and electronic parts, as shown in Fig. 2.2.

The mechanical part includes the base, spindle motor, platters, voice coil motor, head group, and top cover, as shown in Fig. 2.3. All platters are mounted in parallel on the same spindle. Each surface of a platter is equipped with a head. All heads are

2.1 HDDs

Fig. 2.2 Components of an HDD

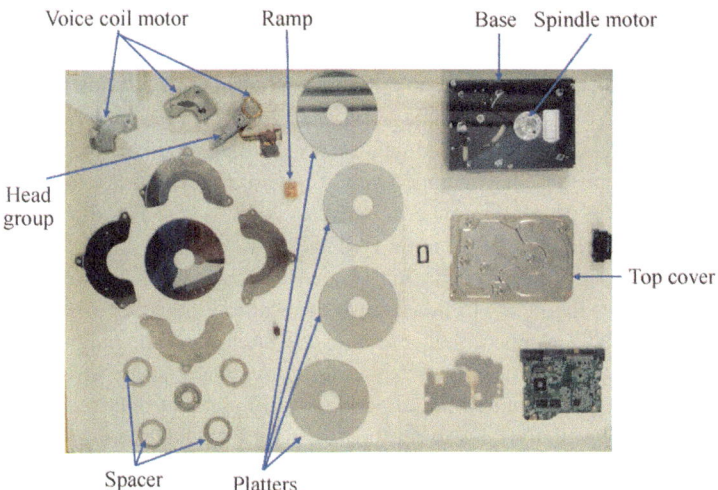

Fig. 2.3 Components of the mechanical part of an HDD

associated with the same head group. A voice coil motor is installed at the tail of the head group to drive the entire head group to move along the radial direction of the platters, coupled with the high-speed rotation of the platters.

The electronic part includes the main control SoC (a monolithic system on a chip), motor driver chip, rotation vibration (RV) sensor, shock sensor, DRAM, and flash read-only memory (ROM), as shown in Fig. 2.4. The RV sensor is typically only used on enterprise-level HDDs.

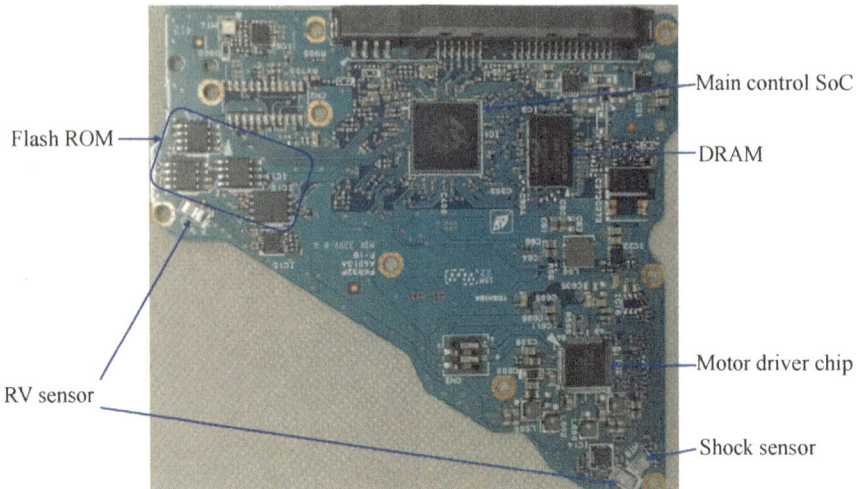

Fig. 2.4 Components of the electronic part of an HDD

2.1.2 Performance

An HDD reads and writes data by using heads to sense and convert the magnetic media on platters. When writing data, a write head of an HDD converts electrical signals into magnetic signals and demagnetizes the magnetic particles on the platter's surface so that they are arranged according to a certain rule. When reading data, a read head of an HDD converts magnetic signals into electrical signals by sensing magnetic field changes to identify data.

When an HDD is working, the high-speed rotation of the platters generates enough force to lift the heads to float above the platters. Each head is a read-write head. The read head is a sensor that is sensitive to changes in the magnetic field. To read data, the read head senses the change of the magnetic field above the rotating platter and outputs the changed voltage signal to implement data reading, that is, magnetoelectric conversion. The write head is a coil wound around the magnetic core. It uses a current to change the magnetization direction of magnetic particles. More specifically, during data writing, a "0" or "1" data stream passes through the write head coil in the form of a pulse current to change the magnetization direction of a magnetic medium layer on a platter below the head. The data write is done via electromagnetic conversion.

An HDD's read performance mainly depends on its internal structure. Its access latency is largely determined by its mechanical components. Figure 2.5 shows the HDD structure. HDDs read data at a sector granularity through two mechanical operations: rotation and seek. The two mechanical operations are related to two parts of an HDD: platter and head.

In terms of platters, they rotate horizontally around a spindle in parallel. The rotation operation refers to the way the spindle rotates to spin the platters. The top

2.1 HDDs

Fig. 2.5 HDD structure

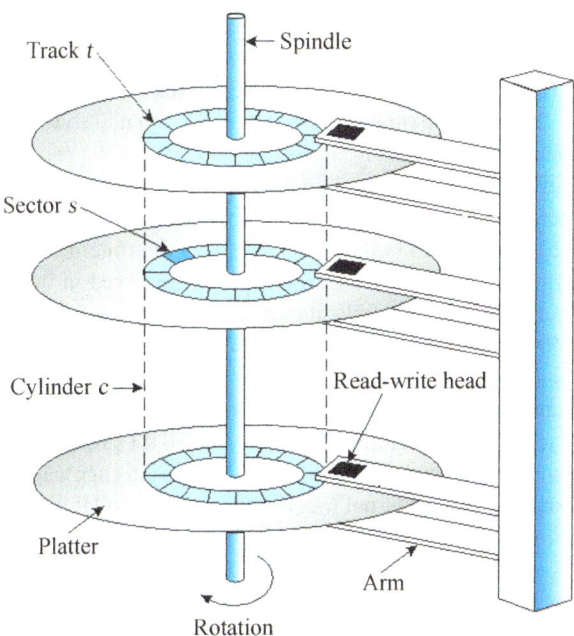

and bottom sides of each platter are called surfaces, and magnetic media are attached to each surface to record data. The surfaces are divided into concentric rings called tracks, which are further divided into sectors of a fixed size, representing the minimum access unit of an HDD.

In terms of heads, each surface corresponds to a head, and the head reads and writes the corresponding surface. Heads for different surfaces form a head group. Each head is driven by a corresponding arm to move across different tracks on a surface, which is known as a seek operation. For example, an HDD's heads have a flying height less than 10 nm, a flying speed close to 200 km/h, a seek precision about 50 nm, and a seek speed at the ms level.

In addition to rotation and seek operations, HDD access latency includes data transfer time and controller processing time. The HDD access latency is calculated using formula 2.1:

$$T_{\text{disk service time}} = T_{\text{seek time}} + T_{\text{rotation time}} + T_{\text{data transfer time}} + T_{\text{controller time}} \quad (2.1)$$

$T_{\text{disk service time}}$ is the estimated data access time, $T_{\text{seek time}}$ is seek time, $T_{\text{rotation time}}$ is rotation time, $T_{\text{data transfer time}}$ is data transfer time, and $T_{\text{controller time}}$ is controller processing time.

Seek time refers to the time taken by the head to locate a specific track. Seek time can be expressed in different ways, including maximum seek time ($T_{\text{full stroke time}}$), average seek time ($T_{\text{average time}}$), and track-to-track seek time ($T_{\text{track-to-track time}}$). The maximum seek time is the time for a head to move from the outermost track to the innermost track on a surface, representing the maximum radial movement distance on a surface. The average seek time is the average time for a head to move to any random track on a surface. The track-to-track seek time is the time for a head to move from one track to an adjacent track.

The rotation time is the duration for a platter to rotate to the location of the sector that the head requests. This time depends on the rotational speed of the spindle and can be calculated based on the rotational speed quoted in the HDD's specifications. Rotational speed is measured in revolutions per minute (rpm), typically at 5400 rpm, 7200 rpm, or 15,000 rpm. $T_{\text{rotation time}}$ is the time it takes for the platter to rotate 180 degrees, completing half a turn. It is measured in ms and calculated using the formula $T_{\text{rotation time}} = 0.5/(\text{rotational speed}/60,000)$. For example, the $T_{\text{rotation time}}$ for a 5400 rpm HDD is 5.6 ms, and for a 7200 rpm HDD, it is 4.2 ms.

The data transfer time is the total duration required to transfer data from the magnetic media to the host, encompassing both internal and external transfer times. Internal transfer time is linked to the HDD's rotational speed and can be estimated using the formula data transfer volume/(storage capacity of a single track × rotational speed). External transfer time relates to HDD interfaces and can be estimated using the formula data transfer volume/interface bandwidth. For example, if the theoretical bandwidth of SATA 2.0 interfaces is 300 MB/s, the external transfer time for 1 MB data is estimated to be 3.33 ms.

The controller processing time mainly refers to the data processing latency of electronic components. This is relatively small compared to the processing latency of mechanical components and can generally be ignored.

When an HDD is working, its read/write request latency is also related to its loads. Their relationship is $T_{\text{avg. response time}} = T_{\text{disk service time}}/(1 - \text{HDD utilization})$. HDD utilization is the ratio of the number of current read and write requests to the maximum HDD throughput. Generally, when an HDD's utilization reaches 70%, its read and write latency increases significantly. Figure 2.6 shows the relationship between HDD latency and loads.

According to formula 2.1 for calculating HDD access latency, HDD performance depends on the rotational speed of platters, seek speed of heads, and the amount of data being transferred. There are two additional factors that also impact performance and should therefore be taken into account:

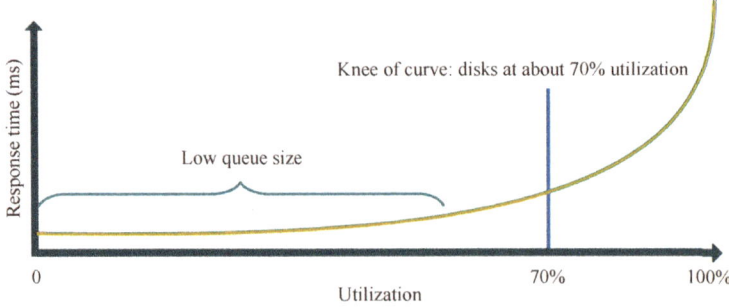

Fig. 2.6 Relationship between HDD latency and loads [1]

1. Sequential data access (locality): When accessing data, HDDs seek out tracks and rotate to locate requested data blocks for read and write. If the locations are random, seek and rotation operations need to be performed for each access, and heads may have to move back and forth a lot. This causes latencies and less effective data read and write. The more sequential the data that needs to be accessed, the better the HDD performance will be.
2. Data access block size: In data access, the seek time, rotation time, and controller processing time are relatively fixed, whereas the transfer time varies depending on the data volume. If the data volume is large, the transfer time accounts for a high proportion of the total time. If the data volume is small, the transfer time accounts for a low proportion of the total time. As most of the time is spent on operations such as seek and rotation, the effective bandwidth for data access is low. To sum up, the larger the size of the data block that needs to be accessed, the better the HDD performance will be.

It is crucial to take these two factors into account when designing storage software. For example, common technologies for improving file system designs include increasing the size of a data access block and improving the locality, prefetching, and caching of data access. These all take advantage of the impact these key factors have on HDD performance.

2.1.3 Firmware

The electronic components of an HDD include the main control SoC chip, which is used for signal processing, error detection and correction, and motor driving. Controllers also run the corresponding firmware, used for address mapping, request queuing and scheduling, and caching. Below, the firmware's address mapping, caching, and scheduling functions for HDDs are explained in more detail.

2.1.3.1 Address Mapping

HDDs export a linear, one-dimensional address space to the operating system, called the logical block address (LBA). Generally, an HDD uses the cylinder, head, and sector (CHS) for the physical block address (PBA) to locate a specific sector. The HDD firmware will translate linear, one-dimensional addresses into sector addresses for the HDD. This is how HDD firmware maps LBAs to PBAs.

A simple method of mapping from LBA to CHS can be expressed as follows:

$$\text{Cylinder No.}(\text{Cylinder}\#) = \frac{\text{LBA}}{\text{Number of sectors in a cylinder}(\text{Cylinder}\_\text{Sectors})} \quad (2.2)$$

$$\text{Head No.}(\text{Head}\#) = \frac{\text{LBA}\%\text{Cylinder\_Sectors}}{\text{Number of sectors in a track}(\text{Track\_Sectors})} \quad (2.3)$$

$$\text{Sector No.}(\text{Sector}\#) = \text{LBA}\%\text{Track\_Sectors} \quad (2.4)$$

where **Cylinder#**, **Head#**, and **Sector#** indicate the cylinder, head, and sector number, respectively. **Cylinder_Sectors** and **Track_Sectors** indicate the number of sectors in each cylinder and track, respectively.

In practice, HDD address mapping is more complex. To optimize storage density and improve access performance, address mapping also includes common technologies such as zones, remappings, and skews.

Zones

An HDD's tracks often have different circumferences. Two approaches can be taken to divide the sectors of a single track: a fixed number of sectors and a fixed bit density (i.e., giving each sector a fixed length). The simple mapping relationship mentioned before assumes that each track has a fixed number of sectors. This is often inefficient and can compromise the storage density of relatively long tracks (such as outer tracks). A fixed bit density, however, makes CHS calculations more complicated. So, the two different zoning approaches are often combined.

Tracks with different radii on HDD surfaces are divided into multiple zones. Each zone is divided into sectors based on a fixed number of sectors per track. The number of sectors per track varies with different zones to obtain the highest possible bit density. This helps achieve a balance between storage density and address mapping complexity.

Once the zones are defined, LBA to CHS translation only requires the start addresses of several zones. The formula is as follows:

$$\text{Zone\_LBA} = \text{LBA} - \text{Zone\_StartLBA} \quad (2.5)$$

$$\text{Cylinder}\# = \text{Zone\_StartCylinder} + \frac{\text{Zone\_LBA}}{\text{Cylinder\_Sectors}} \quad (2.6)$$

$$\text{Head}\# = \frac{\text{Zone\_LBA}\%\text{Cylinder\_Sectors}}{\text{Track\_Sectors}} \quad (2.7)$$

$$\text{Sector}\# = \text{Zone\_LBA}\%\text{Track\_Sectors} \quad (2.8)$$

where **Zone_LBA** indicates the LBA skew in the current zone. **Zone_StartLBA** and **Zone_StartCylinder** indicate the start LBA and start cylinder number of the current zone.

2.1 HDDs

Remappings

Address mapping also needs to cope with sector failures. If a sector is faulty, the HDD needs to remap the faulty sector to a new sector for data read and write. In an HDD, one or more sectors are usually reserved on each track for sector remapping.

When sector remapping occurs, the HDD firmware creates a remapping table that records the mapping from the faulty sector's address to the new sector's address. During CHS address translation, the remapping table is then looked up to determine whether a sector has been remapped. If one has, the new address of the remapped sector is obtained.

Skews

Platters continuously rotate when a head moves to switch tracks. When the head switches between tracks, it generally does not read the data on an adjacent sector. Skews help address this. As shown in Fig. 2.7, the numbering of sectors in the next accessed track is optimized based on the track switching time. Track switching takes time, and surface and cylinder switching also do. So, the skew method can also be used for surface or cylinder switching.

Once skews are used, they are calculated during the translation from logical to physical addresses and factored into the existing calculation result.

2.1.3.2 HDD Cache

A memory of a certain size is embedded in an HDD for data caching. The typical HDD cache size ranges from 2 MB to 64 MB (the HDD cache in the main memory used in the operating system is different from that described in this section). This

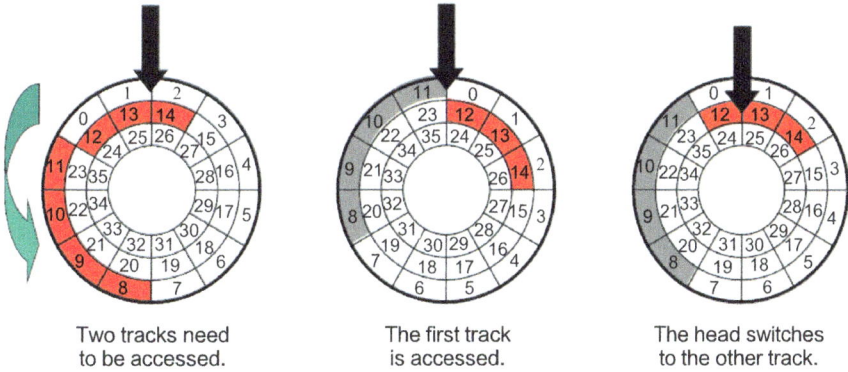

Two tracks need to be accessed.   The first track is accessed.   The head switches to the other track.

Fig. 2.7 Block address mapping with skews

HDD cache functions similarly to most caches and includes the write buffer and read cache.

Write Buffer

When a host sends data to an HDD, the HDD buffers the data in the HDD cache and returns a completion notification to the host. Then, the data is written to the HDD media in the background. A write buffer significantly improves the HDD write performance for hosts. However, a system power failure can cause data loss because the HDD cache is volatile. To prevent data loss, hosts can disable the HDD cache or enable synchronization operations such as explicit sync.

Read Cache and Prefetch

Since the interface performance between hosts and HDDs is usually higher than the internal access performance of HDD media, an HDD cache can also act as a read cache. With prefetching, an HDD can first read data from media to the cache. When a host sends a read request to the HDD, the read cache increases the probability that data is hit in the HDD cache, thereby reducing the read latency.

2.1.3.3 HDD Scheduling

Addresses in the read and write requests received by a disk are typically not contiguous. HDD scheduling algorithms optimize the sequence of these requests to reduce their average access time, improve the overall HDD throughput, and enhance fairness among different requests.

FCFS Algorithm

The First Come First Serve (FCFS) algorithm does not change the sequence of read and write requests but serves them in the sequence they arrive. This algorithm is fair and delivers relatively low request latency but results in an overall low throughput for HDDs.

SSTF Algorithm

The Shortest Seek Time First (SSTF) algorithm first serves the read/write request closest to the current track positioned by the head. This algorithm reduces the HDD seek time and improves overall HDD throughput. However, fairness cannot be ensured, and some requests may be "starved," meaning these request will be continuously ignored and not be responded to.

SCAN Algorithm

With the SCAN or elevator algorithm, the head starts from one end of the surface and moves toward the other end, serving read/write requests along the way. Once the head reaches the other end, it reverses the direction and moves to the start end, serving new requests along the way. The advantage of this algorithm is that the overall HDD throughput is high, and no starvation occurs.

The circular SCAN (C-SCAN) algorithm is similar to the SCAN algorithm. The difference is that once the head reaches to the other end, it immediately returns to the start end without serving any requests on the return trip. Then, it moves in the same direction and continues serving new requests along the way.

LOOK Algorithm

The LOOK algorithm is similar to the SCAN algorithm. The difference is that with the LOOK algorithm, as a head moves from one end to the other end, it stops as soon as it completes the request with the largest track number from all requests, instead of moving all the way to the end of the surface. This algorithm results in less unnecessary head movement than the SCAN algorithm.

The C-LOOK algorithm is a variant of the LOOK algorithm. Similar to the C-SCAN algorithm, once the head completes the last request at one end, it immediately returns to the new request with the smallest track number without serving any requests on the return trip. Then, it processes the remaining requests in the same direction as before.

HDD technologies are still evolving, and many are exploring how to improve the density of HDDs. Examples of promising new HDD technologies include shingled magnetic recording (SMR), heat-assisted magnetic recording (HAMR), and microwave-assisted magnetic recording (MAMR).

2.2 SSDs

An SSD uses nonvolatile storage chips to store data. Currently, SSDs mainly use flash memory or other nonvolatile memory chips like phase-change memory.

Flash memory is an electrically erasable programmable read-only memory (EEPROM). It was proposed by the Japanese company Toshiba in 1984 and has gained popularity since 2005. Flash memory includes not-or (NOR) flash memory and not-and (NAND) flash memory. Currently, NAND flash memory is more widely used. Unless otherwise specified, flash memory mentioned in this document refers to NAND flash memory.

SSDs consist of controllers, NAND flash memories, DRAM, power supplies, backup capacitors, connectors, and firmware. Figure 2.8 shows an example of a typical SSD. The connector is the physical interface an SSD uses to interact with a host. At the front end, the controller provides the Serial Advanced Technology Attachment (SATA), Serial Attached SCSI (SAS), Peripheral Component Interconnect Express (PCIe), or Non-Volatile Memory Express (NVMe) interface to interact with the host for protocol parsing and data transfer. Internally, it is responsible for data assembly and status management. At the back end, it provides multiple channels to mount multiple flash memories, is responsible for data access and reliability management of flash memories, and functions as a DRAM controller to provide cache read and write interfaces. The power supply module converts any power supplied by the host into the different voltages required by the internal components of the SSD. In addition, it works with the backup capacitor to provide backup power in the case of a power failure. Generally, enterprise-level SSDs have backup capacitors while consumer-level SSDs do not. This means consumer-level SSDs typically have no power failure protection. Firmware manages and controls SSD resources. Its front end interacts with the host based on protocols. Internally, it manages the flash memories and provides address mapping, garbage collection, and wear leveling functions. In addition, it provides reliability functions through temperature and voltage monitoring. Flash memories are the main storage components of SSDs.

Fig. 2.8 Structure of an SSD

2.2.1 Flash Memory Cells and Structure

Unlike a conventional complementary metal-oxide-semiconductor (CMOS) cell, a flash memory cell has an additional layer of the floating gate (FG), as shown in Fig. 2.9. Between the FG and substrate is an oxide insulating layer called a tunneling layer. The flash memory cell injects electrons into the FG by applying a voltage. The tunneling layer traps electrons, and the FG stably maintains their states to represent flash memory cells' states.

A flash memory cell reads and writes data by sensing and changing the amount of charge in the FG. Data writing operations inject charge into the FG to form a potential well, which is the "0" data state. An uncharged FG represents the "1" data state. Data reading operations identify "0" and "1" by sensing higher or lower electrical levels on the bit line.

Based on the number of bits stored in each storage cell, flash memory cells are classified into single-level cell (SLC), multi-level cell (MLC), triple-level cell (TLC), and quad-level cell (QLC), as shown in Fig. 2.10. SLC means that a cell stores only 1 bit. It only needs to determine whether a certain amount of charge is stored on the FG. MLC means that a cell stores 2 bits. It needs to determine whether and how much charge is stored in the cell and control the amount of charge programmed for the FG. TLC means that a cell stores 3 bits and QLC means that a cell stores 4 bits. Taking TLC as an example, bits stored in a TLC are classified into a lower bit, upper bit, and extra bit. Reading the lower bit requires only one read voltage, but reading the upper or extra bit requires multiple read voltages.

A flash memory has the following characteristics:

Erase-before-write: In flash memory, the programming of a cell is unidirectional, meaning it only supports writing from "1" to "0," not "0" to "1." Before rewriting a page, the flash memory performs an erase operation. The flash memory is read and written in the unit of a page and erased in the unit of a block.

Different read, write, and erase granularities: The latencies of read, write, and erase operations on the flash memory vary greatly. For a single page, the average latencies for read, write, and erase operations are in the tens of microseconds, hundreds of microseconds, and milliseconds, respectively.

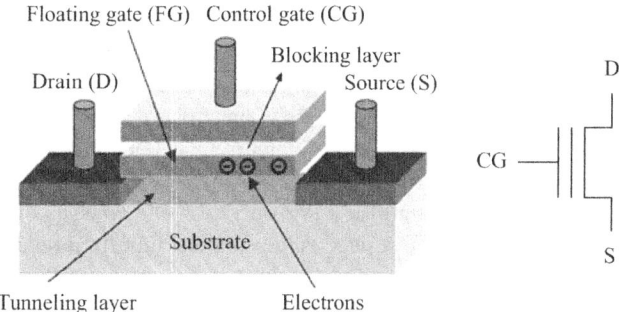

Fig. 2.9 Floating gate memory cell and its schematic symbol [2]

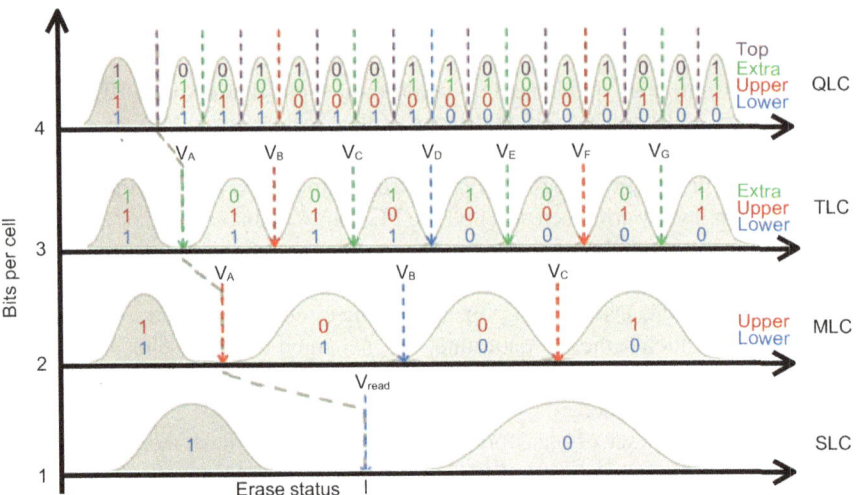

Fig. 2.10 Flash memory cell types [2]

Wearing: A cell can endure a limited number of programming/erase (P/E) cycles, meaning each cell has a limited lifespan. When the number of P/E cycles for a cell is close to the upper limit, the cell cannot reliably store data states. This is called the endurance of flash memory. Although storage density has improved and the price per unit capacity has been reduced, the problem of poor endurance is becoming more and more severe. Each SLC endures 100,000 P/E cycles, each MLC endures 10,000 P/E cycles, and each TLC endures only 1000 P/E cycles.

To prevent latencies caused by P/E operations, flash memory rewrites pages using a remote update policy, meaning it redirects a new page to an idle page and marks the current page as invalid for subsequent reclamation.

Inside an SSD, flash memory chips are connected to flash memory controllers through different channels, as shown in Fig. 2.11. In a flash memory chip, a single chip encapsulates multiple dies, and each die can independently execute instructions. Each die contains multiple planes. Each plane has an independent register and can execute pipeline instructions between planes.

Figure 2.12 shows the internal structure of a NAND chip's target. An encapsulated NAND chip can have multiple targets. Each target is controlled by an independent chip enable signal CE#. Each target can have 1, 2, 4, or 8 logical unit numbers (LUNs)/dies. A LUN is the smallest unit for executing instructions, and different LUNs can execute instructions concurrently. Each LUN can be divided into one or more planes, and each plane corresponds to a group of blocks and a cache. A block is the smallest unit for a P/E operation and consists of several cells controlled by word lines (WLs). A page is the smallest unit for a read/write operation. For the TLC, one WL corresponds to three pages, including data and redundancy (out-of-band data). Different LUNs can concurrently execute instructions, and different planes in the same LUN can also concurrently execute some operations.

2.2 SSDs

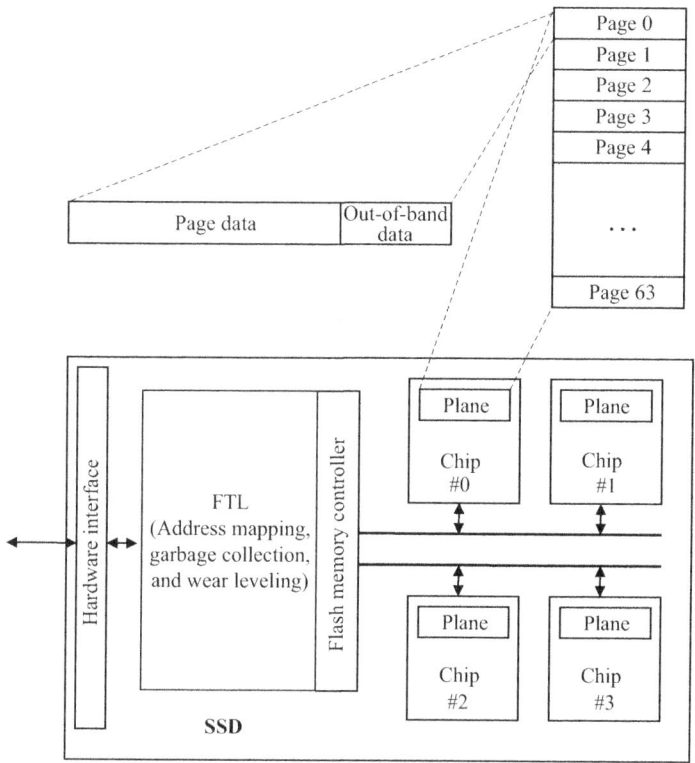

Fig. 2.11 Internal structure of an SSD. (Note: FTL stands for flash translation layer)

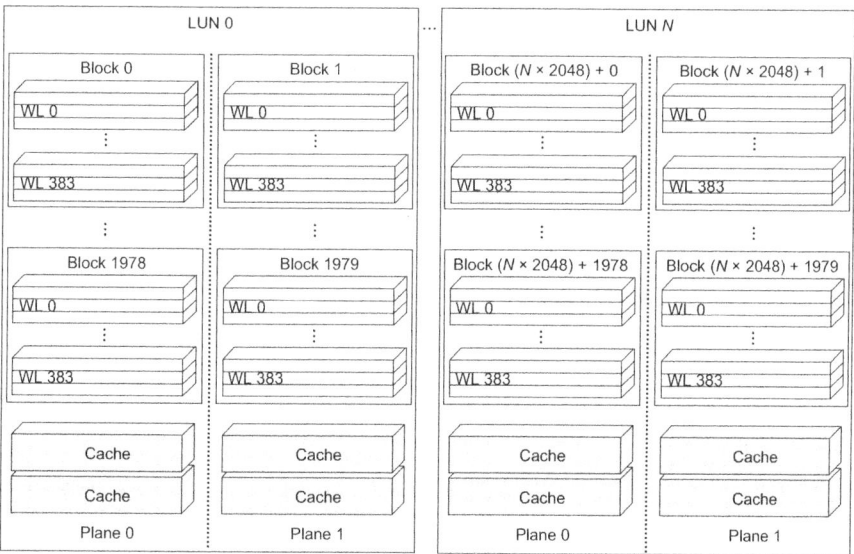

Fig. 2.12 Internal structure of NAND chips

Table 2.1 Performance comparison between SSDs and HDDs [3]

Type	Device model	Read bandwidth/ (MB·s^{-1})	Write bandwidth/ (MB·s^{-1})	Read latency (ms)	Write latency (ms)
HDD	Seagate Savvio	202	202	2.000	2.000
SATA SSD	Intel X25-E	250	170	0.075	0.085
PCI-e SSD	Fusion-io ioDrive octal	6000	4400	0.030	0.030

With different levels of concurrency, SSDs can provide sufficient access bandwidth. This feature is called the internal concurrency of SSDs. Table 2.1 compares the performance between SSDs and HDDs.

2.2.2 FTL

An SSD uses FTL to manage the read, write, and erase operations of flash memory and provides read and write interfaces for software systems. FTL offers functions such as address mapping, garbage collection, and wear leveling. In addition, other functions such as error correcting code (ECC) and bad block management are supported within the SSD.

2.2.2.1 Address Mapping

Address mapping records the mapping between the SSD's logical address and flash memory's physical address and supports remote updating of the flash memory. It is categorized into three types: page-level, block-level, and hybrid address mappings.

Page-Level Address Mapping

This is used to perform remapping at a page granularity, preventing page replication during block merging but requiring a large mapping table.

In page-level address mapping, an entry is created for each logical page number (LPN) and mapped to a physical page number (PPN) of an SSD, as shown in Fig. 2.13. The advantage of this is that each LPN can be mapped to any PPN. However, the overhead of a mapping table is very high. The mapping table's size in this approach is calculated as follows: total SSD capacity/size of a single PPN × size of each mapping.

As typically each page is 4 KB and a mapping needs several bytes to be stored, page-level address mapping may require a mapping table one-thousandth the size of the total flash memory.

2.2 SSDs

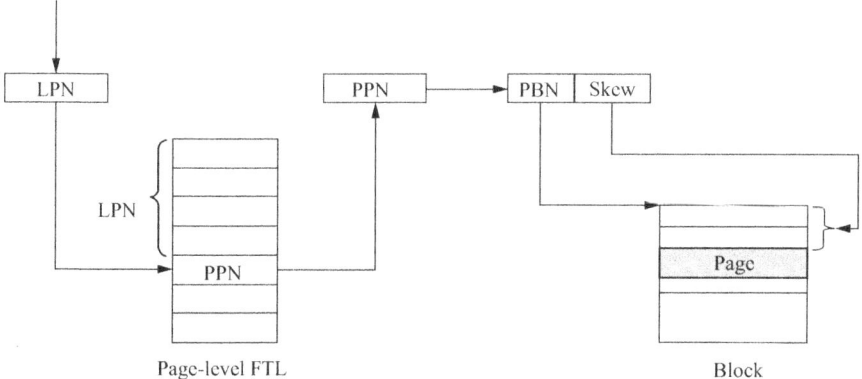

Fig. 2.13 Page-level address mapping

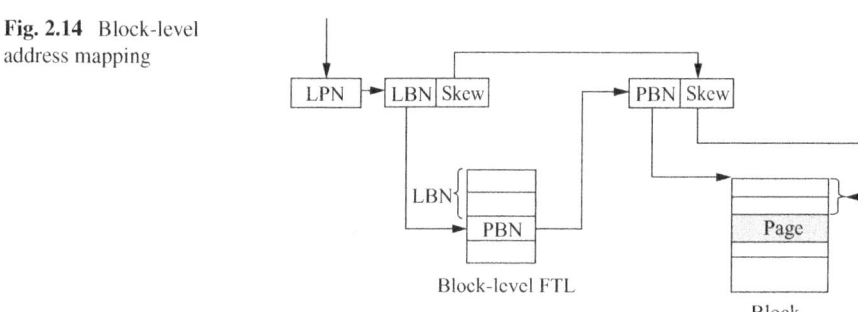

Fig. 2.14 Block-level address mapping

Block-Level Address Mapping

This is used to perform remapping at a block granularity. The advantage of this is that the mapping table is relatively small. However, to keep data in the blocks in the same sequence, the blocks need to be merged, causing unnecessary page replications.

In addition, since mapping is performed at a block granularity, a logical address is divided into a logical block number (LBN) and an in-block skew during address translation. The LBN is translated to a physical block number (PBN) using FTL entries, and then the in-block skew is added to obtain a physical page-level address, as shown in Fig. 2.14. Pages with the same LBN must also be on the same physical block, which causes a large number of pages to be moved when the mapping changes.

Hybrid Address Mapping

It is a compromise between page-level and block-level address mappings and has different hybrid modes. Generally, it uses page-level address mapping to store new or hot data and block-level address mapping to store old or cold data.

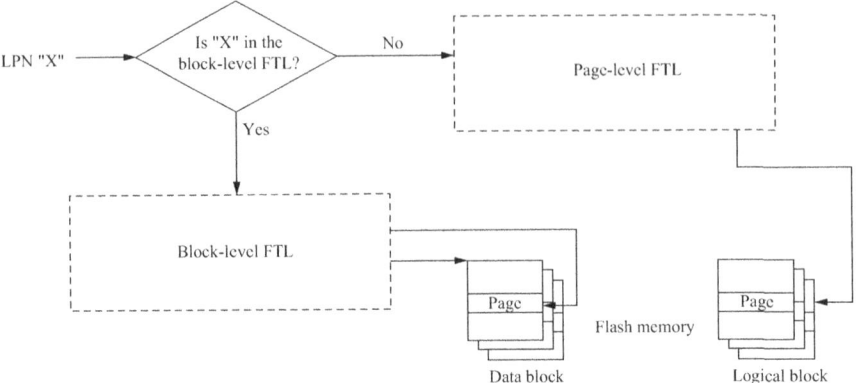

Fig. 2.15 Hybrid address mapping

Data updates are first written into log blocks as logs. When log blocks are used up, valid data in these blocks is merged and then written into data blocks. For frequent read and write operations on some files, the log block design reduces the number of erases and erase overhead. The hybrid mapping mechanism effectively reduces the mapping table's size, as shown in Fig. 2.15.

2.2.2.2 Garbage Collection

This is used for selecting and erasing invalid pages to restore idle states and wait for new data to be written. Garbage collection can be performed in the foreground and background. Foreground garbage collection refers to forcible garbage collection triggered when the idle space of an SSD is less than a specified threshold. Background garbage collection occurs when garbage collection threads are periodically started to erase invalid blocks. Both methods can be implemented simultaneously in a single FTL. During garbage collection, the FTL first selects a block to be erased (victim block), then copies valid pages in the block to the idle pages of another block, and finally erases the block. Moving valid pages causes extra writes inside the SSD, which not only introduces extra latency, affecting flash memory performance, but also increases wear, shortening the flash memory's lifespan. This is also referred to as a write amplification problem of an SSD. To mitigate this problem, when selecting a block to be erased, the FTL chooses one with as few valid pages as possible for garbage collection.

The flash memory is erased at a block granularity, but the overhead of garbage collection exceeds that of erasing a single block. The flash memory is read and written at a page granularity, and a subsequent write operation can only be performed after a page is erased. Since the reads and writes of different pages in the same block are not entirely uniform, a block may contain both valid and invalid pages, and it can be erased after the valid pages are relocated to another block. In this scenario,

2.2 SSDs

the overhead of garbage collection involves moving valid pages and erasing the block.

2.2.2.3 Wear Leveling

Because flash memory has endurance limitations, ensuring the reliable storage of data in an SSD involves marking a page with P/E cycles reaching a specified value as invalid. The lifespan of an SSD refers to how long it can provide sufficient available space. To prolong the SSD's lifespan, the FTL employs a wear leveling policy to distribute P/E cycles evenly across all pages as much as possible. There are two types of wear leveling policies: static and dynamic. Static wear leveling involves selecting blocks with fewer P/E cycles from all blocks (including both idle and used blocks) for space allocation and data writing. Dynamic wear leveling, on the other hand, selects blocks with fewer P/E cycles only from idle blocks for space allocation and data writing.

2.2.2.4 ECC

The BCH algorithm was invented by and named after three mathematicians: Bose, Ray-Chaudhuri, and Hocquenghem. During data writing, the BCH algorithm uses an algebraic formula to sequentially and cyclically encode the original data and store it in NAND media. During data reading, the BCH algorithm applies mathematical principles to cyclically calculate and read data, determining its correctness and correctable errors. The core of the BCH algorithm lies in using the original data to establish a polynomial code *CI*, calculating and generating redundant data *CR*, and establishing a strict verification relationship between the redundant data and encoded data, ensuring that their sum equals 0. When a small number of errors occur in the encoded data, the error data can be recovered by solving polynomial equations.

LDPC is essentially a linear ECC, capable of linearizing encoding/decoding time and code length. It employs a sparse matrix and performs iterative computation to implement information redundancy and error correction. LDPC stands out from other information encoding algorithms in that it uses both hard decision and soft decision schemes. In the LDPC hard decision scheme, a single read voltage is used to determine the likelihood of 0 or 1 for the NAND media, enabling linear data decoding. Conversely, in the soft decision scheme, the received information is represented as a log-likelihood ratio (LLR) sequence. Each real number in the LLR sequence indicates a probability value of the corresponding bit being 0 or 1. A positive number signifies a higher probability value of the bit being 0, while a negative number represents a higher probability value of the bit being 1. The LDPC decoder assigns 0 for positive values or 0 in the LLR sequence and assigns 1 for negative values, generating a sequence comprising 0 s and 1 s. Then, the LDPC decoder multiplies this sequence with the verification matrix. If the resulting sequence consists entirely of 0 s, the codes are deemed correct, and the iteration concludes. Otherwise, the iteration continues until the correct codes are attained or the maximum number of iterations is reached. LDPC soft decision dynamically adjusts the quantization level of the read voltage within the NAND media to obtain multiple possible values such as 0 or 1, maximizing the utilization of the channel's effective soft information. Multiple soft decisions can significantly improve the coding signal-to-noise ratio (SNR) gain, thereby boosting the decoding success rate of the LDPC algorithm. LDPC algorithms based on NAND flash media can enhance media error tolerance, thereby elevating SSD reliability.

2.3 Main Memory

Main memory, or simply memory, is directly connected to the CPU via a high-speed memory bus. The CPU accesses data from memory and writes data onto it using load/store instructions. Currently, main memory commonly uses DRAM media, although other emerging byte-type nonvolatile memories include phase-change memory (PCM), resistive random-access memory (RRAM), and magnetroresistive random-access memory (MRAM). The following information will primarily focus on DRAM and provide a brief overview of nonvolatile memory.

2.3.1 DRAM Components and Structure

DRAM uses a MOS structure to store 1 bit of binary data. Currently, commonly used DRAM uses a 1T1C structure, comprising one transistor and one capacitor, as shown in Fig. 2.16. Data is stored in the parasitic capacitor C of the source of the MOS transistor T. For example, a charged capacitor C represents "1," while an uncharged one represents "0."

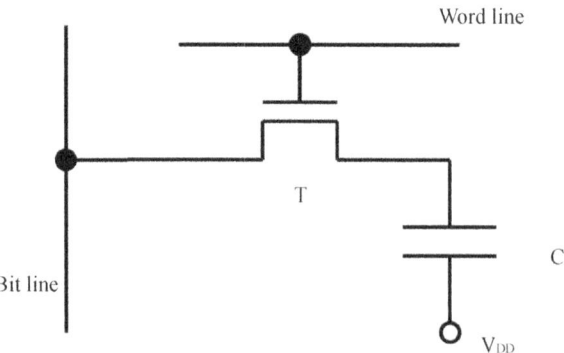

Fig. 2.16 Example of DRAM cell structure

2.3 Main Memory

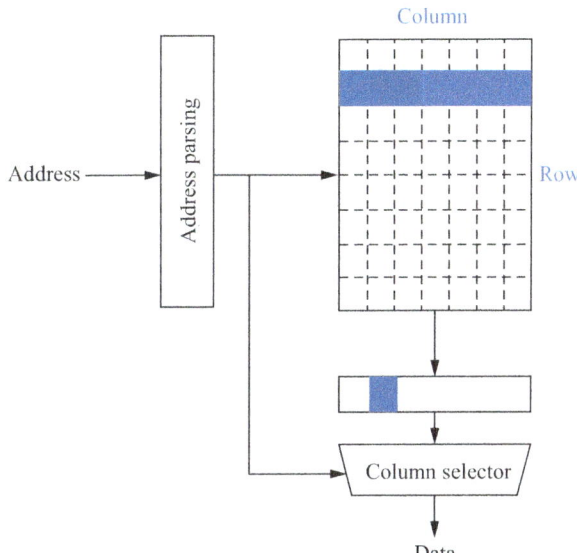

Fig. 2.17 2D row-column structure of DRAM

Each DRAM cell is organized in a 2D row-column structure ($2^m \times 2^n$), as shown in Fig. 2.17. The data address of a DRAM read/write request comprises a row address (m) and a column address (n). Initially, DRAM parses the row address, selecting a row (also known as a page) within the DRAM. Subsequently, it reads data from the row in the structure and stores it in a row buffer via a sense amplifier. Then, the column address is parsed, and the corresponding data block is selected from the row buffer for transmission.

As described above, DRAM components are organized into banks to perform read/write operations. These banks combine to form chips, which in turn constitute ranks, dual in-line memory modules (DIMMs), and channels sequentially. The channels are interconnected with CPUs. Figure 2.18 demonstrates the components and structure of DRAM from the CPU's perspective.

A DRAM chip contains several banks, with different banks in the same chip sharing control lines, address lines, and data lines. Due to limitations in chip design area, the number of chip pins takes precedence. Typically, each chip supports a bit width of 4–16.

To support larger bit widths, DRAM organizes different chips into a rank structure. Different chips within each rank can then perform read and write operations simultaneously, resulting in a larger bit width. Different chips in the same rank share the same address and control lines but provide different data. For example, if a chip has a bit width of 8, DRAM can arrange eight chips in a rank to read and write simultaneously, thus supporting 64-bit wide read and write.

A physical memory module of DRAM, such as a DIMM, incorporates one or more ranks. For example, a DIMM with two sides has eight chips on each side, forming a rank, resulting in the DIMM having two ranks in total.

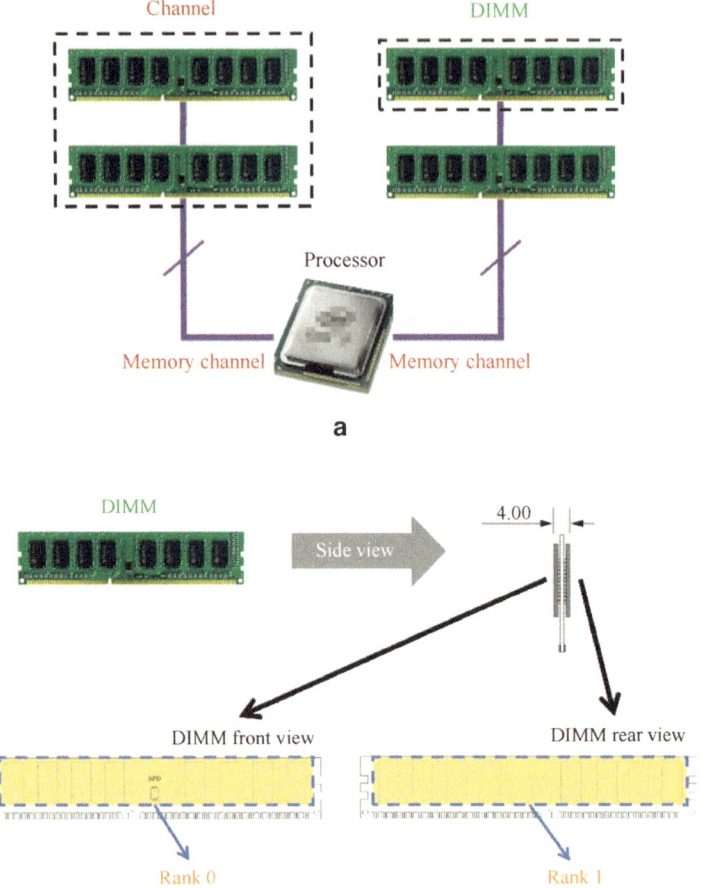

Fig. 2.18 Hierarchical structure of DRAM. (**a**) Channel structure. (**b**) DIMM structure. (**c**) Rank structure. (**d**) Chip structure

2.3 Main Memory

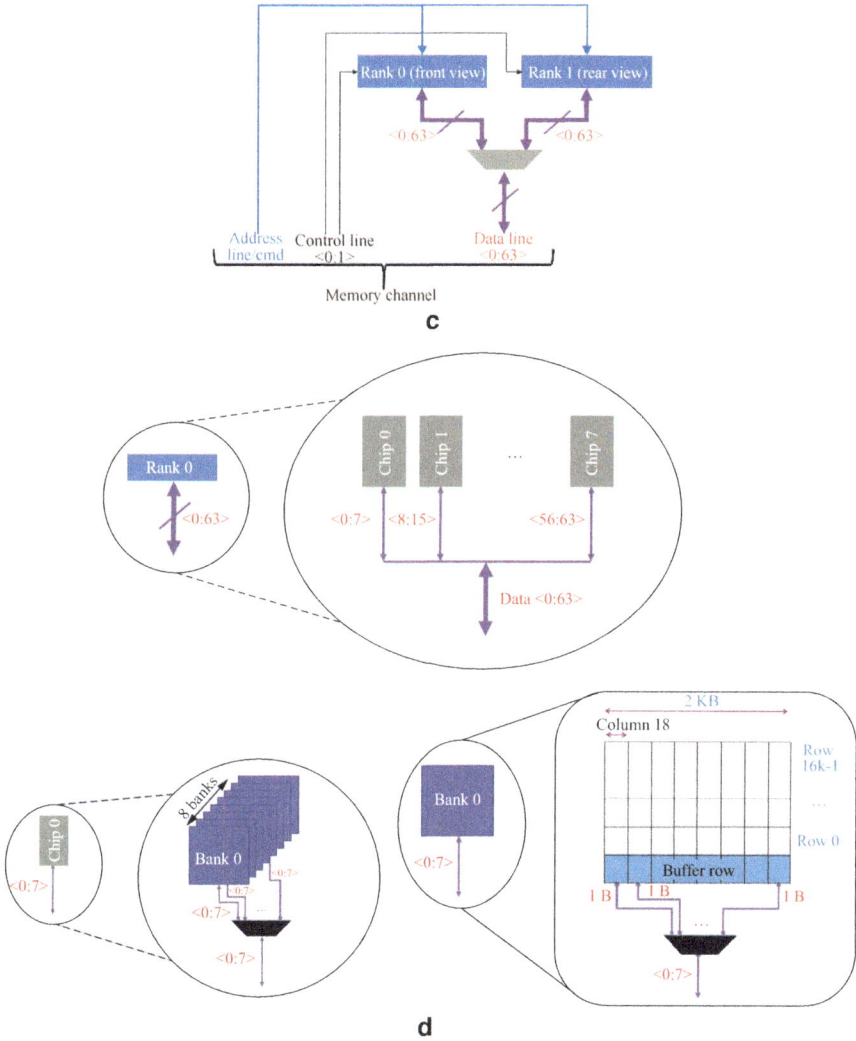

Fig. 2.18 (continued)

A physical memory module is a physical device that can be directly inserted into a motherboard and is connected to a CPU via a bus. To improve memory access performance, a CPU can connect to a physical memory module through multiple buses, each of which is also referred to as a channel.

Currently, connections between CPUs and DRAM mainly use the double data rate (DDR) synchronous memory interface. The data bandwidth of DDR is determined by the frequency and bit width of the data bus. This data bus is used to transfer data between the functional components of a computer. For a data bus, the product of its bit width (bus width) and clock frequency is directly proportional to

Table 2.2 DDR bandwidth

Model	Memory clock frequency (MHz)	I/O bus clock frequency (MHz)	Transfer rate (MT·s^{-1})	Theoretical bandwidth (GB·s^{-1})
DDR-200, PC-1600	100	100	200	1.6
DDR-400, PC-3200	200	200	400	3.2
DDR2–800, PC2–6400	200	400	800	6.4
DDR3–1600, PC3–12800	200	800	1600	12.8
DDR4–2400, PC4–19200	300	1200	2400	19.2
DDR4–3200, PC4–25600	400	1600	3200	25.6
DDR5–4800, PC5–38400	300	2400	4800	38.4
DDR5–6400, PC5–51200	400	3200	6400	51.2

Note: MT·s^{-1} stands for million transfers per second

the maximum data throughput (input/output) capability supported by the bus. Table 2.2 provides a summary of DDR bandwidth.

2.3.2 DRAM Refresh

DRAM records the "0" and "1" states based on the number of electrons stored on the capacitor. However, when DRAM reads a memory cell, electrons on the capacitor in that cell gradually escape over time, a phenomenon referred to as destructive read. Therefore, DRAM must periodically read and then write data into the cell, a process known as refresh.

Destructive read requires DRAM to rewrite the original data after reading it. Within the DRAM, the data held in banks is read into the row buffer through a read amplifier, causing the data in the original DRAM row to be lost. After the read is completed, the DRAM must write data from the row buffer back to the original DRAM row, a procedure known as pre-charge. Then, the previous data is retained in the original row after being read.

The ability of DRAM cells to retain data over time is known as retention. In order to retain data correctly, DRAM periodically reads and rewrites data in DRAM rows via refresh operations. For example, if the retention of DRAM cells generally exceeds 64 ms, DRAM must ensure that each row is refreshed within a 64-ms interval.

DRAM refresh operations include centralized and distributed refreshes, as shown in Fig. 2.19. Centralized refresh means stopping external read/write operations periodically, allowing memory controllers to refresh memory rows sequentially. This method is simple to implement but comes with the disadvantage of high latency in external reads and writes. Conversely, distributed refresh alternates between refreshing memory rows and performing external read/write operations, evenly distributing these operations throughout the refresh interval.

2.3 Main Memory

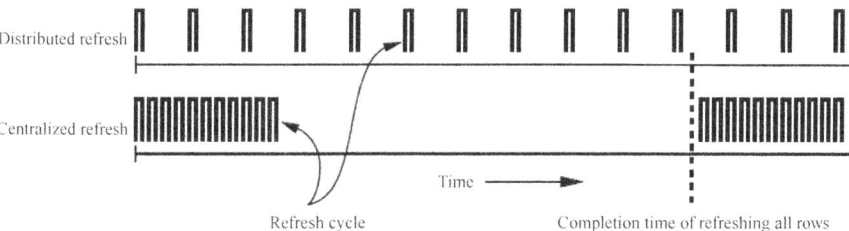

Fig. 2.19 Example of DRAM refresh

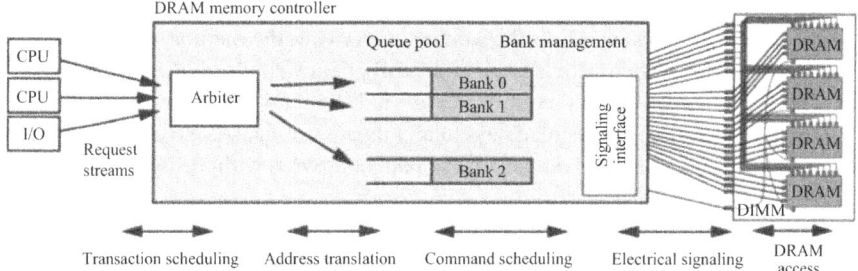

Fig. 2.20 DRAM memory controller structure [1]

2.3.3 Memory Controllers

A memory controller schedules and distributes memory read and write requests sent by a CPU to memory chips. This process involves transaction scheduling, address translation, command scheduling, and electrical signaling, as shown in Fig. 2.20. In addition to external DRAM read and write requests, a DRAM memory controller manages refresh operations.

The memory controller can be deployed either in a memory module or a CPU. Deploying a memory controller in a memory module can support the features of different types of memory media and reduce the energy consumption of a CPU chip. However, the disadvantage is that access latency increases. In addition, due to missing semantic information on the host side, less scheduling information is contained in memory access requests between multiple cores, adversely affecting memory bandwidth. In comparison, deploying a memory controller in a CPU has the opposite advantages and disadvantages.

In a memory controller, the scheduling of memory requests is an important factor that affects memory performance. Memory request scheduling shares similar objectives with all other scheduling policies: to improve throughput, reduce latency, and ensure fairness. However, unlike other policies, this scheduling must consider memory features. Specific factors that must be considered include the row buffer hit or miss state of a request, the request arrival sequence, the request type (prefetch, read, or write), and request priority.

The request's row buffer hit or miss state is a unique factor to be considered during memory scheduling. DRAM has two basic policies for row buffer management: open row and closed row.

With the open row policy, after a memory row is accessed, data is retained in the row buffer and does not need to be written back (i.e., pre-charged) to the memory row. Under this policy, if the next request accesses the same memory row, the row buffer is hit, eliminating the overhead of write-back. If the next request accesses a different row, the row buffer is missed. In this case, write-back is performed, a new row is activated, and a data access operation is performed, resulting in high latency.

With the closed row policy, after each memory access, the row buffer will write the accessed memory row's data back to its original memory row. Under this policy, even if the next request accesses the same memory row, the memory row still needs to be activated first, then data is read, and finally data write-back is executed, causing a high latency. If the next request accesses a different row, the memory row must be activated first, then data is read, and finally data write-back is executed, eliminating the write-back overhead on the critical path compared with the open row policy.

2.3.4 Nonvolatile Memory

In recent years, advancements in materials, devices, and related technologies have fueled the emergence and application of nonvolatile memory, which has influenced the structure and performance of computer systems and storage systems. The following information will briefly describe PCM, RRAM, and MRAM.

2.3.4.1 PCM

PCM is a type of non-volatile memory that uses phase change materials as the storage media. The image on the left in Fig. 2.21 shows the basic structure of a PCM cell. This structure consists of two metal layers, sulfur compounds, and a heater between the layers. The image on the right in Fig. 2.21 shows an example of the cell's structure.

The read and write operations to the PCM are performed by applying a current pulse of specific intensity to the cells for a certain duration. A write operation includes a set (writing "1") and a reset (writing "0"). During a set, the heater applies a low-intensity current pulse to a PCM cell for a long duration (see Fig. 2.22) to raise its temperature between the crystallization point (about 300 °C) and melting point (about 600 °C) of sulfur compounds. This causes part of the programming area of phase-change materials to become crystalline, storing binary data "1." During a reset, the heater applies a high-intensity current pulse to a PCM cell for a short duration to raise its temperature higher than the melting point of sulfur compounds. This causes the programming area to become amorphous, storing binary data "0." Because the resistance of PCM cells in the set state differs

2.3 Main Memory

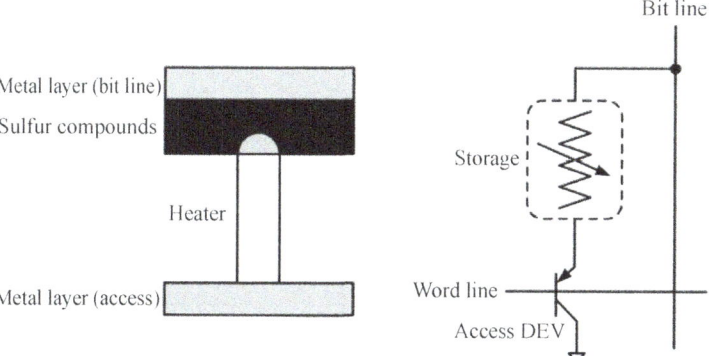

Fig. 2.21 PCM cell and its structure

Fig. 2.22 Current pulse needed for reads/writes on PCM cells

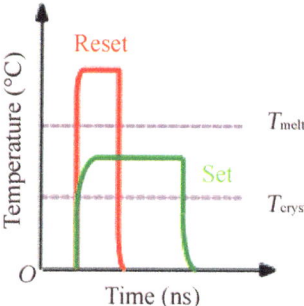

greatly from that in the reset state, binary data can be distinguished by using a low-intensity current pulse. Specifically, during a read operation, the heater applies a low-intensity current pulse to a PCM cell for a short duration to obtain the cell's resistance value. This value is used to distinguish "1" or "0." In addition, to avoid reaching the crystallization or melting point of sulfur compounds, the intensity and duration of the current pulse used for a read operation are less than those of a write operation.

The characteristics of PCM are as follows: high integration, good space scalability, and the ability to continue reducing the cell size below 20-nm technology. It outperforms DRAM in terms of storage density. Furthermore, its access latency is similar to DRAM, but its static power consumption is extremely low.

2.3.4.2 RRAM

RRAM is a type of nonvolatile memory that stores data by changing the resistance of cells. Figure 2.23 shows the basic structure of an RRAM cell, which consists of the top electrode, bottom electrode, and the metal-oxide layer between them. By applying an external voltage, the RRAM cell can switch between low-and

Fig. 2.23 Basic structure of an RRAM cell

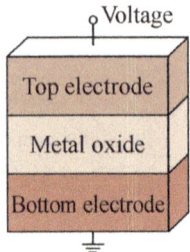

Fig. 2.24 I-V curve of the RRAM cell [4]

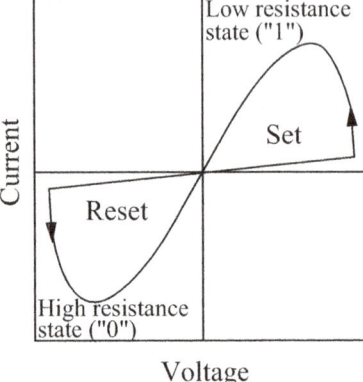

high-resistance states, which are used to represent the stored binary data "1" and "0," respectively.

Figure 2.24 shows the current-voltage (I-V) curve of the RRAM cell. When a positive voltage is applied to a cell, it switches from a high-resistance state (logic "0") to a low-resistance state (logic "1"), known as a set operation. Conversely, when a negative voltage is applied, the cell switches from a low-resistance state (logic "1") to a high-resistance state (logic "0"), known as a reset operation. When data is read from a cell, only a small read voltage, which will not affect the memory state, is applied to detect whether the cell is in a high-or low-resistance state.

RRAM has the advantage of high storage density, as one cell can store 1 bit (SLC) or 2 bits (MLC) of data. In addition, the minimum area of a cell can be 4F2. However, while RRAM has a read speed equivalent to DRAM, its write speed is much slower.

2.3.4.3 MRAM

MRAM is a type of nonvolatile memory that uses magnetic resistance to store data. The image on the left in Fig. 2.25 shows the basic structure of an MRAM cell: a sandwich of two ferromagnetic layers separated by a tunneling barrier layer (an insulating material). One layer is called the reference layer, with its magnetization

2.3 Main Memory 37

direction fixed along the easy axis. The other ferromagnetic layer is called the free layer, with its magnetization direction having two stable orientations: parallel or antiparallel to the reference layer. The image on the right in Fig. 2.25 depicts this structure.

Due to the quantum tunneling effect, a small current can flow through the thin insulating layer of a magnetic memory. When an electron crosses the insulator barrier while keeping its spin direction unchanged, the magnetic moments of the two layers of magnetic materials are parallel (as shown in the upper part of Fig. 2.26), resulting in a low resistance state that represents binary data "0." Conversely, when the magnetic moments are antiparallel (as shown in the lower part of Fig. 2.26), the materials exhibit a high resistance state, representing binary data "1." When data is read, data bit "0" or "1" is determined based on whether the magnetization directions are consistent.

In terms of performance, MRAM has high-speed read/write capabilities similar to static random-access memory (SRAM). As for reliability, MRAM exhibits exceptional robustness due to its inherent radiation resistance, rendering its cells immune

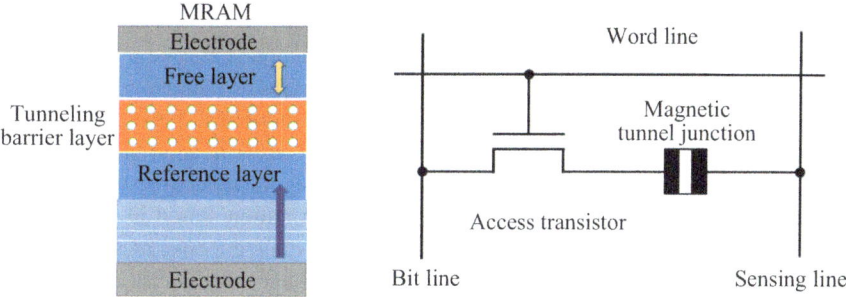

Fig. 2.25 MRAM cell and its structure

Fig. 2.26 Data storage logic of MRAM

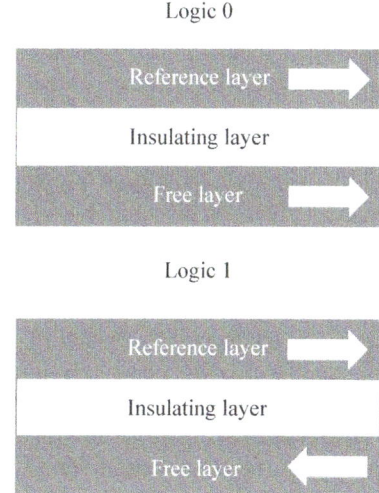

to soft errors. Regarding storage density, MRAM can achieve a level comparable to DRAM. In addition, MRAM consumes less power than DRAM, as it lacks electric leakage and does not require additional energy for refreshing.

2.4 Other Storage Media

2.4.1 Optical Storage

Optical storage uses lasers to irradiate media, inducing physical or chemical changes that alter certain properties of the media to represent different data. Today, optical storage encompasses CDs, DVDs, BDs, and ADs. Optical discs are ideal for single-write and multiple-read operations and are typically used to store data that is infrequently rewritten, such as archived data and cold data. To minimize costs, instead of using CDs and DVDs, storage systems generally use optical discs with high storage density, such as BDs and other high-density optical media types that have emerged in recent years.

The core components of optical storage systems are optical discs and optical disc drives capable of loading optical discs. Optical disc drives use semiconductor lasers and reflected light to write data onto and read data from optical discs, respectively. By using lasers, optical storage enables contactless read/write and random access, as lasers can freely traverse the surface of a disc during the read/write process.

Early on, optical discs were read-only, with information stored in the form of shallow pits embossed on a polymer surface coated with a reflective metal film. The data of an optical disc is read by identifying the changes in the reflected light to determine the presence or absence of pits, representing bit "0" or "1." A read-only optical disc satisfies copyright-compliant distribution needs for both audio and video files but falls short in terms of backup. This led to the emergence of the write once read many (WORM) optical disc. The WORM optical disc reads information from changes in optical properties such as reflectivity and absorptivity. Its writing principle involves using a focused laser beam to create holes in the surface of the optical disc as permanent marks.

Today, optical storage includes CDs, DVDs, BDs, and ADs. CDs appeared in the 1970s, mainly in the format of compact disc read-only memory (CD-ROM). A CD is about 12 cm in diameter, has a capacity of about 150 MB, and provides a read/write bandwidth of about 4.3 Mbit/s. Later, other formats of CDs like compact disc recordable (CD-R) and compact disc rewritable (CD-RW) emerged. A DVD has a higher capacity than a CD, with each DVD layer holding up to 4.7 GB of data and providing a bandwidth of about 11 Mbit/s. DVDs come in several formats, including dual-layer DVDs that nearly double the capacity, double-sided dual-layer DVDs that offer higher capacity, and writable and rewritable DVDs. A BD is about 12 cm in diameter, storing up to 25 GB of data per layer and 50 GB in dual layers and providing a bandwidth of 36 Mbit/s. BDs are in a rewritable optical disc format. An AD is a double-sided storage and has three storage layers for each side, meaning six

2.4 Other Storage Media

layers in total. Each storage layer is in a sandwich structure, consisting of a storage layer sandwiched between two dielectric protection layers. Oxide storage materials have resulted in significant improvements in storage rate and capacity, further extending disc durability. Each storage layer of an optical disc contains a series of concentric spiral grooves and lands, while earlier optical discs stored data only on grooves. To maximize the storage capacity of each layer, ADs use both grooves and lands to store data, doubling the storage density of a single layer and achieving a significant increase in storage capacity. The capacity per disc for a first-generation AD is up to 300 GB. ADs can also withstand temperature and humidity changes and are waterproof and dustproof, ensuring stable data storage for 50 years.

2.4.2 Tapes

A tape is a magnetic, high-capacity storage medium that supports sequential read but not random access.

A tape library is a device consisting of multiple tapes, robotic arms, drives, tape slots, and barcode readers. Robotic arms are used to remove and add tapes, moving them into corresponding slots. A tape library can contain multiple drives and support the concurrent work of multiple servers. A tape drive performs read/write operations for a single tape, and each tape is marked with a barcode label. A barcode reader identifies a single tape by scanning its barcode.

The advantages of tapes are high capacities, low costs per unit capacity, energy efficiency, and high reliability. Currently, a single tape can store about 15 TB of data, and a tape library can store hundreds of TB of data. Once data is written to tapes, no power supply is required as long as data is not being read or written, resulting in low power consumption. In addition, the error rate of tapes is relatively low.

However, tapes do not support random reads and writes. They are suitable for sequential writes of backup data but not for random reads. Tapes used in open environments are susceptible to temperature, humidity, and dust, which can lead to issues such as tape wear. Table 2.3 compares three types of backup media for cold storage.

Table 2.3 Comparison between backup media for cold storage

Item	HDD	Tape	Blu-ray storage
Media cost	$0.0128/GB	$0.013/GB	$0.017/GB
Access latency	10 ms	60 s	1 s
Read/write bandwidth	150 MB/s	300 MB/s	45 MB/s
Storage duration	5 years	10 years	50 years
Storage environment	Air conditioners required	Air conditioners required	No air conditioner required
30-year power consumption for 100 TB of data read/write	108,000 kWh	3500 kWh	3200 kWh

2.5 Summary

In a computer, a storage system virtualizes and unifies different types of storage media according to a storage hierarchy. Throughout the development history of computers, various types of storage media have emerged, ranging from high capacity to high performance, to meet different requirements. Currently, traditional magnetic disk storage continues to evolve toward higher density and capacity, while new nonvolatile memory technologies are advancing toward lower latency and higher bandwidth. Serving as the hardware foundation of storage systems, storage media are driving advancements in both storage and computer systems. Looking forward, the emergence of new storage media will present new opportunities and challenges for the design of new storage systems.

2.6 Practice Questions

1. **The sequential read/write performance of HDDs is much better than that of random read/write. How does the physical structure of the HDDs contribute to this?**
    Answer: An HDD must perform seek and rotation operations to access data. Random reads and writes require an HDD to frequently perform seek and rotation operations, severely affecting the access performance.
2. **What is the biggest disadvantage of the Shortest Seek Time First (SSTF) algorithm for HDD scheduling? How can we overcome this disadvantage?**
    Answer: The biggest disadvantage of SSTF is that it cannot ensure fairness because some requests are ignored and will not receive a response. This disadvantage can be overcome by combining SSTF with FCFS. Specifically, an HDD groups requests according to a request sequence, with each group containing several requests. The SSTF algorithm is used between requests in the same group, and the FCFS algorithm is used between groups.
3. **What are the major differences between NOR and NAND flash memories?**
    Answer: NOR flash memory supports random access. It writes slowly and has a small capacity, but it is very reliable. It is often used to store firmware codes. NAND flash memory is accessed using blocks. It has a high capacity but is not very reliable. It is often used to store large amounts of data.
4. **Flash memory stores more bits in each cell to improve storage density. What are the negative effects of increasing storage density?**
    Answer: If the number of bits stored in each cell increases, the theoretical number of P/E cycles of the flash memory decreases, shortening the flash memory's lifespan. In addition, the read/write performance deteriorates.
5. **Both HDDs and SSDs perform address mapping internally. Why does each of them perform this function?**
    Answer: Both HDDs and SSDs perform address mapping to handle sector failures and improve data access. SSDs also do it to support remote updates, garbage collection, and wear leveling.

6. **Each page in the flash memory has a small spare area called an out-of-band (OOB) area. What are its functions?**

    Answer: The OOB area is used to store error-correcting codes (ECCs) of the pages and metadata required by upper-layer storage software. For example, in a file system, the metadata of the file to which a data block belongs is stored in the OOB.

7. **Assume that a computer contains both SSD and PCM. Could you propose data placement policies and tell us which data should be stored in the SSD and which should be in the PCM?**

    Answer: The first policy would be that frequently accessed (or hot) data is stored in the PCM, and the rest is stored in the SSD. This is because the PCM offers better performance but smaller capacity than the SSD. The second policy would be that small-sized data is stored in the PCM and large-sized data is stored in the SSD because the SSD is accessed at a block-level granularity, whereas the PCM is accessed at byte-level granularity.

8. **Could you suggest a method to reduce the number of refresh operations required during DRAM running?**

    Answer: The DRAM ensures data correctness through refresh operations. However, not all DRAM cells contain valid data. To reduce the number of refresh operations required during DRAM running, we could enable the DRAM to also record data validity and only periodically refresh the rows containing valid data.

References

1. Jacob B, Wang D, Ng S. Memory systems: cache, DRAM, disk. Burlington: Morgan Kaufmann; 2010.
2. Rino M, Crippa L, Marelli A. Inside NAND flash memories. Dordrecht: Springer; 2010.
3. Youyou L, Jiwu S. Survey on flash-based storage systems. J Comput Res Dev. 2013;50(1):49–59.
4. Zahoor F, Azni Zulkifli TZ, Khanday FA. Resistive random access memory (RRAM): an overview of materials, switching mechanism, performance, multilevel cell (MLC) storage, modeling, and applications. Nanoscale Res Lett. 2020;15:1–26.

Chapter 3
Storage Arrays

Storage arrays have evolved with the development of information technology (IT) and new data requirements [1]. The earliest storage devices were the disks in computers and servers. As data volume increased, one or multiple individual disks were unable to meet the capacity requirements of applications. In addition, the high failure rate of disks made them unsuitable for the increasing need for reliability. Therefore, it became standard practice to separate disks from servers and manage them in a centralized pool. This pool serves as a unified storage space and provides data access and storage services for all hosts. Just a bunch of disks (JOBD) was the earliest example of such a storage array [2].

As the main component of an external storage system, a storage array consists of software systems and hardware modules, such as disks, controllers, and interface modules. An engine (controller enclosure unit) includes two or more controllers and shared disk enclosures. The capacity of the storage array can be scaled up by adding disk enclosures, while scaling out controllers can enhance performance and expand specifications. Redundancy technologies such as RAID are used to implement disk fault tolerance and data layout optimization, as shown in Fig. 3.1.

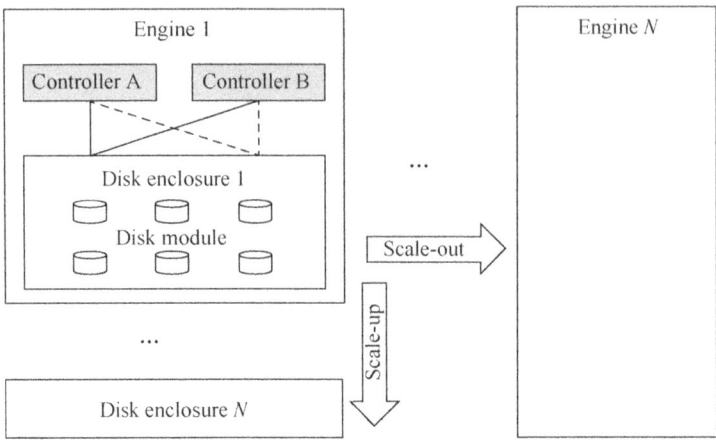

Fig. 3.1 Storage array

3.1 Hardware Architecture

A storage array is a cluster storage system deployed based on a dual-controller redundancy architecture. The cluster storage system must deliver high concurrency for data inputs and outputs; support user access to storage media, such as disks, at any time and place; and provide abundant computing power for various advanced functions and features in the storage array. The development of media technologies has laid the groundwork for the evolution of storage array hardware. Original disk arrays have evolved from HDD-only to hybrid modes (HDDs and SSDs), and recently, all-flash storage arrays have emerged as a prominent alternative. Despite the differences in the underlying media of these storage arrays, their basic hardware architectures are essentially the same.

3.1.1 System Architecture

Storage array hardware consists of several components, including the chassis, interconnection backplane, controller module, interface module, disk enclosure, disk module, heat dissipation module, and power module. As shown in Fig. 3.2, each component has specific functions and is indispensable for the storage array. For example, the power module supports power conversion, the disk module provides storage spaces, and the heat dissipation module ensures efficient heat dissipation for the array.

3.1 Hardware Architecture 45

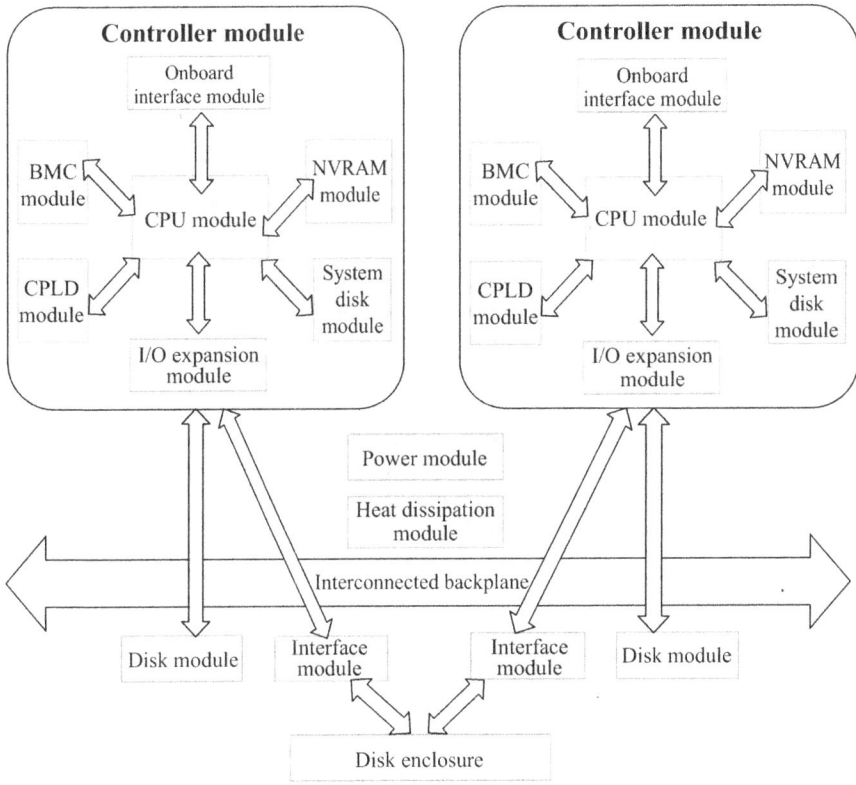

Fig. 3.2 Architecture of the entire system

3.1.2 Controller Module

The controller module is the core component of the storage array, providing computing and management capabilities. It consists of several components, including a central processing unit (CPU) module, input/output (I/O) expansion module, onboard interface module, bubble memory controller (BMC), complex programmable logic device (CPLD), system disk module, and nonvolatile random-access memory (NVRAM) module, among others.

The CPU module contains the CPU and the memory. Currently, CPUs use the x86 or ARM architecture, and Double Data Rate 4 (DDR4) is the most predominant memory type. The interaction between CPUs and memory enables the computing behind high-performance services of storage arrays. Each CPU also provides various I/O interfaces to connect to front-end hosts, back-end disks, and management chips. These interfaces include Peripheral Component Interconnect Express (PCIe) 3.0, PCIe 4.0, Serial-Attached SCSI (SAS) 3.0, Serial Advanced Technology Attachment (SATA) 3.0, Serial Gigabit Media-Independent Interface (SGMII), Serial Peripheral Interface (SPI), local bus, and low pin count (LPC) bus, as well as other bus interfaces.

The I/O expansion module is used to expand I/O interfaces of the controller module. When the CPUs are unable to provide enough PCIe channels, the I/O expansion module can be used to expand the channels, supporting higher system specifications and capabilities.

Both the CPU module and I/O expansion module support the data mirroring function for mutual data backup between the two controller modules in the storage array. If either controller module fails, the other one can take over the services to ensure a high level of system reliability.

The onboard interface module is directly integrated into the controller module.

The BMC module is responsible for the out-of-band management of the controller modules and the entire storage array. The out-of-band management function oversees basic system information, power modules, and heat dissipation modules. It also supports component temperature monitoring, board power-on and power-off, voltage monitoring, error monitoring, and firmware upgrade functions.

The CPLD module provides functions such as power-on and power-off control, reset control, interface conversion, external serial port management, clock monitoring, system watchdog, dual-BIOS switchover, hot swap status management of disk modules, and indicator control. The BMC and CPLD modules are combined to manage the hardware system.

The system disk module is used to store the operating system of the storage array.

The NVRAM module is used to store data when the storage array is powered off unexpectedly. Data in the storage array is transmitted or stored among the CPU cache, memory, and disks. Because the CPU cache and memory are volatile media, an unexpected power outage will cause them to lose all data. To tackle this issue, when the external power supply is abnormal, the storage array can use batteries and capacitors to supply power for the CPU cache and memory, allowing them to write data to the NVRAM module to prevent data loss.

3.1.3 Interface Module

Interface modules connect storage array hardware to other devices and also support special functions, as shown in Fig. 3.3.

Interface modules are classified into front-end, back-end, or switch interface modules. A front-end interface module connects a host to a storage array, facilitating data transmission from the host to the storage array. A back-end interface module connects a storage array to disk modules, enabling storage expansion and increasing the array's storage capacity to store more data. A switch interface module

3.1 Hardware Architecture   47

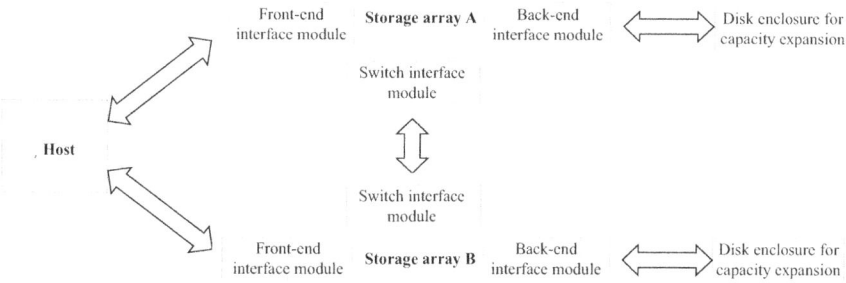

Fig. 3.3 Storage array interface modules

connects multiple storage arrays, enabling expansion of their computing and storage capabilities.

Interface modules may use different protocols, such as Fibre Channel (FC), Ethernet (ETH), and SAS, to interconnect with other hardware. Furthermore, interface modules support both customized and standard physical forms, allowing for flexibility in hardware configuration. For example, standard forms in the industry include both PCIe FHFL (full-height, full-length) and PCIe HHHL (half-height, half-length) interface modules.

As network technologies develop, new interface modules have emerged. These include front-end RDMA over Converged Ethernet (RoCE) and back-end remote direct memory access (RDMA) interface modules, facilitating end-to-end performance of storage arrays. The growing diversity and size of interface modules represent a key competitive advantage for storage arrays.

3.1.4 Disk Enclosure and Disk Module

A disk enclosure consists of disk modules, expansion modules, enclosure-level chips, and enclosure-level heat dissipation modules. Disk modules include HDDs and SSDs and mainly come in two sizes: 2.5 in. and 3.5 in.

Disk interfaces are classified into Advanced Technology Attachment (ATA), SATA, SAS, and NVMe interfaces, with the latter three currently being the most popular. SAS and NVMe interfaces are the main interfaces of SSDs.

A disk enclosure using SSDs (or SSD enclosure) is much more complex than one using HDDs. One reason for this is the emergence of new NVMe interfaces for SSDs. Furthermore, the performance of SSDs is exponentially higher than that of HDDs. Therefore, SSD enclosures need to support NVMe interfaces and provide strong external throughput and high-speed network interfaces to fully utilize SSD performance. The SSD enclosures even need to provide computing power to support service innovation of all-flash storage arrays (all disks are SSDs in the storage array). For example, data-intensive background tasks such as reconstruction can be offloaded to SSD enclosures to further improve system performance. Hence, a brand-new SSD-oriented disk enclosure that integrates enclosure-level computing chips and high-speed network interfaces has emerged, commonly known as a smart disk enclosure within the industry.

In addition, the high-density design of disk enclosures is critical for capacity-intensive scenarios. Higher disk density means higher capacity density, which helps reduce power consumption and required space in data centers. However, this poses new requirements on the heat dissipation and structure of disk enclosures. For example, conventional vertical backplanes feature small heat dissipation windows and large wind resistance, and connectors on both sides interfere with each other. In addition, a conventional 2.5-in. SSD can be as thick as 14.8 mm. These factors limit the number of SSDs supported by a 2 U enclosure in a 19-in. cabinet to no more than 25. Therefore, a horizontal backplane with an orthogonal connection structure has been introduced, as shown in Fig. 3.4. Disk connectors and controller connectors are connected orthogonally, preventing mutual interference and improving disk connector density. Furthermore, the thickness of NVMe SSDs has been reduced to 9.5 mm, allowing a 2 U enclosure to accommodate 36 SSDs. This greatly improves disk density and dissipation capabilities.

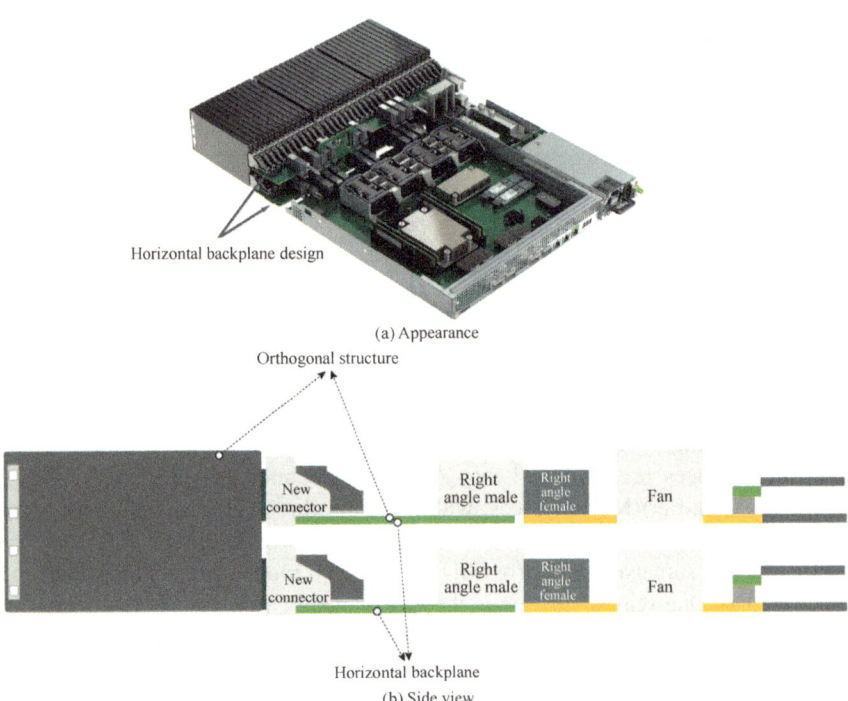

Fig. 3.4 Horizontal backplane with the orthogonal connection structure

3.1.5 Heat Dissipation Module

The heat dissipation module mainly consists of air cooling components such as fans, including centrifugal fans and axial fans. Fan speed directly impacts heat dissipation and system power consumption. Therefore, the fan module must provide basic functions such as fan speed detection and intelligent fan speed adjustment. Fan speed detection serves as the basis for intelligent fan speed adjustment, which involves real-time adjustment based on service workloads and fan speeds to balance the need for heat dissipation and power efficiency.

As energy conservation becomes increasingly important, heat dissipation modules will evolve from air-cooling-only to air and liquid cooling modes. Balancing high performance with low power consumption will be crucial in the heat dissipation designs of next-gen storage arrays.

3.2 Software Architecture

A storage array's software system is designed to provide reliable and high-performance services that are easy to manage. It usually comprises multiple subsystems, among which the RAID subsystem and cache mirroring subsystem are the most important. They have a direct effect on the most critical data read/write process and high-reliability protection mechanism, and they are crucial determinants of system performance, reliability, scalability, and other quality-related attributes.

3.2.1 RAID Subsystem

The RAID subsystem, based on RAID redundancy capability, is the fundamental subsystem of a storage array. It combines a large number of disks to form a reliable local data storage unit, providing unified storage space. Mirroring, one of the earliest data protection methods, writes one copy of the data to two or more disks so that if one disk fails, the data can still be read from a normal copy. New data is also be copied to two or more disks to ensure data reliability.

3.2.1.1 RAID Algorithm Principles

The term "RAID" was proposed by David Patterson, Garth A. Gibson, and Randy Katz, when they were academics at the University of California, Berkeley, in 1988 [3]. It is a storage technology that combines multiple physical storage devices into a large-capacity logical unit to support high concurrent access and high fault tolerance capabilities. RAID technologies can be classified into RAID 0, RAID 1, RAID 2, RAID 3, RAID 4, RAID 5, and RAID 6 based on the operations and the layout of the data. The following section will introduce the configurations and performance characteristics of these categories.

RAID 0: RAID 0 splits files and stripes them across multiple disks. This means that RAID 0 can fully utilize the bandwidth of buses and disks and improve access concurrency. However, RAID 0 has no fault tolerance capabilities. Figure 3.5 illustrates how RAID 0 works. Blocks 1 (A1) to 8 (A8) are cyclically stored on two different disks. The data on the two disks can be accessed concurrently, but data loss will occur if either disk fails. Therefore, RAID 0 should be used in scenarios that prioritize high access performance but do not need fault tolerance.

RAID 1: With RAID 1, also called mirroring, data is written identically to two disks to improve reliability. RAID 1 can tolerate the failure of a single disk, but only half of its storage capacity is effective. In addition, RAID 1 causes unbalanced read and write performance. It improves read performance, as the same data can be read from two disks simultaneously. However, it incurs additional write operations, as the mirrored data needs to be updated synchronously when a copy of data is updated. This means that the slowest disk limits the write performance. Therefore, RAID 1 is usually recommended for scenarios that require fault tolerance and can afford higher capacity overheads. Figure 3.6 illustrates how RAID 1 works.

Fig. 3.5 RAID 0

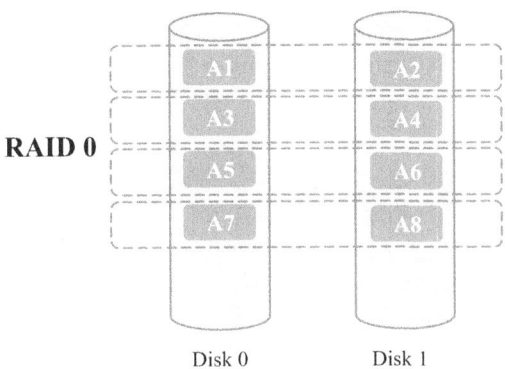

Fig. 3.6 RAID 1

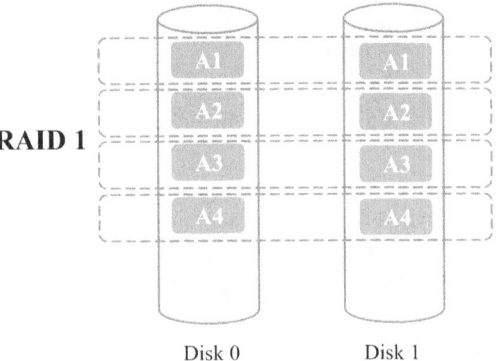

3.2 Software Architecture

RAID 2: Unlike RAID 0, which performs block-level striping, RAID 2 performs bit-level striping. The redundant parity bits are generated using Hamming code, allowing for bit-level error detection and correction. The data bits and parity bits are stored on different disks [4]. To facilitate concurrent data reads, the controller requires all disks to spin at the same angular orientation. It has a built-in ECC function that supports bit-level fault tolerance. However, RAID 2 has complex and strict requirements, so it is rarely used in practice. Figure 3.7 illustrates how RAID 2 works. A1–A4 are data bits and A_{p1}–A_{p3} are parity bits.

RAID 3: RAID 3 performs byte-level striping. Data is stored on multiple disks, and parity bits are stored on a dedicated disk, as shown in Fig. 3.8. This configuration enables RAID 3 to tolerate the failure of any single disk.

RAID 4: RAID 4 performs block-level striping and stores parity blocks to a dedicated disk. This means that RAID 4 can tolerate the failure of any single disk. Since data is striped at the block level and stored on multiple disks, RAID 4 performs well on random reads. However, the maximum write performance is determined by the dedicated disk that stores the parity blocks. This is because whenever a data block is written, the corresponding parity block also needs to be updated. Figure 3.9 illustrates how RAID 4 works. Blocks A1, A2, A3, and A_p are in the same stripe. Disk 3 is a dedicated disk that stores all of the stripe parity blocks.

RAID 5: RAID 5 performs block-level striping with distributed parity, preventing all parity blocks from being stored on the same disk. RAID 5 can

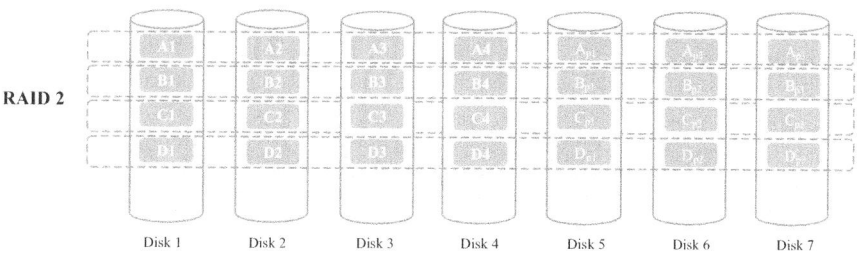

Fig. 3.7 RAID 2

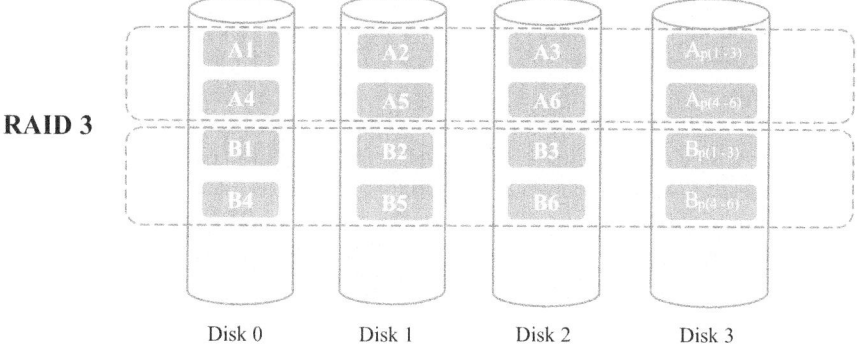

Fig. 3.8 RAID 3

RAID 4

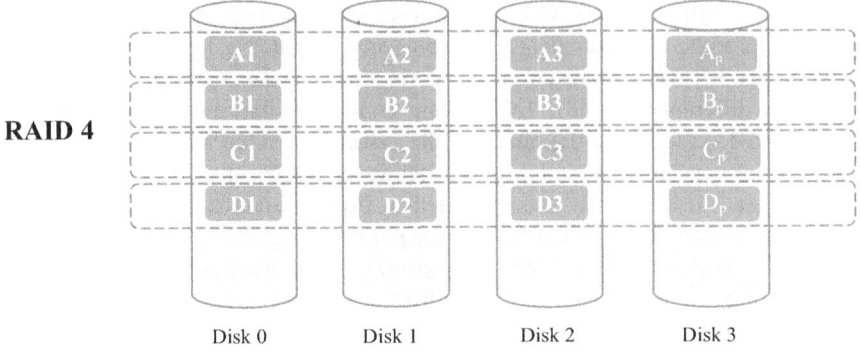

Fig. 3.9 RAID 4

RAID 5

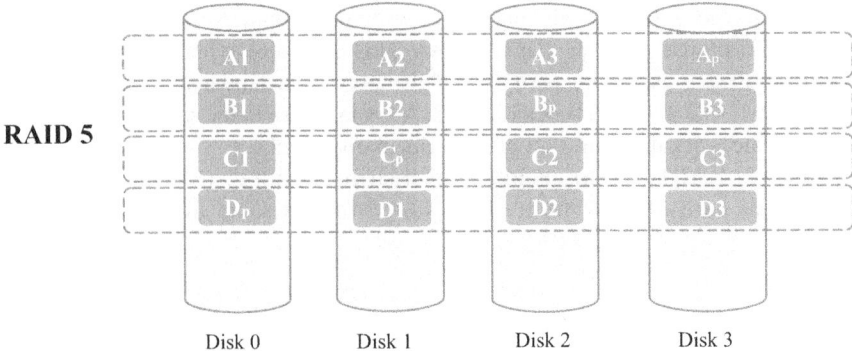

Fig. 3.10 RAID 5

tolerate the failure of a single disk. Since all disks hold parity blocks, the I/O operations for updating parity blocks are distributed across all disks. The RAID 5 algorithm can produce a different layout than other RAID algorithms, depending on the placement policies of data blocks and parity blocks. To evenly distribute parity blocks, as shown in Fig. 3.10, data blocks are written in ascending order from left to right, while parity blocks are written from right to left, starting in the right-most disk and moving one disk to the left with each additional stripe.

RAID 6: RAID 1 to RAID 5 can only tolerate the failure of one disk. RAID 6 adds a parity block to each stripe and stores the blocks within each stripe on different disks, thus allowing it to tolerate the failure of two disks. Various techniques for implementing RAID 6, such as EVENODD coding and RDP coding, have been developed based on different layouts of data blocks and parity blocks, as well as various parity block generation policies. Figure 3.11 illustrates how RAID 6 works with five disks. Each stripe contains two parity blocks, starting from the right-most two disks and moving left with each additional stripe.

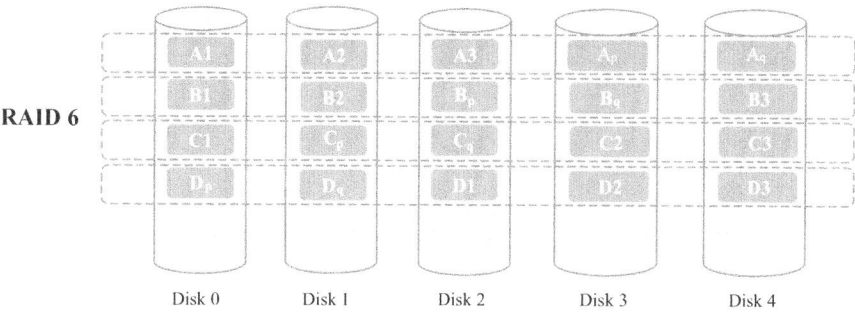

Fig. 3.11 RAID 6

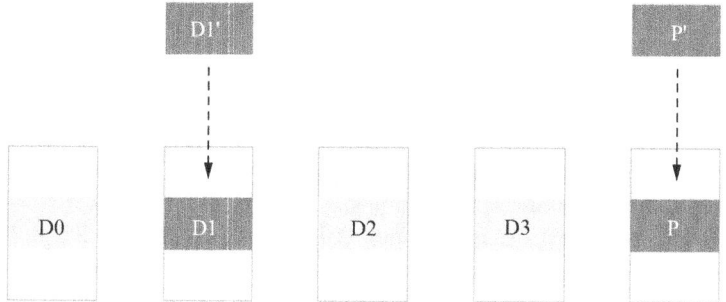

Fig. 3.12 RAID data update

3.2.1.2 RAID Data Update

When data in a RAID group is modified, multiple data blocks are updated to ensure that the RAID stripes remain consistent. Stripe consistency means that multiple data blocks in a stripe adhere to the RAID algorithms. For example, all RAID 1 data copies are the same. In RAID 5, as shown in Fig. 3.12, P is the result calculated by performing an exclusive OR (XOR) procedure on the data. Therefore, it is necessary to update both the modified data blocks and the corresponding parity blocks during data modification.

In Fig. 3.12, D0–D3 and P are legacy data, D1' is new data written by the host, and the data to be written by the system is D1' and P'. P' can be generated in two modes: Read-Modify-Write and Reconstruction-Writes.

For the Read-Modify-Write mode, the calculation formula is $P' = D1 + D1' + P$. This means that D1 and P should be read from disks. For minor modifications on a single disk, RAID 5 will generate two read requests and two write requests on the disk. For the Reconstruction-Writes mode, the formula is $P' = D0 + D' + D2 + D3$. This means that D0, D2, and D3 should be read from disks. This mode can be used when a large amount of data needs to be modified at a time, and data on multiple disks can be overwritten.

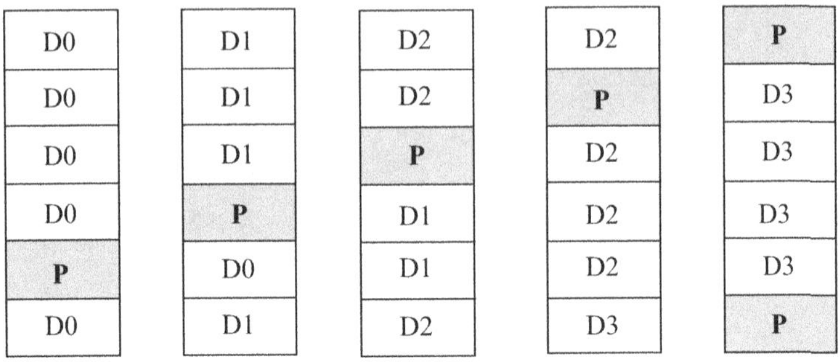

D0	D1	D2	D2	P
D0	D1	D2	P	D3
D0	D1	P	D2	D3
D0	P	D1	D2	D3
P	D0	D1	D2	D3
D0	D1	D2	D3	P

Fig. 3.13 RAID striping

The overwrite operation of RAID 6 is similar to that of RAID 5 and will not be described in this chapter.

The overwrite operation causes I/O amplification. This means that a write request from a user will be amplified into multiple read and write operations on disks. Therefore, the software system of storage arrays has been purposely designed to reduce write amplification. For example, data can be aggregated, sorted, and processed in batches by using cache technology.

Moreover, parity disks may become performance bottlenecks. If independent parity disks are used (like RAID 3 and RAID 4), parity blocks need to be updated synchronously when a copy of data is updated. This leads to frequent access of the parity disks, thus affecting system performance. Therefore, RAID 5 and RAID 6 distribute the parity blocks of each stripe on different disks to balance the workload of member disks, thus eliminating the performance bottleneck, as shown in Fig. 3.13.

The overwrite operation does, however, bring a host of other problems, such as the write hole, which requires storage arrays' software systems to provide a fault tolerance function. The "write hole" effect can happen if an exception, like a power failure, occurs during RAID 5's overwrite operation. In this situation, some columns fail to be written. For example, if a power failure occurs after D1' has been successfully written but P' has not, the data on the disks becomes D0, D1', D2, D3, and P. In this case, stripes are inconsistent. If this inconsistency remains unresolved, the system will restore incorrect data during data reconstruction in the event of disk failures. For example, if a disk storing D2 is faulty, D2 will be restored based on the formula D2 = D0 + D1' + D3 + P. However, this result is incorrect, leading to serious data inconsistency. Hence, storage arrays must be designed to solve the write hole problem. Generally speaking, this problem can be solved by using the NVRAM to back up logs.

3.2.1.3 RAID Space Management

In addition to the basic RAID algorithms, the RAID subsystem, as the most important subsystem of a storage array, needs to provide other key functions, such as efficient space management, and reconstruction and balancing mechanism for disk faults and disk capacity expansion.

3.2 Software Architecture

Regarding the space management mechanism, traditional RAID subsystems directly combine a group of disk modules to form a RAID group. This mechanism uses disks as the minimum unit for space management. However, the evolution of media technologies allows a storage array to accommodate more disks with greater capacities, and conventional RAID groups are insufficient for data reconstruction if these high-capacity disks are faulty. This is because the reconstruction of a conventional RAID group involves only the member disks in this group, which limits the reconstruction rate and prolongs the reconstruction period. Taking a 4 TB 7200 rpm disk as an example, if RAID 5 (8D + 1P) is used, the reconstruction time is about 40 h, which can pose significant risks in operations and maintenance (O&M). Moreover, although the number of disks is increasing, the LUN space provided by a storage array to a host can be allocated only from the RAID group. Therefore, the performance of a single LUN may be limited by the number of member disks in the RAID group. This means that the increasing number of disks cannot bring more benefits to the performance.

Therefore, optimizing the RAID space management mechanism is becoming increasingly urgent in the industry. Some vendors have made gradual improvements. At the underlying layer, they still use disk modules as the space management unit, while at the upper layer, they aggregate multiple RAID groups into a space pool and divide the space pool into small-granularity management units (e.g, 8 MB/ unit). These units are then recombined and mapped to LUNs that are visible to users. This method allows the distribution of single LUN space to multiple disks to overcome LUN performance bottlenecks. However, only a few underlying disks are used for reconstruction, and the reconstruction time is not reduced.

Some other vendors have completely redesigned and optimized the mechanism by adopting a two-layer virtualization management mode for underlying disks and upper-layer resources, as shown in Fig. 3.14. In terms of the underlying layer virtualization, disks in a storage pool are divided into small-granularity data blocks before a RAID group is created. RAID groups are created based on these data blocks instead of the entire disks. This means data is evenly distributed to all disks. The upper-layer virtualization is similar to the design mentioned above. Before

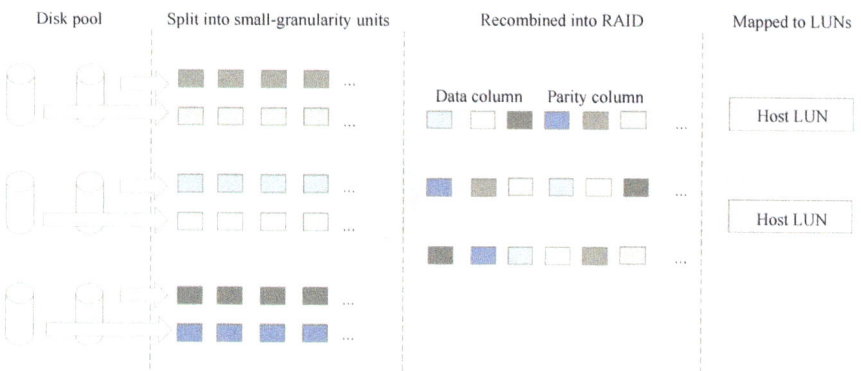

Fig. 3.14 Two-layer virtualization

RAID groups are mapped to LUN space, fine-grained partitioning is performed to divide the space of multiple RAID groups into small-granularity units, which are then recombined and mapped to LUN space. This design solves the two weaknesses of conventional RAID groups. If a disk is faulty, almost all disks can be used for reconstruction simultaneously, significantly improving the reconstruction speed. It also breaks the LUN performance bottlenecks to achieve balanced distribution on all disks. Currently, most emerging storage arrays adopt the two-layer virtualization design for the space management of the RAID subsystem.

Media development is critical for evolving the RAID subsystem. Alongside the evolving RAID space management mechanism, the emergence of new flash storage media has led to the gradual incorporation of a large number of flash-oriented designs being into the RAID subsystem. First of all, considering the high cost of flash storage, the RAID subsystem must support data reduction technologies such as deduplication and compression, which pose new challenges to the RAID space management granularity and internal indexing technology. Second, as data erasure would influence the lifespan of SSDs, technologies such as wear leveling, anti-wear leveling, and system-level garbage collection should also be used in the RAID subsystem.

3.2.2 Cache Mirroring Subsystem

The cache mirroring subsystem provides an efficient cache acceleration layer for storage arrays and enables highly reliable cache mirroring between two or more storage controllers. Cache technology is widely used in IT systems. When it comes to storage arrays, the cache should not only accelerate read and write operations but also provide mirroring and power failure protection mechanisms to ensure data security and reliability.

A cache is typically used to buffer data between modules with different rates to reduce the coordination overheads caused by different rates between the modules and improve overall performance. Common caches include the L1/L2 CPU cache, the page cache of Linux Virtual File System (VFS), and the VFS inode cache. In a storage array, the cache is used to equilibrate host processing and media response, as they usually perform at different rates. In particular, early disks perform poorly in terms of I/O throughput and response latency. Therefore, cache is essential for storage arrays to support efficient data access.

3.2.2.1 How Does the Cache Mirroring Subsystem Improve Storage Array Performance?

The cache mirroring subsystem usually builds a cache layer based on the DRAM media and uses the principle of locality to improve performance. The principle of locality covers temporal and spatial locality [5]. Temporal locality refers to reusing

3.2 Software Architecture

specific data within a relatively short time frame. For example, some hotspot data in a storage array may be frequently accessed. Spatial locality means that when this data is accessed, its adjacent data may also be accessed very soon afterward, such as during a full-disk scanning of the storage array. Based on the temporal locality principle, the cache mirroring subsystem employs cache technology to enhance storage efficiency. Based on the spatial locality principle, the cache mirroring subsystem can use content prefetching technology to read data into the cache layer in advance.

The advantage of cache layers is that they optimize storage performance indicators by utilizing a limited amount of high-performance and costly media based on the principle of locality. These performance indicators include input/output operations per second (IOPS), bandwidth (the amount of data that can be read and written per second), and latency (measured from the time a read/write request is initiated to the time a response is returned).

To accelerate read and write requests, caches are classified into read caches and write caches. The write cache leverages the low latency of DRAM to enable immediate response to a host's write I/O request as soon as the request is received. This reduces the write latency from more than 10 ms (when writing data directly to disks) to less than 1 ms. The read cache, which also uses DRAM, monitors and collects statistics on the read/write frequency and time interval of each I/O operation. Based on the statistics and a dedicated algorithm, the read cache retains frequently accessed hot data and evicts seldom accessed cold data. This ensures that hot data can be accessed at the cache layer instead of being read from underlying disks, which improves the overall IOPS performance.

Moreover, the read cache can provide content prefetching to improve bandwidth performance, as shown in Fig. 3.15. When the read cache identifies that the current data access model is a sequential model, meaning that N consecutive read requests access consecutive logical block addresses (LBAs), it proactively reads a large segment of data from the next LBA to be accessed. When a read request for this data

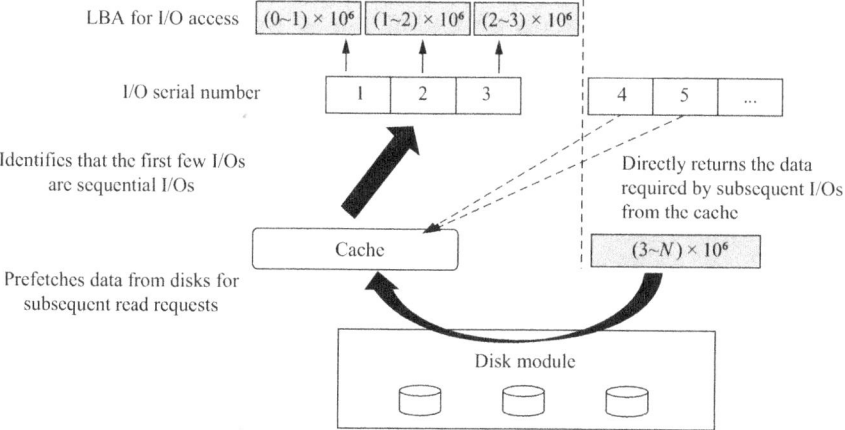

Fig. 3.15 Content prefetching technology

arrives, the data can be directly obtained from the cache instead of being read from disks. This is called content prefetching, and it improves bandwidth performance and reduces latency.

3.2.2.2 How Does the Cache Mirroring Subsystem Improve the Reliability of Storage Arrays?

Storage arrays usually adopt a dual-controller or multi-controller design to maximize reliability and redundancy. After data is written to the cache of a controller, a success message will be returned to the host. However, if this controller is faulty before the data is written to disks, the other controller will not be able to access the correct data even if it takes over services. This can result in serious data access issues. The cache mirroring technology is used to solve this problem. The temporary data in the write cache must be mirrored to the write cache of another controller for backup before a response message is returned to the host. The key to cache mirroring is to efficiently synchronize data between controllers without affecting system performance. This is why early cache mirroring usually uses PCIe—a high-speed serial bus used in computers—to ensure efficient data transmission, as shown in Fig. 3.16.

Although the DRAM-based cache layer provides robust performance acceleration capabilities, its volatile memory can compromise system reliability. In the event of an unexpected power outage, cache data that has not been written to disks will be lost. Therefore, power failure protection technology is crucial for the cache mirroring subsystem. The controller enclosures of storage arrays are equipped with highly redundant batteries. If a power failure occurs, the batteries allow controllers to store the data that has not been written to disks into coffer disks to prevent data loss. Coffer disks are dedicated disks that store storage system logs, configuration information, and cache data in the event of power failures. Storage arrays have built-in coffer disks, which are equipped for each controller. They can also select several external disks as external coffer disks. Coffer disks do not require much space.

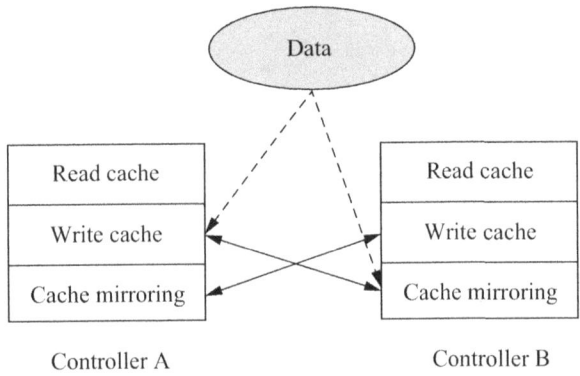

Fig. 3.16 Cache mirroring

Therefore, the external type only uses a small amount of space on the selected disks, and the remaining space is still available storage capacity for users.

3.2.2.3 New Trends in Cache Mirroring Subsystems

With the development of network, media, and cluster technologies, cache mirroring subsystems are also evolving. For example, the data channel of cache mirroring has evolved from a PCIe-based design to a high-speed network based on RDMA such as RoCE and InfiniBand (IB) for improved scalability and flexibility. The cluster scale of storage arrays has been expanded from 2 controllers to 8, 16, or even more controllers, which has improved system reliability. The multi-copy mirroring technology for the write cache is being used across multiple controllers to increase reliability and tolerate the concurrent failure of multiple controllers. To take advantage of the low latency of flash media, the way that data is organized at the cache layer has developed from an LBA-based memory page to a log-based cache structure. This has reduced the write cache latency, ensuring the stable and low latency performance of all-flash storage arrays.

3.3 High Performance and Reliability Design

3.3.1 Application Scenarios

Storage arrays are important for storing critical data in IT systems and are integral to the core production systems of finance sectors, governments, enterprises, carriers, and other large-scale organizations. Designed for mission-critical applications sensitive to latency and throughput, storage arrays prioritize service continuity. They boast high performance and reliability across multiple dimensions, including components, software, and the system as a whole.

Storage arrays primarily serve workload scenarios sensitive to response time and IOPS. Typical applications include mainstream databases and transactional applications, providing persistent storage for enterprise platforms like Oracle databases and virtualization environments.

3.3.1.1 Database

Database serves as the core service system of an enterprise, demanding 24/7 high availability and low latency for online transaction processing (OLTP) applications like enterprise resource planning (ERP) and customer relationship management (CRM) to ensure service continuity. Therefore, priorities lie in performance, reliability, and security. Storage arrays have proven to be the optimal storage platform for commercial databases such as Oracle and DB2.

Typical database workloads require high IOPS/throughput and low latency. Therefore, storage arrays and application servers connect via Fibre Channel (FC) networks, with storage space mapped to hosts as volumes to improve data transmission efficiency and processing speeds. Furthermore, during database deployment, storage arrays' backup and disaster recovery features, such as snapshot, replication, and active-active configurations, safeguard database volumes, preventing huge business losses caused by service interruptions (Fig. 3.17).

3.3.1.2 Virtualization Platform

An increasing number of IT organizations are adopting virtualization in constructing their data centers, with VMware emerging as a prominent and extensively utilized virtualization platform. Enterprises often build virtualization platforms with VMware as the foundation, where they run various development and testing applications, as shown in Fig. 3.18. Leveraging VMware's powerful management platform, enterprises effortlessly virtualize their data centers and establish private cloud systems. Therefore, VMware has become one of the most important applications in enterprise data centers, with Virtual Server Infrastructure (VSI) and Virtual Desktop Infrastructure (VDI) serving as typical examples.

The primary workload on virtualization platforms comprises random small and medium I/O access, alongside performance-intensive scenarios like boot storms and login storms. In these scenarios, the virtualization platform generates a large number of I/O requests, which requires the storage platform to deliver optimal performance to manage the service burst effectively. Furthermore, a large amount of duplicate data in the system volumes and data volumes of VMs necessitates storage arrays equipped with technologies such as deduplication and compression to help enterprises in reducing the required storage space. With features like thin

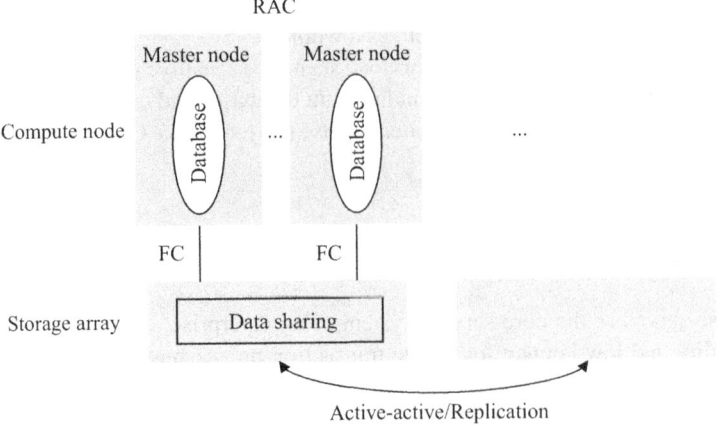

Fig. 3.17 Database scenario. Note: RAC stands for Real Application Cluster

3.3 High Performance and Reliability Design

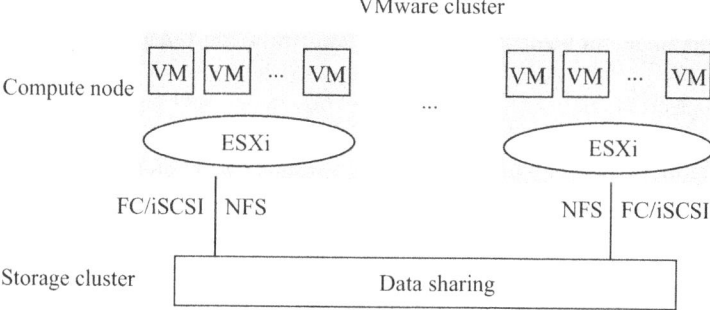

Fig. 3.18 Virtualization scenario. Note: VM stands for virtual machine; iSCSI stands for Internet Small Computer System Interface; NFS stands for Network File System

provisioning, clone, and snapshot, storage arrays greatly reduce the total cost of ownership (TCO) while enhancing operational efficiency.

3.3.2 High-Reliability Subsystems for Redundant Failover

Controllers serve as the core hardware units in storage arrays, housing key components such as CPUs, memory modules, and interface modules. Therefore, protection against controller failure stands as a top priority in ensuring the high reliability of storage arrays.

The high-reliability design of storage arrays should not only offer high fault tolerance capability to ensure service continuity but also deliver optimal device performance and alarm capability to optimize user experience in the event of a failure.

There are four levels of tolerance for controller failure:

Level 1: Services remain uninterrupted if a single controller fails.
Level 2: Services remain uninterrupted if any two controllers in a multi-controller cluster fail.
Level 3: Services remain uninterrupted if any engine in a multi-engine cluster fails. An engine essentially constitutes a controller enclosure typically housing two or four controllers. Storage arrays can be scaled out by adding engines, with failover primarily executed within an engine. Therefore, in a multi-engine cluster, one of the criteria for achieving high reliability is maintaining uninterrupted services in the event of a single engine failure.
Level 4: Services remain uninterrupted even if the controllers in a multi-controller cluster fail one by one until only one functioning controller remains.

There are three levels of user experience in the event of a failure:

Level 1: I/O performance deterioration or suspension lasts less than 30 s. A link disconnection alarm is reported. The host performs a multi-path switchover.

Level 2: I/O performance deterioration or suspension lasts less than 5 s. A link disconnection alarm is reported. The host performs a multi-path switchover.

Level 3: I/O performance deterioration or suspension lasts less than 1 s. No link disconnection alarm is reported. The host does not perform a multi-path switchover.

To ensure the high reliability and service continuity of mission-critical production systems in the event of controller failures, a storage array should deliver Level 2 or better fault tolerance capability and Level 3 user experience.

To achieve these levels, storage arrays can utilize a variety of technologies to develop a dedicated design geared toward ensuring reliability in the event of controller failures.

3.3.2.1 Fast Controller Failure Detection

The underlying hardware acceleration technology enables the storage array OS to quickly detect controller failures and report them to system components within 200 ms. When the OS detects a controller failure, it disables the front-end and back-end interface modules and the expansion modules before the controller resets. Then, it informs other nodes of the issue, achieving fast detection and response to controller failures.

3.3.2.2 Cache Mirroring Technology (Including Continuous Mirroring)

Storage arrays need to be equipped with cache mirroring technology so that, in the event of a controller failure, other controllers can seamlessly take over services. Cache mirroring is especially important for the write cache. Write cache technology is designed to provide write-back capability, meaning the I/O interface will return a write success message to the user as soon as it writes data to the cache. While some users may tolerate the loss of cache data in certain cases, in most snecarios, highly reliable cache is required.

Cache subsystems can be enhanced in two ways to ensure high reliability. First, redundant copies can be added through data mirroring. Second, power failure protection for memory is used to reduce the risk of data loss in the event of a power failure.

During data mirroring, a pair of controllers in a cluster is selected to form a mirroring relationship. This means that when data is written to the cache of one controller, it will be mirrored to the cache of the other controller through a mirroring channel (such as the backplane channel or another communication channel). Once the data is written to the caches of both controllers, a write success response message is returned to the host.

In addition to common data mirroring mechanisms, cutting-edge storage arrays support continuous mirroring technology to further increase reliability. Figure 3.19

3.3 High Performance and Reliability Design

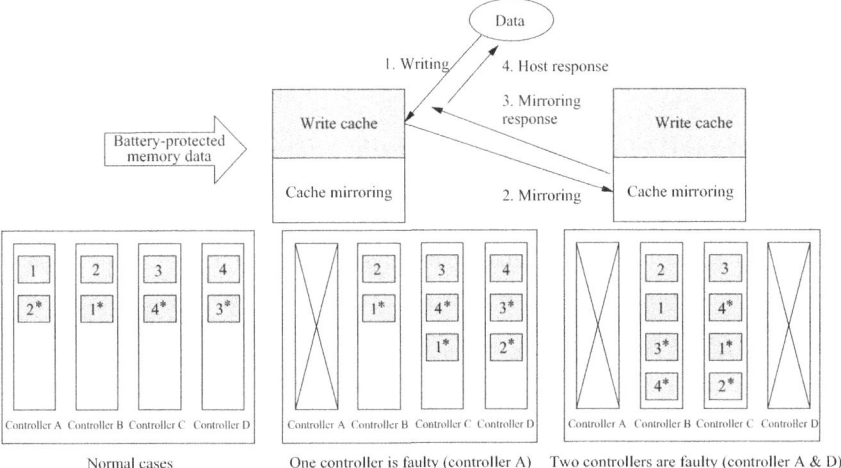

Fig. 3.19 Continuous mirroring technology

shows how continuous mirroring works. Data blocks 1 and 2^* on controller A and data blocks 1^* and 2 on controller B are mirror copies of each other. If controller A fails, blocks 1^* and 2 lose their mirror copies on controller A. In this case, they will be mirrored to both controllers C and D to ensure dual-copy redundancy. If controller D fails, the data blocks on controllers B and C will be mirrored to each other to ensure dual-copy redundancy of the cached data.

Continuous mirroring is preferentially performed between controllers within a controller enclosure. If the controller enclosure has only one functioning controller left, it will mirror cache data to controllers of other controller enclosures. This process continues until the whole storage array has only one functioning controller. The continuous mirroring technology is essential for achieving Level 4 controller failure tolerance because it can prevent data loss and service interruption even if the user fails to replace the faulty controller and more controllers fail in sequence, thus maximizing service continuity.

3.3.2.3 Fast Controller Failover

In addition to fast controller failure detection and cache mirroring technologies, storage arrays need to have fast controller failover capabilities. If the system detects a controller failure, it needs to assign another controller to take over the affected services as soon as possible to deliver a Level 3 user experience. Because the metadata of service instances in the system is saved on disks, some key metadata involved in system takeover can be mirrored and cached on paired controllers in advance to accelerate the takeover procedure. In the event of a failure, services are preferentially taken over by the node where the metadata is mirrored, and the metadata is directly restored from the cache. This reduces the system's dependency on reading data from disks and enables fast service takeover.

3.3.3 High-Performance Cluster Subsystems

In terms of system performance balancing, the cluster subsystems of the storage arrays can be categorized into balanced and unbalanced architectures. The balanced architecture is further classified into partially symmetric architecture and fully symmetric architecture.

The main difference between an unbalanced and a balanced architecture lies in whether services can be evenly distributed throughout the entire cluster and whether a volume can flexibly utilize the performance of the entire cluster rather than being restricted by a resource bottleneck of a single controller node in the cluster.

3.3.3.1 Unbalanced Architecture

The unbalanced architecture has been widely used by many vendors. It is also called active-passive (A-P) access mode. With this architecture, a volume or file system is owned by one specific controller, and other controllers serve as highly available backup, regardless of the number of controllers that share the back-end disk enclosures. If a fault occurs, the volume or file system is switched to another backup controller. If the services of a volume or file system are delivered to a nonowner controller, the storage system will first forward the services to the owner controller and then perform related operations such as reading and writing data.

There are two problems with this architecture. First, the networking is complex. This is because each controller must be connected to all hosts that use its owning volume or file system to prevent services from being delivered to a nonowner controller. Otherwise, a large number of I/Os are forwarded within the storage system, which wastes forwarding channel resources and seriously affects performance. Second, the performance of a volume or file system is limited by the processing capability of a single controller. Idle resources, such as CPU resources, on nonowner controllers cannot be utilized to boost the volume or file system performance.

3.3.3.2 Balanced Architecture

The core production systems of some large enterprises require large-scale storage systems capable of delivering ultra-large capacity and excellent performance. Therefore, storage arrays need to support scale-out of controllers. As the cluster scale and the number of controllers increase, it becomes urgent to solve the problems in the unbalanced architecture so that the performance of individual volumes

3.3 High Performance and Reliability Design

and file systems can increase linearly with the number of controllers. Consequently, balanced architectures were developed and have since been widely used in core fields that require superb performance and high reliability.

In a balanced architecture, any volume or file system is not owned by any specific controller. Instead, it is distributed to every controller in a cluster based on a finer granularity such as address segments, data slices, or directories. As a result, all services are evenly distributed across the entire cluster, and storage volumes or file systems can achieve cluster-level performance that is not limited by the resource bottleneck of a single controller node, regardless of the interface where they are accessed. This meets the ever-increasing performance requirements for the flexible expansion of host applications. A balanced architecture is also called the active-active (A-A) access mode or A-A balanced architecture.

There are two types of balanced architecture: partially symmetric architecture and fully symmetric architecture. It is possible to achieve balanced distribution on three layers: front-end networks, controllers, and back-end disk enclosures. A partially symmetric architecture only supports balanced distribution at the controller layer, which means that data from volumes or file systems can be distributed to all controllers at a certain granularity, and services are also evenly distributed on each controller. However, services may not be evenly distributed on front-end interface modules or back-end disk enclosures. For example, if a front-end interface module is owned by a specific controller, and a back-end disk enclosure is owned by a specific engine, services cannot be evenly distributed on the front-end network and back-end disk enclosure. Services received by a front-end interface can only be processed by the controller that owns the front-end interface because the interface cannot interconnect with other controllers. Therefore, to achieve full balance, a complex cross-connected network must be established across all volumes and all front-end interfaces. Similarly, an engine's ownership of a back-end disk enclosure may also compromise reliability. For example, if the entire engine is faulty, the back-end disk enclosure will be inaccessible, and services will be interrupted.

A fully symmetric architecture, on the other hand, allows for services to be evenly distributed on all three of the aforementioned layers. Front-end interface modules are not owned by a single controller, and back-end disk enclosures are also not limited to a single controller or engine. Therefore, services can be distributed from any interface to all controllers and all disks to achieve a fully symmetric balanced distribution. This not only enhances performance but also improves ease of use and reliability.

A balanced architecture cannot be implemented without the collaboration of hardware and software. It requires either loose coupling or tight coupling hardware and the balanced distribution algorithm.

The balanced distribution algorithm is the key to building a symmetric balanced architecture. A well-performed balanced distribution algorithm can ensure that data is well-balanced across all components of the system, as shown in Fig. 3.20. As volumes or file systems are no longer owned by any controllers, users only need to specify the total storage capacity and their performance requirements, without

Fig. 3.20 Balanced distribution algorithm. Note: CIFS stands for Common Internet File System

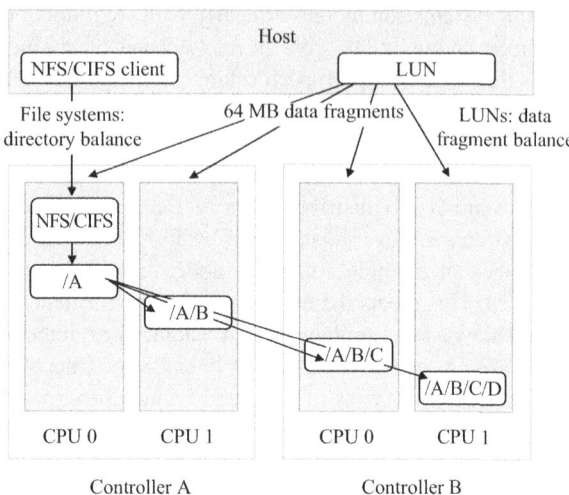

worrying about the distribution of capacity and performance in the storage system. This simplifies storage system resource planning.

The common balanced distribution algorithm for volumes is to slice their LBAs based on a fixed granularity (e.g., 64 MB) and then distribute the slices to each controller processing unit using the hash algorithm. For file systems, balancing is achieved by distributing directories or files to each controller processing unit.

In addition to ensuring good balance, the balanced distribution algorithm needs to ensure fast failover and takeover in the event of a fault. Therefore, global view management, partition management, and metadata design are all essential parts of the algorithm.

3.3.4 Redirect-on-Write and Garbage Collection Technologies

As storage media continues to evolve, traditional HDDs are gradually being replaced by SSDs. The redirect-on-write (ROW) mechanism is designed to resolve the write penalty problem caused by the RAID overwrite operation on all-flash storage arrays.

With the ROW mechanism, all new data can be written to new blocks to avoid write penalties caused by the traditional RAID write process. This greatly reduces overheads on controller CPUs and the read and write latency on SSDs caused by write operations. Compared with the write-in-place mode of traditional RAID, the ROW's full-stripe write mode can improve the overall performance at various RAID levels.

As shown in Fig. 3.21, the system uses RAID 6 (4 + 2) and writes new data blocks 1, 2, 3, and 4 to modify the existing data.

In the traditional write-in-place mode, a storage system must modify every RAID group in which the data blocks reside. For example, when writing data block 3 to

3.3 High Performance and Reliability Design

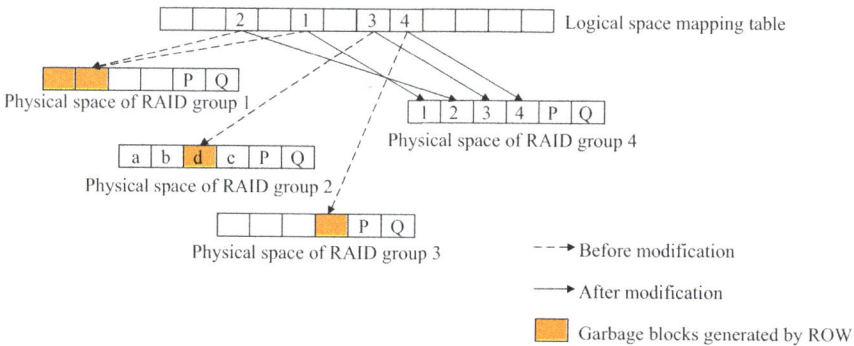

Fig. 3.21 ROW mechanism

RAID group 2, the system must first read the original data block D and the parity blocks P and Q. Then, it must calculate the new parity blocks P′ and Q′ based on the redundancy algorithm and write P′, Q′, and data block 3 to RAID group 2. In the ROW full-stripe write mode, shown in Fig. 3.21, the system uses data blocks 1, 2, 3, and 4 to calculate P and Q and writes them to disks in a new RAID group. Then it modifies the LBA pointer to point to the new RAID group. There is no need to read any existing data during this process.

For traditional RAID, such as RAID 6, if the data is updated, the system must first read the original data blocks and parity blocks P and Q before writing new data in new data blocks and parity blocks P′ and Q′. Therefore, three read I/O requests and three write I/O requests are generated. Generally, the read and write amplifications of small random I/Os using traditional RAID ($xD + yP$) is $y + 1$.

However, ROW causes a new problem. If the modified data is redirected to a new space, "garbage" is produced. "Garbage" refers to the original data that has become invalid but still occupies physical storage space. Therefore, a garbage collection (GC) mechanism is developed to reclaim the space occupied by garbage data. Because it is not possible to predict when GC will be triggered on each SSD, GC's impact on system performance also becomes unpredictable. If there is a large amount of garbage data on SSDs, GC may be triggered on multiple SSDs at the same time, which would cause a significant drop in performance.

The global GC technology is a common solution to this problem. It allows the system to periodically check the ratio of garbage data in each RAID group and transfer valid data from a RAID group with a large amount of garbage to a new RAID group. After all valid data has been transferred, the SSD is instructed to erase the original blocks to reclaim the original RAID group. This reduces the amount of data that needs to be moved during GC and the impact of SSD GC on system performance. This is one of the key factors behind the exceptionally stable performance of some new high-end all-flash storage arrays.

To further reduce the impact of GC on system performance, some all-flash storage arrays provide technologies that can separate hot and cold data. An example of hot data is the metadata of a system, which is updated frequently and is more likely

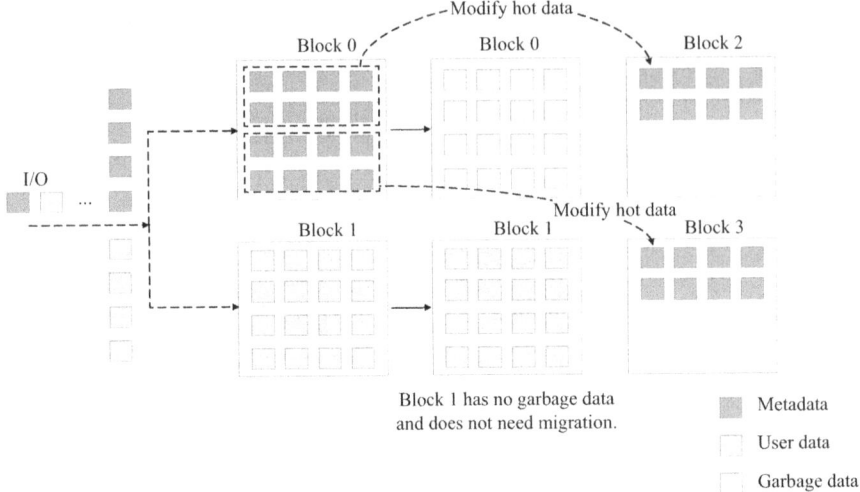

Fig. 3.22 Hot and cold data separation technology

to generate garbage data. An example of cold data, on the other hand, is user data that is rarely modified, which is unlikely to generate garbage data. The industry's mainstream storage arrays use multistreaming technology to enable collaboration between disks and controllers so that hot and cold data can be stored in different blocks. This increases the probability that all data in a block is invalid, reduces the amount of valid data to be migrated during GC, and improves SSD performance and reliability. As shown in Fig. 3.22, separating hot and cold data greatly reduces the amount of garbage data that needs to be migrated.

3.4 Summary

Storage arrays are a critical component in the development of IT infrastructure. They can meet the growing demand for information acquisition and efficient data storage, making them ideal for use in cross-era applications such as databases and virtualization, injecting new momentum into the digital economy. The evolution of storage arrays toward all-flash and all-IP is being driven by new media such as flash memory and emerging network technologies like RoCE. Despite the emergence of diverse new storage media, storage arrays still offer obvious advantages. The unique combination of software and hardware, along with their abundant enterprise features, enables them to provide optimized performance, reliability, and efficiency. Storage arrays not only play an important role in conventional core production systems but have also found utility in emerging containers and cloud-based applications. The ongoing evolution of storage arrays remains one of the main driving forces of data infrastructure construction.

3.5 Practice Questions

1. **Which core modules form the controller module of a storage array? What are the main functions of these modules?**

   Answer: The controller module comprises a central processing unit (CPU) module, an input/output (I/O) expansion module, an onboard interface module, a bubble memory controller (BMC), a complex programmable logic device (CPLD), a system disk module, and a nonvolatile random-access memory (NVRAM) module, among other components.

   1. The CPU module, housing the CPU and memory, executes computing operations to support the high-performance services of the storage array.
   2. The I/O expansion module expands the I/O interfaces of the controller module.
   3. The onboard interface module integrates the interface module directly into the controller module.
   4. The BMC module manages out-of-band management for controller modules and the entire storage array.
   5. The CPLD module handles functions such as power-on and power-off control, external serial port management, clock monitoring, and indicator control, often collaborating with the BMC module to manage hardware systems.
   6. The system disk module stores the operating system of the storage array.
   7. The NVRAM module stores data when the storage array experiences unexpected power loss.

2. **Why do the RAID data updates cause obvious disk I/O amplification issues? What kind of write mechanism can be designed for a storage system to effectively overcome the write penalty issues?**

   Answer:

   1. When a single piece of data within a RAID group is modified, it necessitates simultaneous modifications to parity data or other data within the RAID stripe to maintain consistency, leading to additional write amplification issues.
   2. The ROW full-stripe write mechanism allows modified data to be written to any physical space, and RAID full-stripe write can be ensured through I/O aggregation. This approach helps reduce write amplification. However, it introduces new issues such as GC.

3. **What are the weaknesses of space management when using conventional RAID algorithms? How can we improve the design to address these weaknesses?**

   Answer:

   1. Conventional RAID algorithms suffer from several weaknesses in space management. One such weakness is that conventional RAID group reconstruction involves only a few member disks within the group, limiting the reconstruction rate and prolonging the reconstruction period. Moreover, despite an increase in the number of disks, the storage array alocates LUN

space to a host solely from the RAID group. Therefore, the performance of a single LUN may be limited by the number of member disks in the RAID group, meaning that the increasing number of disks may not significantly enhance performance.
2. To address these weaknesses, the design can be improved in the following ways: One approach is to use disk modules as the space management unit at the underlying layer. At the upper layer, multiple RAID groups are aggregated into a space pool, which is then divided into small-granularity management units (e.g., 8 MB/unit). These units are recombined and mapped to LUNs visible to users. This method allows the distribution of single LUN space to multiple disks to overcome LUN performance bottlenecks. However, since only a few underlying disks are used for reconstruction, the reconstruction time is not reduced significantly. Another way is to adopt a two-layer virtualization management mode for underlying disks and upper-layer resources. For underlying layer virtualization, disks in a storage pool are divided into small-granularity data blocks before a RAID group is created. RAID groups are then created based on these data blocks instead of the entire disks, ensuring even data distribution across all disks. For upper-layer virtualization, fine-grained partitioning is performed to divide the space of multiple RAID groups into small-granularity units before mapping them to LUN space. These units are then recombined and mapped to LUN space. This approach resolves issues related to both reconstruction rate and LUN performance bottlenecks.

4. **How does read/write cache improve storage array performance? What issues are introduced by the cache system while improving performance? How can we solve these issues?**

    Answer:

    1. Cache uses the principle of locality to enhance performance. Caching technology leverages the temporal locality principle in the data access model to improve efficiency, while content prefetching technology benefits from the spatial locality principle.
    2. Although caching can improve performance, it also introduces data reliability issues. Upon writing data to a controller's cache, a success message is sent to the host. However, if the controller fails before data is written to disks, accessing the correct data becomes problematic for the other controller even if it assumes service. This issue necessitates cache mirroring technology. To mitigate this, the temporary data in the write cache should be mirrored to another controller's write cache for backup before sending a response message to the host.

5. **How can we evaluate the high reliability design of controllers in a storage array?**

    Answer: The evaluation of high reliability design in storage arrays encompasses not only ensuring high fault tolerance capability for uninterrupted service

3.5 Practice Questions 71

continuity but also optimizing device performance and alarm capability to enhance user experience during failures.

There are four levels of tolerance for controller failure:

1. Level 1: Services remain uninterrupted if a single controller fails.
2. Level 2: Services remain uninterrupted if any two controllers in a multi-controller cluster fail.
3. Level 3: Services remain uninterrupted even if any engine in a multi-engine cluster fails. An engine, typically housing two or four controllers, forms the core of a controller enclosure. Scaling out storage arrays involves adding engines, with failover primarily occurring within an engine. Therefore, maintaining continuous services in the event of a single engine failure is crucial for achieving high reliability in multi-engine clusters.
4. Level 4: Services remain uninterrupted even if the controllers in a multi-controller cluster fail sequentially until only one functioning controller remains.

There are three levels of user experience in the event of a failure:

1. Level 1: I/O performance deterioration or suspension lasts less than 30 s. A link disconnection alarm is reported, and the host executes a multi-path switchover.
2. Level 2: I/O performance deterioration or suspension lasts less than 5 s. A link disconnection alarm is reported, and the host executes a multi-path switchover.
3. Level 3: I/O performance deterioration or suspension lasts less than 1 s. No link disconnection alarm is reported, and the host does not execute a multi-path switchover.
4. To ensure the high reliability and service continuity of mission-critical production systems during controller failures, a storage array should provide fault tolerance capability of at least Level 2 and a user experience of Level 3 or better.

6. **What are the problems with the unbalanced architecture? What are the key differences between a fully symmetric balanced architecture and a partially symmetric balanced architecture?**
   Answer:

   1. There are two problems with the unbalanced architecture. First, the networking is complex. This is because each controller must be connected to all hosts that use its owning volume or file system to prevent services from being delivered to a nonowner controller. Otherwise, a large number of I/Os are forwarded within the storage system, which wastes forwarding channel resources and seriously affects performance. Second, the performance of a volume or file system is limited by the processing capability of a single controller. Idle resources, such as CPU resources, on nonowner controllers cannot be utilized to boost the volume or file system performance.

2. Achieving balanced distribution across three layers—front-end networks, controllers, and back-end disk enclosures—is feasible. In a partially symmetric architecture, balanced distribution occurs mainly at the controller layer. This means data from volumes or file systems can be evenly distributed among all controllers, ensuring services are evenly distributed across each controller. However, this architecture may not ensure uniform distribution across front-end interface modules or back-end disk enclosures. Conversely, a fully symmetric architecture ensures services are evenly distributed across all three layers. Front-end interface modules are not restricted to a single controller, and back-end disk enclosures are not confined to one controller or engine. Therefore, services can be distributed from any interface to all controllers and disks, achieving a fully-symmetric balanced distribution. This not only enhances performance but also improves usability and reliability.

7. **In RAID 6 (8+2) scenarios, what are the write amplification values of the conventional write-in-place mechanism and ROW mechanism, respectively?**
   Answer:
   1. Conventional write-in-place mechanism: In the Read-Modify-Write mode, if one piece of data is modified, one piece of old data and two parity bits need to be read. After the new parity bits are calculated, the new data and two parity bits are written onto the disks simultaneously. Therefore, three read I/Os and three write I/Os are generated, and the read/write amplification is 6.
   2. ROW mechanism: As ROW supports full-stripe write, only full-stripe data needs to be aggregated. Two parity bits are calculated and written to disks. Write amplification = (number of data blocks + number of parity blocks)/ number of data blocks = (8 + 2)/8 = 1.25.

8. **How does the multistreaming technology that enables collaboration between disks and controllers optimize GC performance?**
   Answer: Data in a storage system can be separated into cold and hot data. An example of hot data is the metadata of a system, which is updated frequently and is more likely to generate garbage data. An example of cold data, on the other hand, is user data that is rarely modified, which is unlikely to generate garbage data. The industry's mainstream storage arrays use the multistreaming technology to enable collaboration between disks and controllers so that hot and cold data can be stored in different blocks. In the blocks where hot data is stored, a large amount of data becomes invalid in a short period of time, thus generating a large amount of garbage and reducing the amount of valid data to be migrated during GC. In contrast, the blocks where cold data is stored are seldom modified, minimizing the need for data migration during GC. This improves GC performance.

References

1. Maier D, Vance B. A call to order. In: Beeri C, editor. Proceedings of the twelfth ACM SIGACT-SIGMOD-SIGART symposium on principles of database systems (PODS '93). New York: Association for Computing Machinery; 1993. p. 1–16.

References

2. Wilkes J, Golding R, Staelin C, et al. The HP AutoRAID hierarchical storage system. ACM Trans Comput Syst. 1996;14(1):108–36.
3. Chen P, Lee E, Gibson G, et al. RAID: high-performance, reliable secondary storage. ACM Comput Surv. 1994;26(2):145–85.
4. Katz R, Chen P. RAID-II: Design and implementation of a large scale disk array controller[R/OL]. (1992-01-01) [2023-04-10].
5. Denning P. The locality principle. Commun ACM. 2005;48(7):19–24.

Chapter 4
Storage Protocols

Storage protocols are used to connect storage devices to hosts for the purposes of data communication and exchange. The current computer storage architecture primarily uses storage block protocols to access data on storage devices based on a multiple of a fixed data block size. Small Computer System Interface (SCSI) and non-volatile memory express (NVMe) are typical storage block protocols, which can be developed into dedicated storage protocols based on the types of bearer links between hosts and storage devices. This chapter will elucidate the models and key command sets of the SCSI and NVMe protocols, along with their respective link bearer protocols, as well as the Compute Express Link (CXL) protocol.

4.1 SCSI Protocol

The SCSI protocol is a set of standards for connecting and transmitting data between hosts and peripheral devices. It has undergone three generations of iterative evolution, known as SCSI-1, SCSI-2, and SCSI-3 [1]. SCSI enables point-to-point connections between multiple hosts and peripheral devices and defines the command set, communication protocol, electrical model, and communication interface required for such connections. In theory, SCSI allows a host to communicate and exchange data with any device. However, for commercial purposes, it is primarily used for communication and data exchange with storage devices.

4.1.1 SCSI Protocol Overview

SCSI has a huge protocol architecture, as seen in Fig. 4.1. From top to bottom, the SCSI standards architecture defines four layers: SCSI device command set, SCSI Primary Commands (SPC), SCSI Architectural Model (SAM), and SCSI physical link mapping.

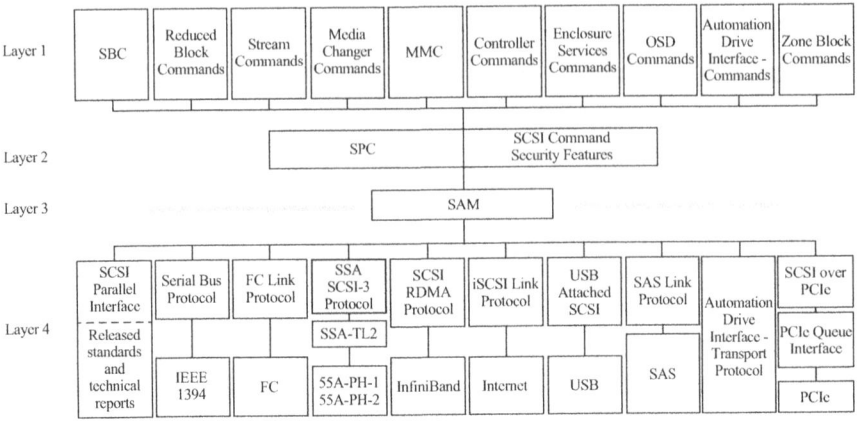

Fig. 4.1 SCSI standards architecture [1]

At the first layer, the SCSI device command set defines the necessary command sets for connecting different types of devices, including SCSI Block Commands (SBC) for block devices, Multimedia Commands (MMC) for multimedia devices, and Object-Based Storage Devices (OSD) for object storage devices.

At the second layer, the SCSI shared command set defines the SPC and security features for the communication models of all SCSI device types.

At the third layer, SAM provides an abstract view of the SCSI architecture and defines the common standards and specifications for communication between different SCSI devices.

At the fourth layer, SCSI link mapping specifies how the SCSI protocol is implemented over bearer links of different protocols, including Fibre Channel (FC), Serial-Attached SCSI (SAS), and Internet Small Computer System Interface (iSCSI).

4.1.2 SCSI Service Model

The SCSI service uses the traditional client/server (C/S) communication model. A client, which is the initiator, sends a request to a server, which is the target. Upon receiving a response from the target, the initiator establishes a request/response model with the target. Figure 4.2 shows a basic distributed service model using the SCSI protocol.

A dotted arrow represents a request/response transaction of a single command between the client and server, while a solid arrow represents the physical communication implemented by the command between the client and server using the service delivery subsystem.

A communication transaction of the SCSI service model can be described as a procedure call. The client sends a procedure call service request, and the server

4.1 SCSI Protocol

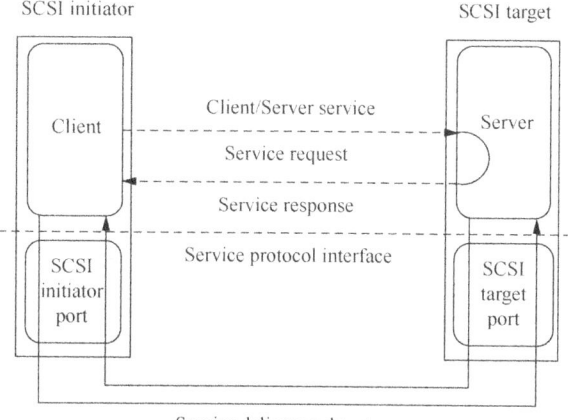

Fig. 4.2 SCSI distributed service model [1]

returns output data and a procedure call response status. The detailed process is as follows:

- Step 1: The client sends a service request to the server through the SCSI initiator port.
- Step 2: The service delivery subsystem transmits the service request.
- Step 3: The server receives the service request through the SCSI target port and executes the requested service.
- Step 4: The server returns the response information and result to the client.
- Step 5: The client receives the service request response result or failure message from the server. A failure message means the current server has a fault in responding to the client.

In a traditional computer storage system, a host and a storage device function as a client and a server, respectively. They communicate and exchange data using the SCSI service model.

The SCSI protocol enables point-to-point communication between multiple hosts and devices. Therefore, the SCSI service model can be further extended to a client/server model. In Fig. 4.3, a SCSI initiator device provides multiple application clients, and a SCSI target device provides multiple logical units. Each logical unit includes a device server and a task manager. The detailed service process is as follows:

- Step 1: An application client encapsulates a device service request or task management request into a procedure call.
- Step 2: The application client sends the device service request or task management request to a logical unit of the SCSI target device.
- Step 3: The logical unit of the SCSI target device receives the request and performs command processing or task management.
- Step 4: The logical unit of the SCSI target device returns the device service response or task management response to the application client.

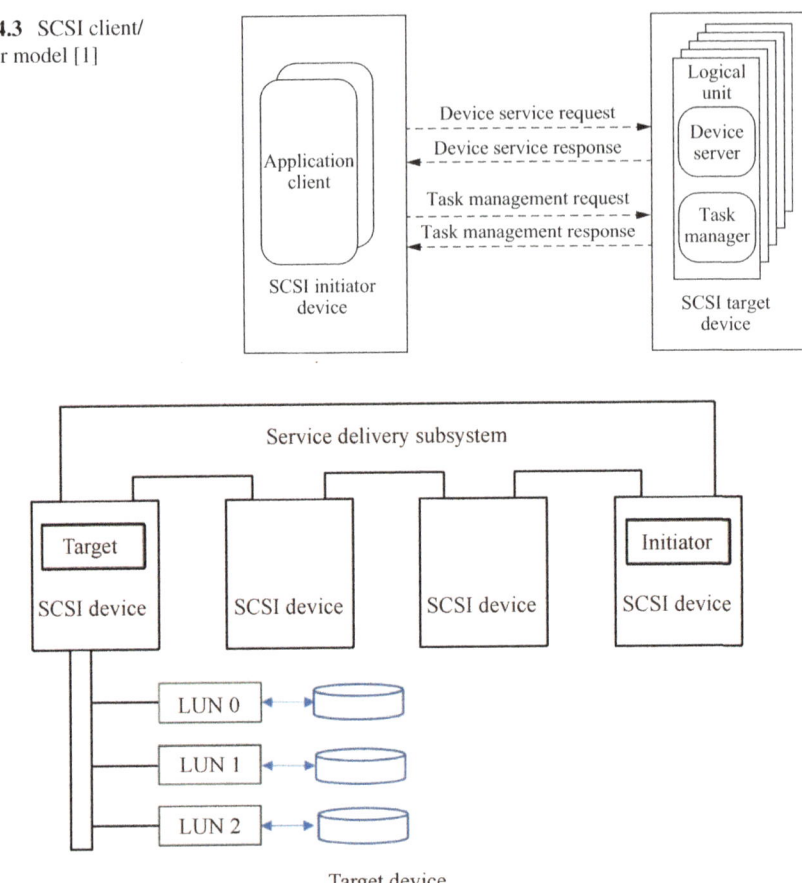

Fig. 4.3 SCSI client/server model [1]

Fig. 4.4 SCSI domain model [1]

The communication interaction process of the SCSI client/server model depends on the SCSI domain model. In Fig. 4.4, multiple SCSI devices are interconnected by the service delivery subsystem. The SCSI device can be instantiated as a client initiator device, a server target device, or relevant infrastructure. The service delivery subsystem transmits information such as commands, data, and task management functions. After a SCSI device is instantiated as a target device, it can include multiple logical units, and each logical unit maintains its own target device.

To facilitate device addressing when there are multiple SCSI devices, the SCSI protocol uses three levels of addresses in the service delivery subsystem: bus ID, SCSI device ID, and logical unit number (LUN) ID. The bus ID identifies each SCSI bus in the service delivery subsystem; the SCSI device ID identifies each SCSI device connected to the SCSI bus; and the LUN ID identifies each target device on a SCSI device. This device addressing mechanism of the SCSI protocol allows different SCSI devices to effectively communicate and exchange data.

4.1.3 SCSI Command Sets

The SCSI standards architecture encompasses a series of command sets, mainly the SCSI device command set and SCSI shared command set [1], for communication and data exchange between SCSI devices.

The SCSI device command set mainly includes the SBC and SCSI Stream Commands (SSC), used for connecting different types of devices. The SBC defines a set of interfaces for accessing block devices (such as HDDs, SSDs, and CD-ROMs) through logical blocks. For example, the FORMAT UNIT command sets a device to a specified logical block format; the READ command reads the data of several logical blocks from specified locations in a device; the WRITE command writes the data of several logical blocks to specified locations in a device; and the COMPARE AND WRITE command performs atomic operations of comparison and writing. The SSC defines a set of interfaces based on the sequential access model for stream devices (such as tapes). For example, the ERASE command erases data from some or all media, and the LOCATE command is used to locate specified media.

The SCSI shared command set defines commands that are relevant to devices. For example, the INQUIRY command queries information about target devices and logical units; the REPORT LUNS command obtains the logical unit list; and the TEST UNIT READY command checks the logical unit status.

4.1.4 SCSI Read/Write Process Analysis

This section further describes the basic read/write process of SCSI. In a SCSI read/write process, the SCSI initiator and target devices must first establish a session for transmitting read/write commands and data.

Figure 4.5 (a) illustrates the steps in a SCSI data read process:

1. The application sends a SCSI read command to the SCSI initiator.
2. The SCSI initiator obtains the right to use the bus, selects the target device, and performs addressing.
3. The SCSI initiator sends Command Descriptor Block (CDB) information to the SCSI target device.
4. After receiving the CDB information, the SCSI target device reads the data and prepares for data transmission.
5. The SCSI target device transmits the data.
6. Upon completion of data transmission, the target device returns a completion command.
7. The SCSI initiator returns a read completion command to the application.

Figure 4.5 (b) illustrates the steps in a SCSI data write process:

1. The application sends a SCSI write command to the SCSI initiator.
2. The SCSI initiator obtains the right to use the bus, selects the target device, and performs addressing.

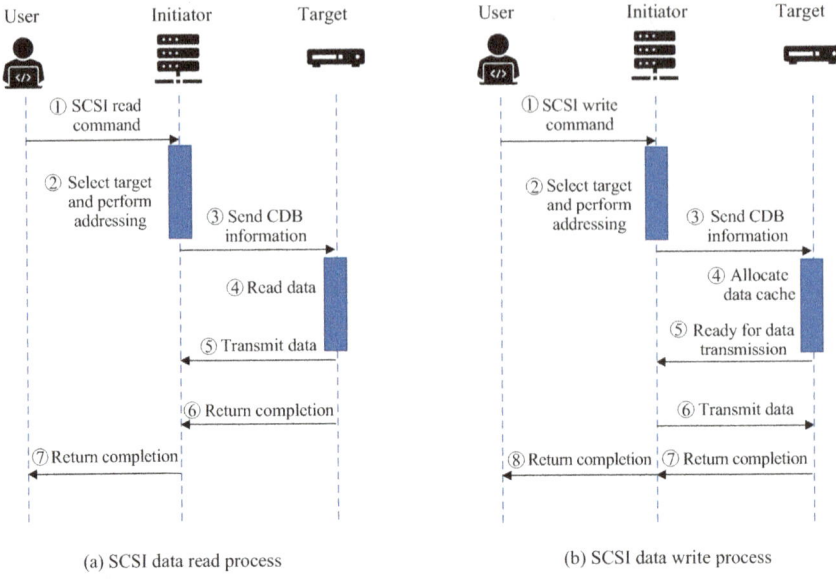

Fig. 4.5 SCSI read/write process

3. The SCSI initiator sends CDB information to the SCSI target device.
4. After receiving the CDB information, the SCSI target device allocates data cache and prepares to receive data.
5. The SCSI target device notifies the SCSI initiator of data transmission.
6. The SCSI initiator transmits the data.
7. Upon completion of data transmission, the SCSI target device returns a completion command.
8. The SCSI initiator returns a write completion command to the application.

4.2 SCSI Link Bearer Protocols

4.2.1 SAS Protocol

SAS is a point-to-point serial SCSI technology that uses standard SCSI command sets and is compatible with Serial Advanced Technology Attachment (SATA) devices. The SAS protocol establishes a dedicated communication channel between two devices, eliminating the need to determine the channel status of the parallel bus used in the conventional SCSI protocol. This results in improved data transmission bandwidth and efficiency.

Figure 4.6 illustrates the SAS protocol architecture, which consists of the SAS application layer, transport layer, port layer, link layer, PHY layer, and physical layer from top to bottom.

4.2 SCSI Link Bearer Protocols

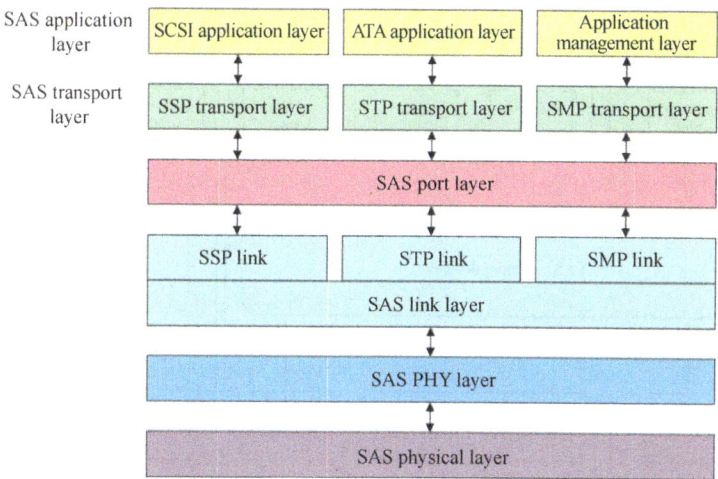

Fig. 4.6 SAS protocol layers [1]

Application layer maintains the use of the SAS protocol with all application software above the transport layer, including applications, file system drivers, SCSI drivers, and microport drivers. The application layer sends requests to the transport layer, specifies the protocol frame format and command type, and receives responses from the transport layer.

Transport layer receives requests from the application layer, encapsulates the requests into protocol frames, and sends the protocol frames to the port layer. It also receives and parses protocol frames from the port layer and sends them to the application layer. The transport layer supports three protocol types: Serial SCSI Protocol (SSP), SATA Tunneling Protocol (STP), and Serial Management Protocol (SMP). The SSP bears the SCSI protocol, the STP is compatible with the ATA command set, and the SMP manages communication between SAS devices.

Port layer provides transmission interfaces for the link layer and transport layer and determines how to establish and terminate links.

Link layer establishes and manages links between devices, which includes encapsulating and parsing SAS frames. The link layer supports SSP, STP, and SMP links according to the protocol types of the transport layer.

PHY layer encodes data transmission bit streams, manages clock offset, and processes out-of-band signals. The PHY layer is primarily designed for bottom-layer physical connections and blocks the physical connection details from upper layers.

Physical layer describes the specifications of SAS physical connection lines, interfaces, and transceivers.

Based on the SAS protocol layers, SAS devices can establish network communication with one another. Figure 4.7 shows the SAS networking topology, which

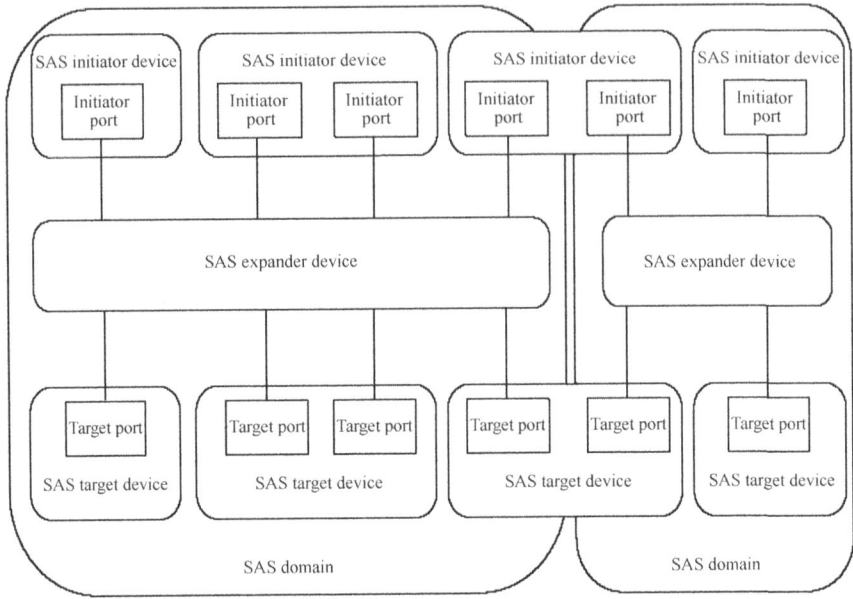

Fig. 4.7 SAS networking topology [1]

consists of SAS initiator devices, SAS expander devices, SAS target devices, and SAS domains.

The SAS initiator and target devices transmit and receive data through the SSP/STP/SMP interfaces. The SAS expander devices expand the SAS network by enabling the connection of more SAS devices. The SAS domain includes multiple SAS devices, physical links, and SAS expanders.

4.2.2 Fibre Channel Protocol

To ensure a solid foundation for covering the Fibre Channel Protocol (FCP), it is essential to understand the fundamental concepts of storage area network (SAN) [2]. SAN is a network-centric data storage architecture that connects external storage devices to servers over networks. Early SANs were mainly carried by FC connections, so the FCP was formulated to transmit SCSI commands over FC networks [3]. The FCP was proposed by the ANSI X3T9 task group in 1988 to provide high-performance and reliable data transmission for SANs.

Figure 4.8 illustrates the FCP architecture, which consists of FC-0 (physical layer), FC-1 (transmission protocol layer), FC-2 (signaling protocol layer), FC-3 (common service layer), and FC-4 (protocol mapping layer).

FC-0 defines the media and transmitter/receiver interfaces of physical transmission links.

4.2 SCSI Link Bearer Protocols

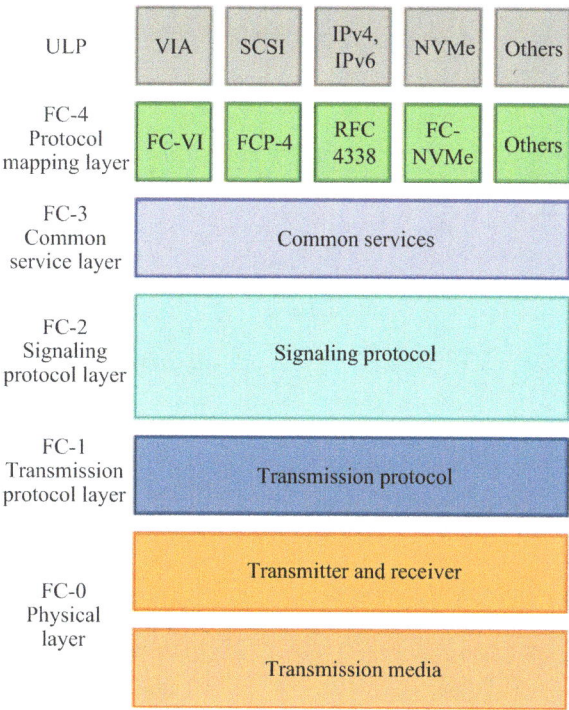

Fig. 4.8 FCP layers [3]

FC-1 defines data coding and the link transmission protocol, including the coding/decoding mode and error control.

FC-2 defines the signaling protocol (including the fundamental rules and mechanism for data transmission), data frame format, flow control mechanism, and QoS.

FC-3 provides common service capabilities for applications, including broadcast, encryption, and compression.

FC-4 specifies the interface for interaction between FC and the upper-layer protocol (ULP), such as the SCSI protocol.

The current FCP for network storage supports three topologies: point-to-point (P2P), arbitrated loop, and switched fabric.

The P2P network is the simplest and most efficient among these topologies. It enables communication between the initiator and target devices by directly connecting their node ports (N_ports) with a fiber cable.

The arbitrated loop network is an early topology. During communication, two node loop ports (NL_ports) exclusively occupy a loop, causing poor communication efficiency.

The switched fabric uses switches to forward datagrams and allocate device addresses. This topology, capable of supporting up to 224 nodes, is widely employed in large-scale data centers. The performance of network transmission is heavily

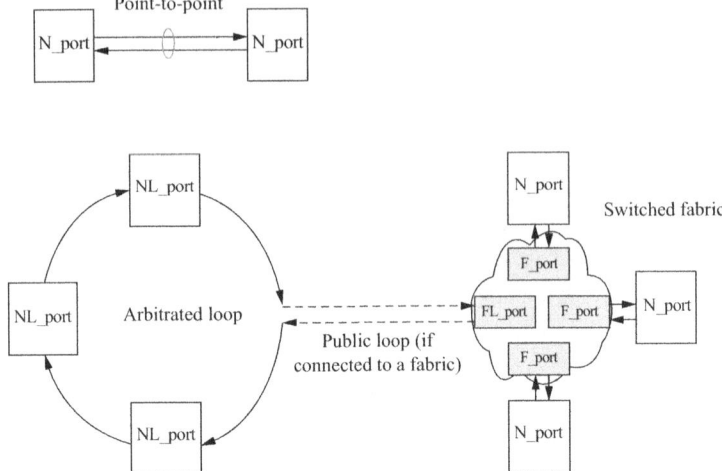

Fig. 4.9 FC network topology [3]

reliant on effective network planning. A well-crafted network plan is essential for maximizing the performance of both the host (initiator) and storage (target) systems while also preventing congestion issues. The switched fabric provides the best performance and scalability among the three topologies. Figure 4.9 shows a node accessing the switched fabric by connecting an N_port to a fabric port (F_port). In addition, the switched fabric offers fabric loop ports (FL_ports) for interconnection with arbitrated loop networks.

4.2.3 iSCSI Protocol

In addition to FC, SCSI can also be mapped to TCP/IP networks. This section provides an overview of the iSCSI protocol, a TCP/IP-based SCSI protocol that enables device interconnection over Ethernet [4]. The iSCSI protocol uses the C/S architecture, wherein the client, typically deployed on the host, serves as the iSCSI initiator device, while the server, typically deployed on the storage device, functions as the iSCSI target device.

The communication process with the iSCSI protocol is as follows: After receiving a SCSI command from the SCSI client, the iSCSI initiator encapsulates the command into an iSCSI packet and sends it to the iSCSI target device over Ethernet. Upon receiving the iSCSI packet, the target device parses and executes the SCSI command, encapsulates the execution result into another iSCSI packet, and returns the packet to the initiator. Finally, the initiator relays the execution result back to the SCSI client. Fig. 4.10 depicts the iSCSI protocol architecture.

4.2 SCSI Link Bearer Protocols

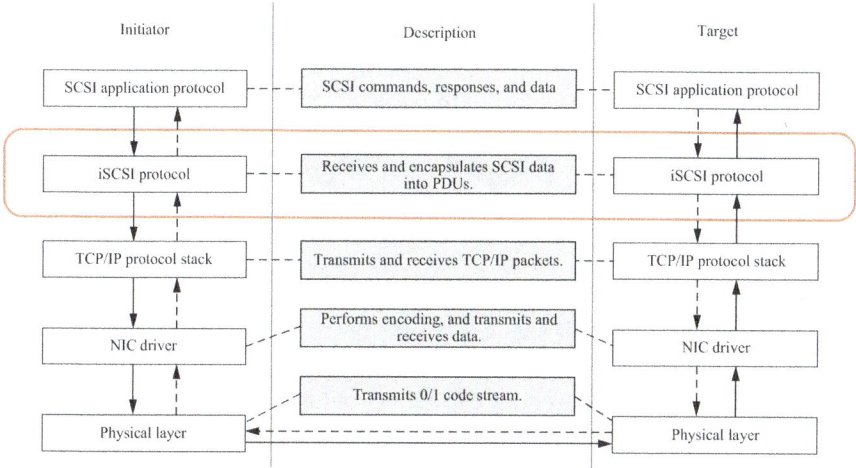

Fig. 4.10 iSCSI protocol architecture [4]

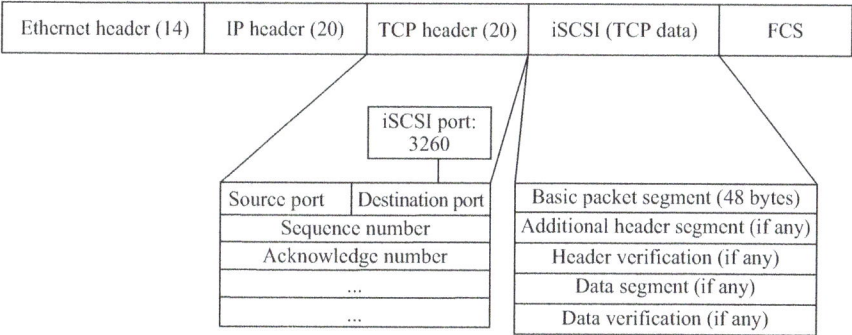

Fig. 4.11 iSCSI-based TCP/IP packet structure [4]

Because the iSCSI protocol is carried by Ethernet, the iSCSI protocol data unit (PDU) is encapsulated as TCP packet data for transmission via TCP/IP. Fig. 4.11 shows the structure of an iSCSI-based TCP/IP packet. The destination port number in the TCP packet header is the port number of the iSCSI destination host. The iSCSI packet is encapsulated as TCP data.

Figure 4.12 shows the iSCSI network structure. The iSCSI network structure fully utilizes existing Ethernet infrastructure to enable communication and data exchange between iSCSI initiator and target devices. By reusing the TCP/IP protocol within the Ethernet infrastructure for underlying network connections, the iSCSI protocol minimizes the development and application costs associated with its implementation.

The iSCSI protocol has gained widespread support from storage vendors and has been widely implemented in server operating systems due to the strong

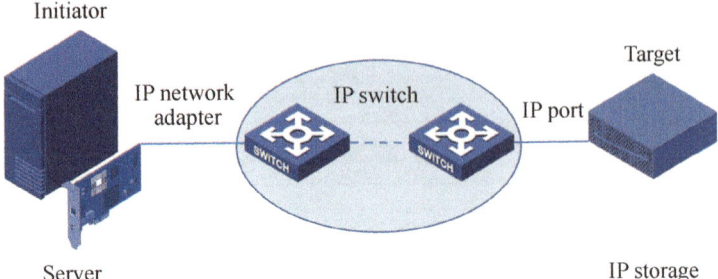

Fig. 4.12 iSCSI network structure [4]

compatibility and robust networking capabilities of Ethernet, which provide support for the fundamental link connections of the iSCSI protocol. Additionally, the high scalability of TCP/IP allows for seamless integration of the iSCSI protocol into transmission networks, realizing cost-effective remote data transmission.

4.3 NVMe Protocol

The NVMe protocol is a storage access and transport protocol defined for SSDs. Early SSDs were designed to use traditional SATA/SAS protocols to minimize changes to the legacy HDD-based storage systems. However, these protocols were never intended for high-speed storage media like NAND flash. Now that SSDs have much lower media access latencies, traditional SATA and SAS protocols make up a higher proportion of overhead costs in links, protocol stacks, and software stacks. Consequently, the NVMe protocol was developed to address the incompatibility of traditional protocols with high-speed storage media.

The NVMe protocol uses the Peripheral Component Interconnect Express (PCIe) bus to access SSDs [5]. A storage device is directly connected to the CPU via a high-speed bus to reduce the latencies of the controller and software interfaces. Unlike the traditional mechanism that uses a single command queue, the PCIe bus supports tens of thousands of command queues for parallel processing. To improve performance and reduce latency, the NVMe protocol outlines a set of efficient command interaction mechanisms that are suitable for OS device drivers running in either interrupt or polling mode [5]. Currently, PCIe 3.0 links can transmit data more than twice as fast as SATA links. Since its widespread adoption, the NVMe over PCIe protocol has been standardized to ensure compatibility across manufacturers. The NVMe protocol leverages nonvolatile memory in diverse compute environments, offering a forward-looking and scalable solution that can be seamlessly integrated with future persistent memory technologies.

The latest NVMe 2.0 protocol was released on June 3, 2021, which includes specifications for the NVMe base, transport, command set, and management interfaces. Figure 4.13 shows how they are related. The NVMe base specification defines

4.3 NVMe Protocol

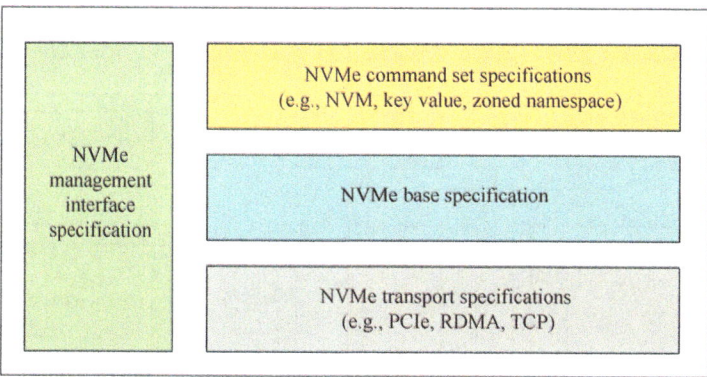

Fig. 4.13 NVMe protocol specifications [5]

a protocol for host software to communicate with nonvolatile memory subsystems over a variety of memory-based transports and message-based transports. The NVMe transport specifications define the binding of the NVMe protocol (including controller properties) to a specific transport protocol (e.g., transport specifications of PCIe, RDMA, TCP, and FC). The NVMe command set specifications define the data structures, features, log pages, commands, and status values (e.g., NVM, key value, and zoned namespace) that extend the NVMe base specification. The NVMe management interface (NVMe-MI) specification defines an optional management interface for all NVMe subsystems [5].

4.3.1 NVMe Device Model

The NVMe storage device model includes the following objects:

Subsystem: This contains multiple domains and controllers, zero or more namespaces, and multiple ports. An NVM subsystem may include a nonvolatile memory storage medium and an interface between the controller(s) in the NVM subsystem and nonvolatile memory storage medium.

Doman: A domain is the smallest indivisible unit that shares state (e.g., power state and capacity information). The boundaries between domains are generally communication boundaries (e.g., fault boundaries and management boundaries).

Endurance group: This manages the NVM(s) in the NVM subsystem based on endurance. An endurance group contains one or more NVM sets. It is used to manage wear leveling, different types of media (hybrid media), etc.

NVM set: An NVM set is logically (and sometimes physically) separate from other sets. It works with the predictable latency mode (PLM) feature to achieve a consistent I/O latency. One or more namespaces can be created for an NVM set.

Namespace: This is a formatted nonvolatile memory space accessible to hosts. A namespace is equivalent to a SCSI LUN. Each namespace belongs to an NVM set.

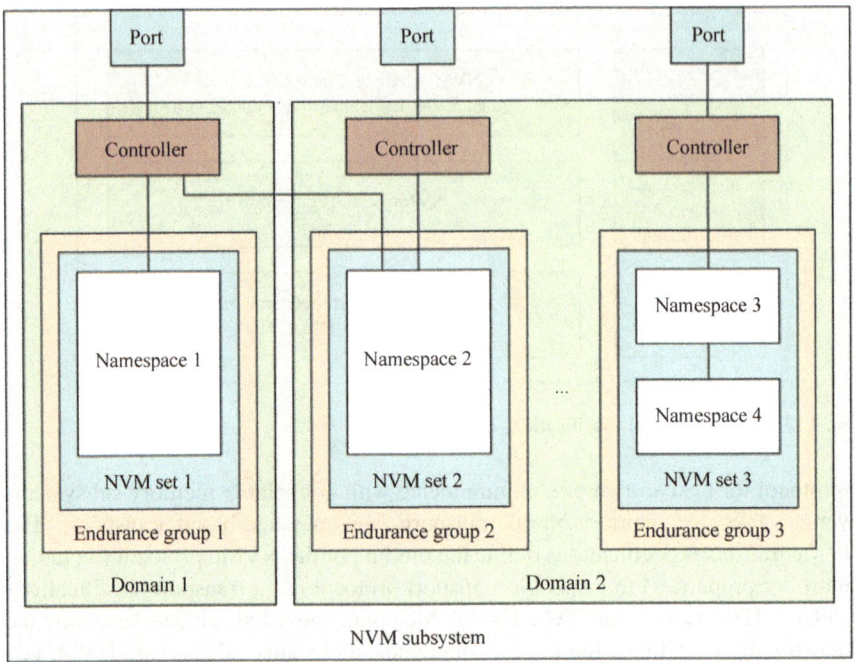

Fig. 4.14 NVM subsystem [5]

Figure 4.14 illustrates the relationship between these objects in an NVM subsystem with multiple domains.

Figure 4.15 shows the device model of a dual-port NVMe SSD. The NVMe SSD functions as a comprehensive NVM subsystem, which has two PCIe ports that connect to two NVMe controllers respectively to access the namespaces within the SSD. The SSD has two endurance groups (Y and Z), accommodating TLC and SLC flash media, respectively. Endurance group Z, which contains SLC media, delivers superior access performance and is ideal for storing hot data. Endurance group Y, which contains TLC media, is divided into NVM sets A and B based on flash memory channels, offering two storage areas with different access latencies. Each NVM set is further partitioned into multiple logical areas, i.e., namespaces, which can be formatted into sectors of different sizes and with different data consistency protection policies.

4.3.2 NVMe Queue Model

Multi-queue technology is crucial for improving NVMe performance. It allows NVMe to manage and allocate queues based on tasks, scheduling priorities, and CPU core loads, enabling high-performance access to storage systems.

4.3 NVMe Protocol

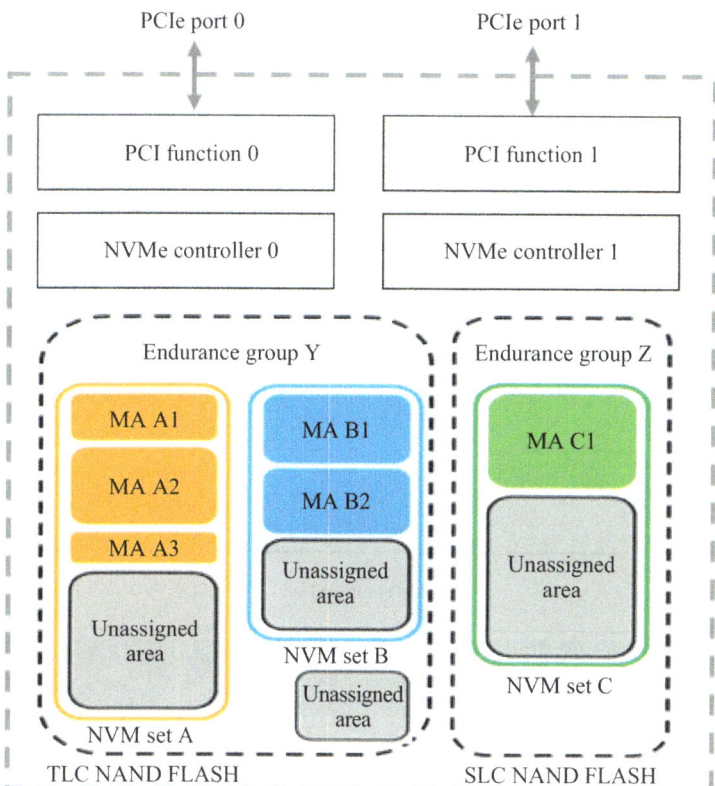

Fig. 4.15 NVMe SSD device model [5]

Figure 4.16 shows the NVMe multi-queue model, in which a group of NVMe queues facilitate command exchange between a host and a device (NVMe controller). These queues are paired Submission Queues (SQs) and Completion Queues (CQs). Each SQ or CQ is a segment of memory, organized as a circular buffer. Each queue has a head pointer and a tail pointer. When the two pointers point to the same address, the queue is empty. Host software places commands into the SQ. The controller fetches the commands from the SQ in order and places completions into the associated CQ.

In the memory-based transmission queue model, multiple SQs may utilize the same CQ. In the message-based transmission queue model, each SQ is mapped to a CQ. For I/O queues, a CQ can be utilized by one or more SQs; for Admin queues, SQs and CQs must be a 1:1 correspondence. Only commands that are part of the Admin or Fabrics command set may be issued to the Admin SQ.

An NVMe controller supports one Admin queue and 65,535 I/O queues, with a queue depth of up to 65,535 commands per queue. In contrast, each SATA or SAS queue can only support 32 or 256 commands, respectively. The NVMe design of queue depth increases the number of channels between computers and storage devices, leading to a significant enhancement of the IOPS of SSDs.

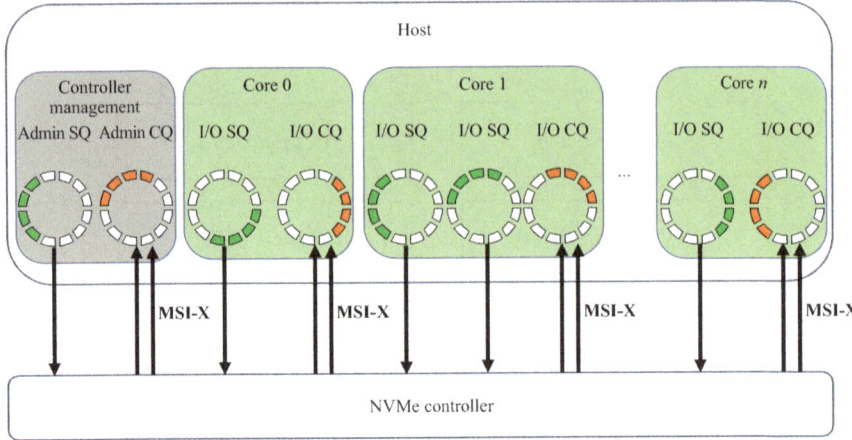

Fig. 4.16 NVMe multi-queue model [5]

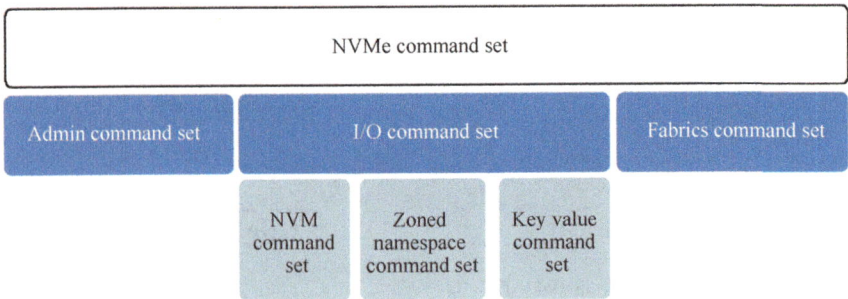

Fig. 4.17 NVMe command set [5]

4.3.3 NVMe Command Set

As shown in Fig. 4.17, NVMe defines three types of command sets: Admin, I/O, and Fabrics. The Admin command set exists for creation and deletion of I/O SQs and CQs, aborting commands, etc. An I/O command set is used with an I/O queue pair and includes the NVM command set, key-value command set, and zoned namespace command set. The Fabrics command set defines commands associated with NVMe over Fabrics operations, e.g., establishing connections, authenticating, and obtaining or setting properties. All Fabrics commands can be issued to the Admin SQ, and some of them can also be issued to the I/O SQ. Unlike Admin and I/O commands, Fabrics commands are processed by the controller.

The following analyzes the command data structure, using the read command as an example. A read command is an I/O command that reads data and metadata from the I/O controller for the LBAs indicated. This command may specify the protection information to be checked as part of the read operation. If the command uses

Physical Region Pages (PRPs) for data transfer, then the Metadata Pointer, PRP Entry 1, and PRP Entry 2 fields are used. If the command uses scatter/gather lists (SGLs) for data transfer, then the Metadata SGL Segment Pointer and SGL Entry 1 fields are used.

4.3.4 NVMe over PCIe

The NVMe over PCIe protocol uses PCIe for NVMe transport. PCIe provides reliable channels for memory-mapped data transfer of Admin and I/O commands. Like most general-purpose PCIe transports, NVMe over PCIe uses common PCIe capabilities, such as memory-mapped I/O for data transfer and register access, PCIe configuration space, and PCIe Message Signaled Interrupts (MSI) or Extended Message Signaled Interrupts (MSI-X).

Figure 4.18 shows the following eight-step NVMe over PCIe command interaction process:

1. The host software (driver software) submits one or more commands to the SQ entry that the SQ tail pointer indicates.
2. The host software writes the updated SQ tail pointer to the Doorbell register associated with the SQ and notifies the NVMe controller of a new I/O.
3. The NVMe controller fetches several commands from the SQ header for execution. This may be done using different algorithms like round-robin or weighted round-robin.

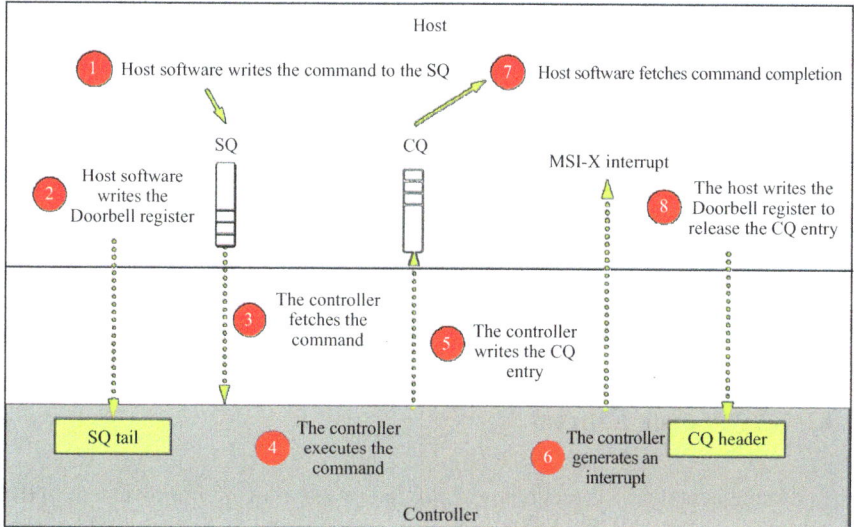

Fig. 4.18 NVMe over PCIe command interaction process [6]

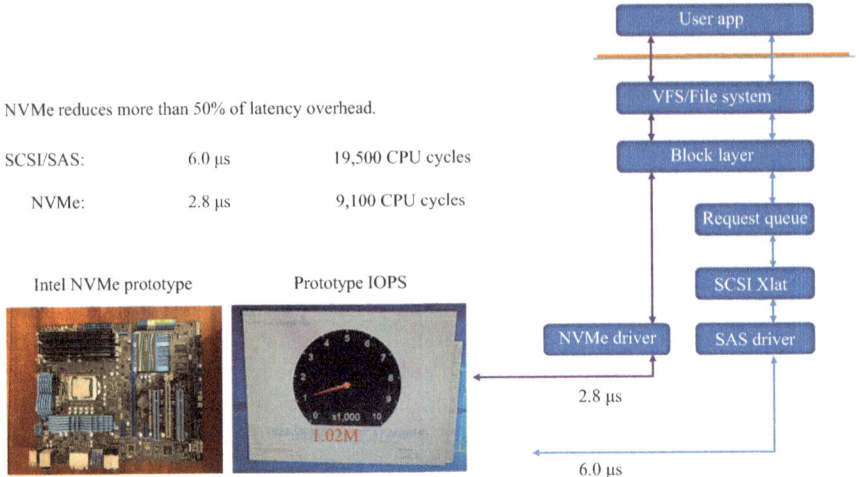

Fig. 4.19 A comparison of NVMe and SCSI software stacks [7]

4. The NVMe controller executes the commands. Generally, the direct memory access (DMA) engine on the PCIe device is used to read and write data.
5. The NVMe controller writes the command completion into the CQ entry that the CQ tail pointer indicates.
6. The NVMe controller generates an MSI/MSI-X interrupt to notify the host software.
7. The host software fetches the command completion from the CQ header.
8. The host writes the updated CQ header to the Doorbell register associated with the CQ and notifies the NVMe controller of releasing the CQ entry.

During this process, the host only needs to submit commands to the SQ and process the command completion in the CQ. The NVMe controller automatically completes the rest of work.

NVMe changes the software stack when used for SSD interfaces. Fig. 4.19 compares the NVMe and SCSI software stacks. The NVMe driver directly connects to the block layer in the Linux kernel, eliminating the software overheads associated with SCSI. In addition, NVMe's multi-queue design optimizes the utilization of multiple CPU cores and minimizes resource contention on I/O paths, significantly improving IOPS and reducing latency.

4.4 NVMe over Fabrics

The advancements in SSD technology have led to a significant improvement in the performance of local NVMe SSDs. However, conventional SCSI-based SANs continue to rely on the FC and iSCSI protocols. This means that hosts continue to use

4.4 NVMe over Fabrics

the SCSI protocol and miss out on the benefits of NVMe's high concurrency and simplified command structure. To address this issue, the storage industry introduced the NVMe over Fabrics (NVMe-oF) protocol in 2016. It extends NVMe access by using a network fabric and is fully compatible with the NVMe architecture, so there is no need for NVMe-to-SCSI conversion. This has drastically improved storage access performance [7].

NVMe-oF defines a general-purpose architecture for accessing NVMe block storage devices over various transport networks. The protocol's front-end ports can extend large-scale networks from data centers to remote NVMe devices and NVM subsystems. NVMe-oF supports multiple transport protocols, including remote direct memory access (RDMA), FC, and TCP, as shown in Fig. 4.20.

The NVMe over RDMA protocol is further categorized into NVMe over RoCE, InfiniBand (IB), and Internet Wide Area RDMA Protocol (iWARP). NVMe over RoCE carries the NVMe protocol using the RDMA over Converged Ethernet (RoCE) technology. NVMe over RDMA maps an NVMe device's I/O queues to the RDMA queue pairs (QPs) and completes I/O interactions using the RDMA SEND, RDMA WRITE, and RDMA READ semantics.

Both NVMe over FC (FC-NVMe) and FC-SCSI are built upon the FCP and utilize exchanges for their I/O interactions. FC-NVMe utilizes existing FC network infrastructure to carry the NVMe protocol for better transmission performance.

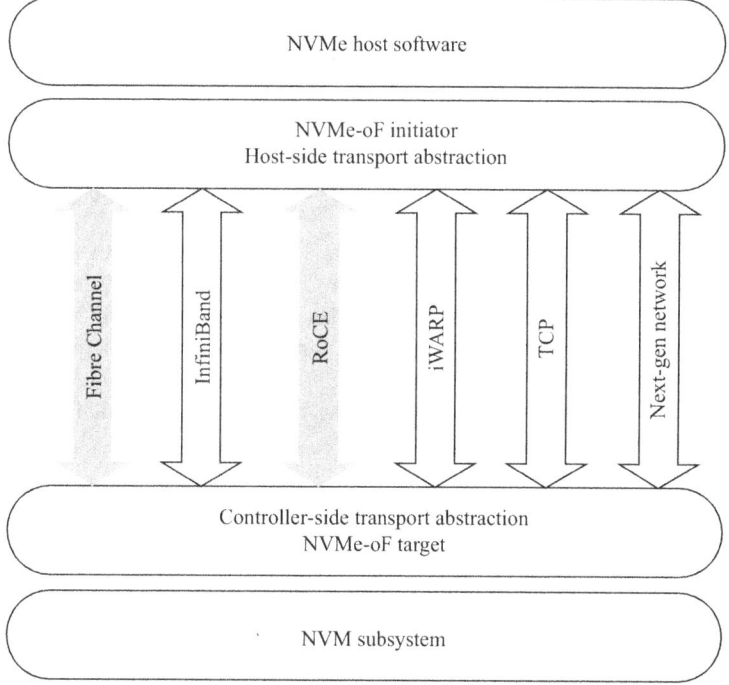

Fig. 4.20 NVMe over Fabrics transport layer [8]

NVMe over TCP leverages existing IP networks and the TCP protocol to transmit NVMe commands, enabling seamless implementation of end-to-end NVMe without changing the network infrastructure.

Table 4.1 compares the transport layer protocols.

Currently, these transport layer protocols for NVMe-oF coexist and complement each other. In practice, the selection of protocols is determined by specific application scenarios and requirements. From the available options, NVMe over RoCE is the optimal transport protocol for NVMe-oF because it offers higher bandwidth and lower latency than FC and also supports all-Ethernet and all-IP. These advantages make it the prevailing technology for NVMe-oF in the industry.

4.4.1 NVMe over RDMA

RDMA utilizes hardware and network technologies to directly access data from the memory of a remote computer without OS intervention, featuring high bandwidth, low latency, and low resource consumption. RoCE makes it possible to deploy RDMA on the most widely used Ethernet to achieve high speed, ultra-low latency, and minimal CPU overhead. This section uses NVMe over RoCE as an example to describe the concept of NVMe over RDMA.

Figure 4.21 shows the RDMA network protocol stack, including RoCE v1, RoCE v2, IB, and iWARP. NVMe over RoCE carries NVMe-oF services over a lossless IP network. RoCE v2 is a network layer protocol that introduces IP to solve scalability issues and implement routing across Layer 2 networks. The service and storage networks are integrated into the IP network to facilitate management and O&M. Because RoCE requires lossless Ethernet, the IP SAN can be planned independently or integrated with the service network. Currently, the majority of RoCE applications are compatible with RoCE v2.

NVMe over RDMA implements multi-queue I/O processing for NVMe in the RDMA queues. Each RDMA SQ/CQ maps to an NVMe SQ/CQ for end-to-end multi-queue processing. NVMe protocol packets are directly transmitted as RDMA data without the need to change the packet format. Figure 4.22 shows the queue mapping of NVMe over RoCE v2.

Figure 4.23 shows the advantages of RDMA over the traditional TCP/IP protocol stack:

- Kernel bypass: User-mode applications directly operate network device interfaces without system invocation, eliminating the overhead cost associated with switching between kernel mode and user mode.
- CPU offload: The CPU does not participate in data processing. The NICs complete the data transmission process without involving software or consuming CPU resources.
- Zero replication: The processing time is reduced by eliminating data replication between the buffers of each layer.

4.4 NVMe over Fabrics

Table 4.1 A comparison of the NVMe-oF transport layer protocols

Protocol	Network	Performance	Application scenario
FC-NVMe (NVMe over FC)	Traditional FC network with upgraded software	Medium. The transport layer does not change. FC-NVMe uses multiple queues and simplifies the host protocol stack to provide slightly better performance than FC-SCSI	This protocol is used to support NVMe-oF on the FC network infrastructure
NVMe over RoCE	Lossless Ethernet	High. RDMA is used to improve performance	This protocol is used to support NVMe-oF on the lossless Ethernet in IP data centers
NVMe over IB	IB network	High. RDMA is used to improve performance	This protocol is used to support NVMe-oF by upgrading the IB networks, e.g., for high-performance computing
NVMe over TCP	Common Ethernet	Low. The TCP transport layer does not change. NVMe over TCP uses multiple queues and simplifies the host protocol stack to provide slightly better performance than iSCSI. TCP can be offloaded to the NIC chipset to improve performance	This protocol is used to support NVMe-oF on the legacy IP networks

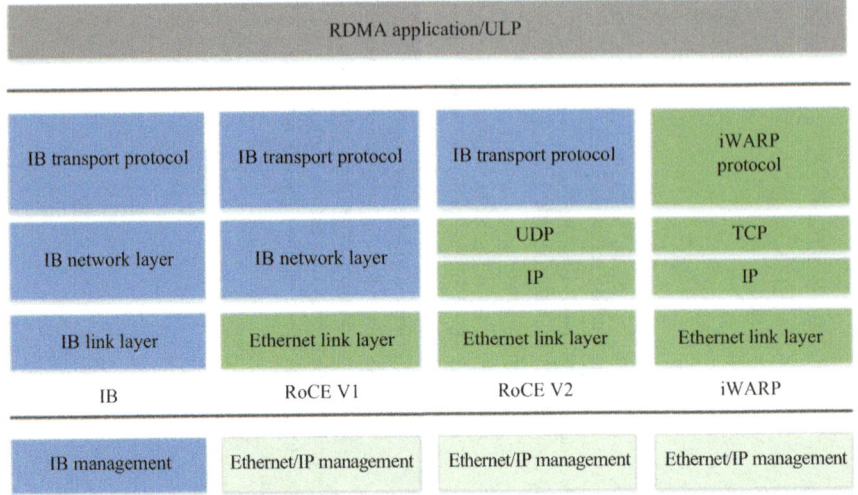

Fig. 4.21 RDMA network protocol stack [9]

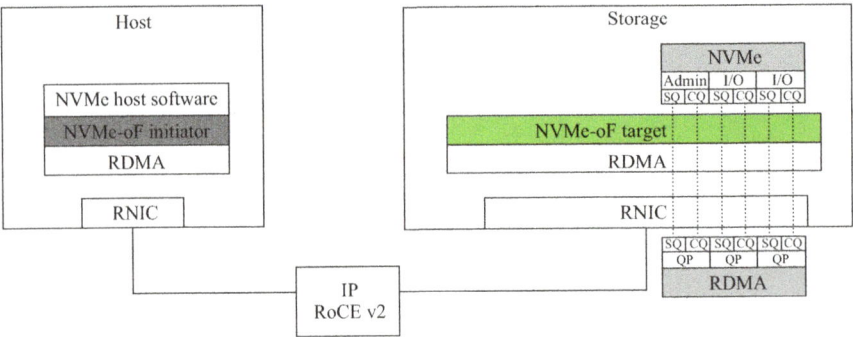

Fig. 4.22 NVMe over RoCE v2 queue mapping. Note: RNIC refers to RDMA network interface controller

Figure 4.24 shows the NVMe over RDMA read I/O process. The steps are as follows:

1. The host builds and places an NVMe-oF read command in an RDMA queue.
2. The target device fetches the read command from the RDMA queue, builds an NVMe read command and fills it in an SQ on the SSD, and then notifies the controller.
3. The SSD executes the read command, writes the data to the buffer of the target device through DMA, and returns a CQ entry (CQE) to notify the target device. Then, the target device initiates an RDMA write operation to transfer the data to the host.

4.4 NVMe over Fabrics

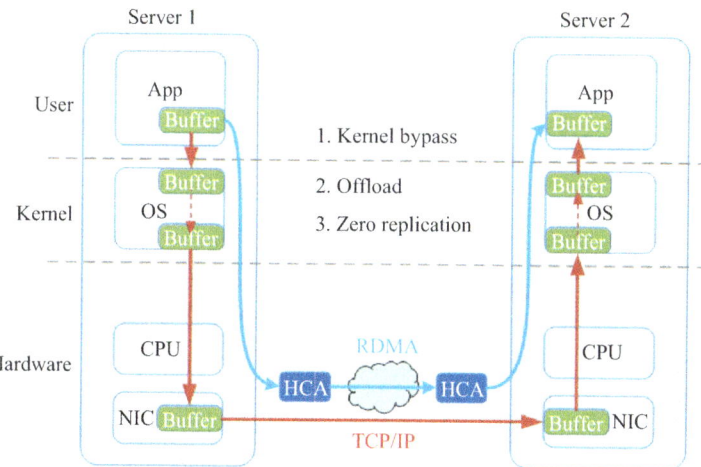

Fig. 4.23 RDMA data flow offload [10]. Note: HCA refers to host channel adapter; NIC refers to network interface controller

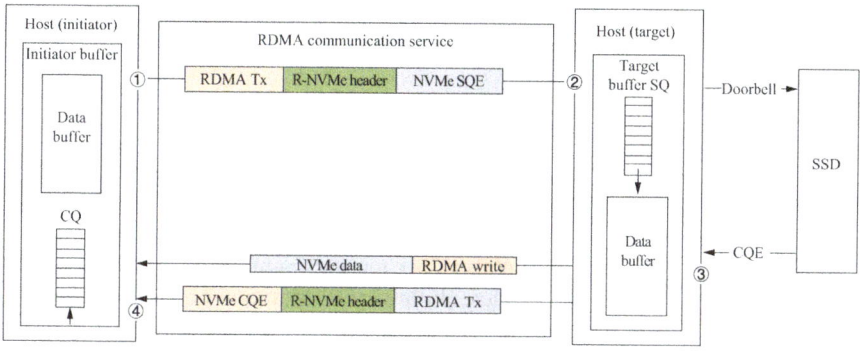

Fig. 4.24 NVMe over RDMA read I/O process [10]

4. The target device builds a CQE, writes it to a CQ within the host, and notifies the host of read completion.

Since data replication between data buffers is handled by NICs, the CPU achieves zero replication.

Figure 4.25 shows the NVMe over RDMA write I/O process. The steps are as follows:

1. The host builds and places an NVMe-oF write command in an RDMA queue, with the option to immediately transfer small data (e.g., less than 8 KB) with the command.
2. The target device fetches the NVMe-oF write command from the RDMA queue, translates it into an NVMe command, writes it to an SQ on the SSD, and then

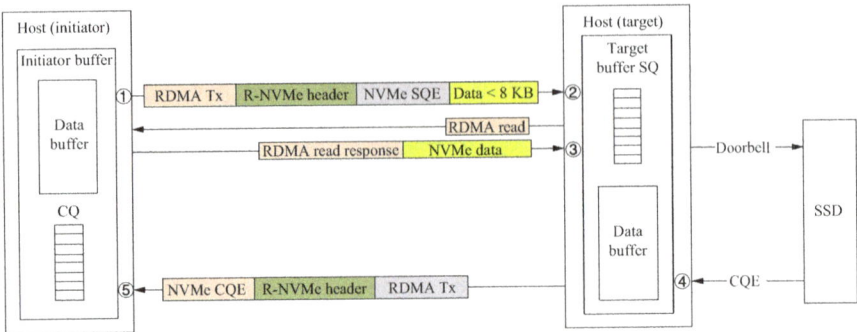

Fig. 4.25 NVMe over RDMA write I/O process [11]

notifies the controller. If the command is accompanied with immediate data, the data is received.
3. The target device initiates an RDMA read operation to fetch data from the host. Once the data is received, the target device writes it to the SQ on the SSD and notifies the controller.
4. The SSD writes the data and returns a CQE.
5. The target device builds a CQE, writes it to a CQ within the host, and notifies the host of write completion.

4.4.2 NVMe over TCP

The NVMe over TCP protocol supports only message semantics. Like the FC protocol, TCP has no memory semantics. With the incorporation of TCP support, the NVMe system can become equivalent to a SCSI system, making it compatible with almost all networks. Figure 4.26 depicts the NVMe over TCP software stack. NVMe over TCP leverages existing IP networks and the native TCP protocol stack in Linux, eliminating the need for any changes to system hardware or software.

NVMe over TCP facilitates communication between storage arrays by providing high bandwidth, low latency, and secure physical isolation. It transmits data over common TCP switching networks without modifying the underlying network infrastructure. Its simplicity and high efficiency make it suitable for use in ultra-large data centers. Currently, NVMe over TCP is widely used in ultra-large flash storage environments, especially for rapid access to large data volumes over existing high-bandwidth switching networks while maintaining a low latency. NVMe over TCP supports fast read access and allows data distribution to multiple data centers. Furthermore, it offers advantages in terms of power consumption, cooling efficiency, and local high availability while avoiding any additional costs associated with expanding fiber networks.

Figure 4.27 illustrates how NVMe over TCP sets up a connection. The host and the controller first set up a standard TCP connection and then an NVMe connection.

4.4 NVMe over Fabrics

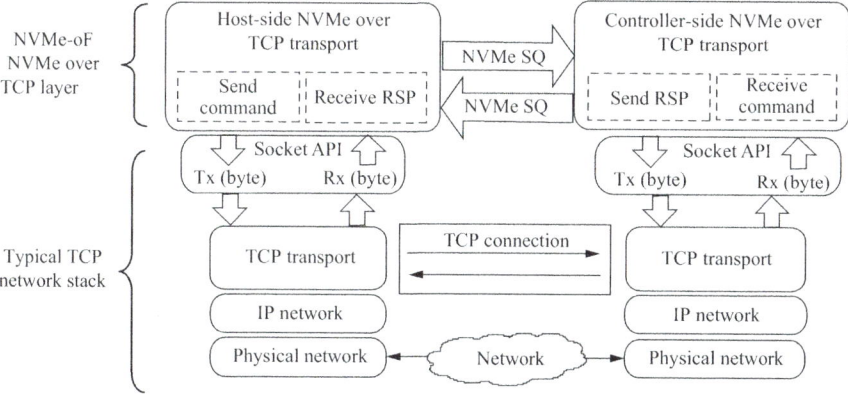

Fig. 4.26 NVMe over TCP software stack [8]

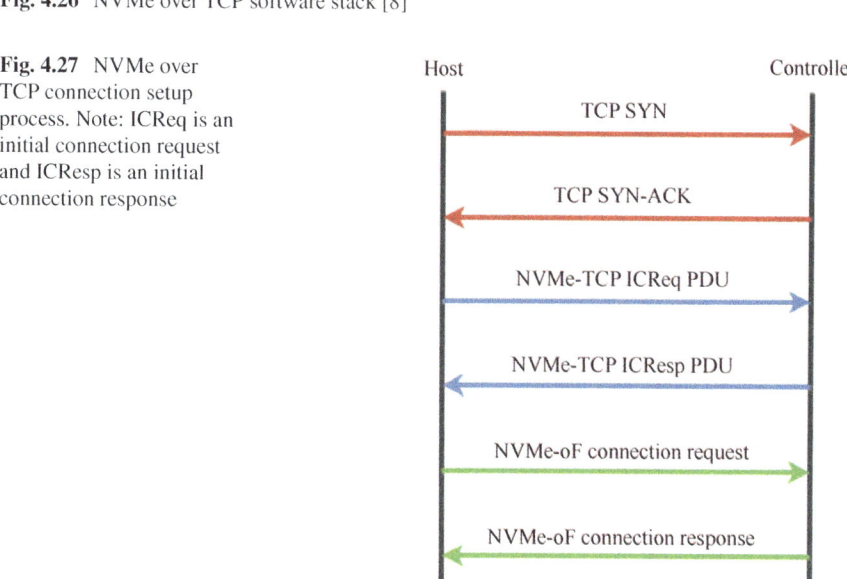

Fig. 4.27 NVMe over TCP connection setup process. Note: ICReq is an initial connection request and ICResp is an initial connection response

During this process, multiple parallel NVMe I/O queues are injectively mapped to an equal number of parallel TCP/IP connections. This mapping between NVMe and TCP enables a simple end-to-end parallel architecture.

4.4.3 NVMe over FC

The NVMe over FC protocol is implemented through the FC-NVMe standard. FC-NVMe directly encapsulates NVMe packets into FCP packets and simplifies the NVMe command set by converting it into basic FCP commands.

FC-NVMe makes full use of existing FC infrastructure, including switches and NICs. Users can simply upgrade the firmware and drivers to support FC-NVMe and improve performance. FC-NVMe provides the same structure, predictability, and reliability features as FC-SCSI. Additionally, one FC network fabric can handle both NVMe-oF traffic and SCSI traffic, allowing for a seamless transition without the need for extensive replacement of the existing network infrastructure. Figure 4.28 shows an NVMe over FC network architecture that supports both FC-SCSI and FC-NVMe. Despite its advantages, FC-NVMe encounters several challenges, such as closed network technologies, limited bandwidth, and complex O&M processes. FC can only support a maximum bandwidth of 64 Gbit/s, which can cause a bottleneck when accessing storage at a network rate of 100–200 Gbit/s. Moreover, the O&M of FC networks heavily relies on the support of the original manufacturer, who may lack O&M personnel and respond to requests slowly.

Figure 4.29 depicts the FC-NVMe protocol layers. Two nodes communicate and exchange data over an FC network. A single node consists of the NVMe host software layer, NVMe-oF layer, FC-NVMe layer, and virtual node port layer. The NVMe host software layer specifies the software and subsystem specifications of the NVMe nodes. The NVMe-oF layer specifies an NVMe protocol extension framework based on network transport. The FC-NVMe layer further defines interface and transport standards for FC and finally connects to the FC network through node ports.

As shown in Fig. 4.29, the FC-NVMe protocol leverages mature FC network technologies and uses physical FC networks to maximize the performance advantages offered by the NVMe protocol.

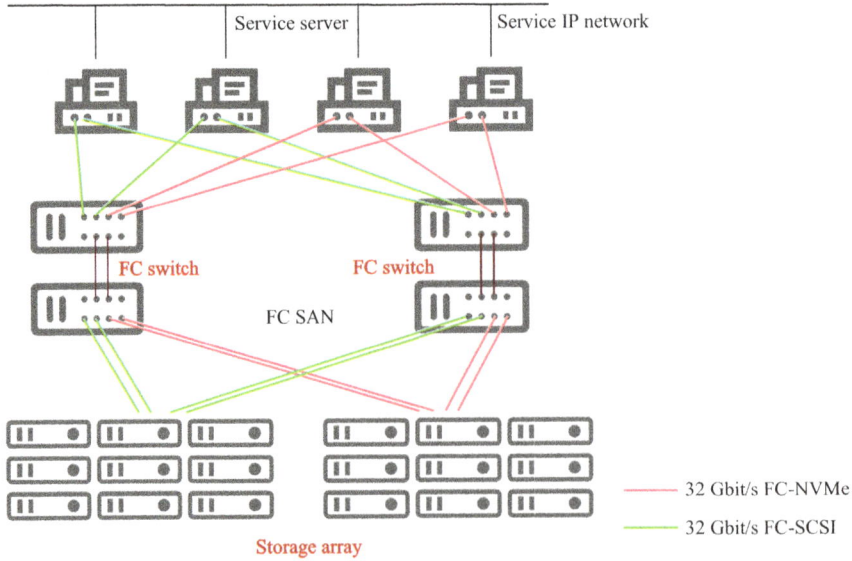

Fig. 4.28 NVMe over FC network architecture

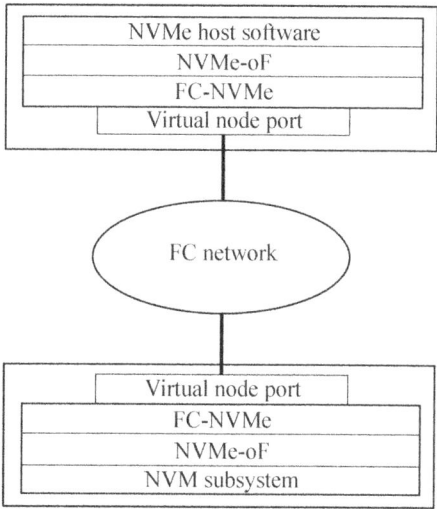

Fig. 4.29 FC-NVMe protocol layers [8]

4.5 Compute Express Link Protocol

With the exponential growth of data, the global semiconductor industry is undergoing a significant transformation poised to deliver breakthroughs in performance, efficiency, and cost-effectiveness for data centers. Currently, mainstream server architectures are transitioning toward a disaggregated pooling model. In this model, hardware resources like memory and GPUs within a data center are interconnected via a cache-coherent interconnect bus, creating a resource pool that facilitates independent expansion of each resource. This pooling approach enables flexible, on-demand resource allocation and expansion, thereby enhancing resource utilization and reducing costs. Such advancements are expected to yield significant benefits for applications experiencing growing data demands, including big data analytics, graph computation, and artificial intelligence.

Sharing hardware resources between servers on traditional architectures presents challenges. In Fig. 4.30, the server on the left struggles to utilize the idle resources of the server on the right. To meet peak hour demands, data centers often need to configure excessive hardware resources on physical servers, leading to low resource utilization [11], as shown in Fig. 4.31. Moreover, traditional interfaces impose limitations during server maintenance and upgrades, often requiring coarse-grained hardware replacement or even complete system replacement. Resource disaggregation addresses these issues by enabling on-demand resource provisioning and offering a comprehensive array of compute and memory instances, thereby reducing maintenance and upgrade costs.

The industry has conducted extensive research on hardware disaggregation, pooling, system interconnects, and even special architectures like blade servers [13]. However, limitations of internet bandwidth and latency have hindered further development of these technologies. The emergence of high-speed network

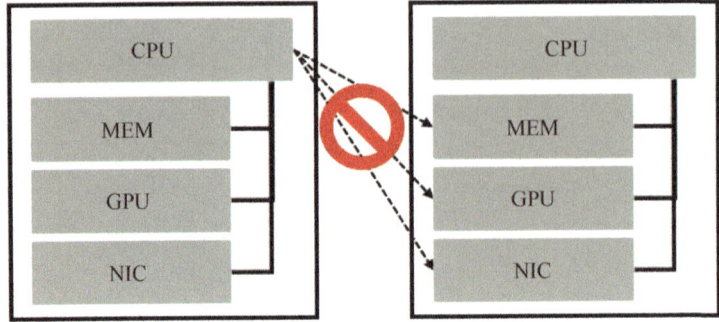

Fig. 4.30 An illustration of how traditional architecture may struggle with resource sharing

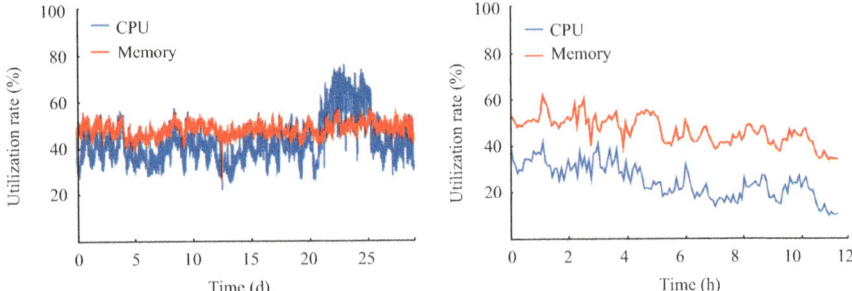

Fig. 4.31 Data center resource utilization [12]

technologies, such as RDMA, has renewed interest in disaggregation technologies within the industry [14, 15]. Take disaggregated memory as an example. Most existing research leverages the virtual memory subsystem, which uses a page fault trap to operate the RDMA software stack for data fetching, prefetching, and eviction of a memory page between different local and remote memory spaces. A similar process is required to share other resources: The OS and device driver coordinate to maintain data consistency between the local CPU memory, device-attached memory, and remote memory.

Such resource-sharing solutions driven by high-speed networks do not change the server architecture and are well compatible with new hardware. However, they add microsecond-level latency overheads to key paths such as system interrupts, context switching, software stacks, and network data transmission. The hardware disaggregation solution powered by the cache-coherent interconnect aims to transform the server architecture so that compute devices such as processors can directly access remote data without involving dedicated drivers on key paths, and this would have the added benefit of reducing latency to the sub-microsecond level.

Despite its advantages, the adoption of hardware disaggregation based on cache-coherent interconnect remains uncommon in the industry. Yet, industry vendors have reached a consensus of using Compute Express Link (CXL) as the cache-coherent interconnect for devices.

4.5 Compute Express Link Protocol

4.5.1 CXL Overview

The CXL specification, now at version 3.0, enables low-latency and coherent data access between devices. Currently, CXL uses the same physical layer as PCIe (as shown in Fig. 4.32) to minimize changes to hardware and facilitate seamless integration of CXL with the mature PCIe ecosystem. However, CXL and PCIe have very different external interfaces. CXL defines three types of protocol interfaces: CXL.io, CXL.cache, and CXL.mem. CXL.io is a non-coherent I/O interface used to initialize, discover, and enumerate devices and access device registers. It functions in the same way and follows the same device interaction process as PCIe. As the foundational CXL protocol on the control plane, CXL.io applies to all CXL devices.

CXL.cache allows devices to efficiently access and cache host memory for enhanced performance. For example, by using CXL.io and CXL.cache, a NIC can directly cache CPU memory. This means that in the case of an RDMA write request, the host can write the data to the NIC cache without involving the CPU or bus. This avoids frequent invocation of software drivers and helps maintain consistency. CXL.mem allows a host to directly access device-attached memory using load/store commands, thereby avoiding the software stack overhead caused by memory paging.

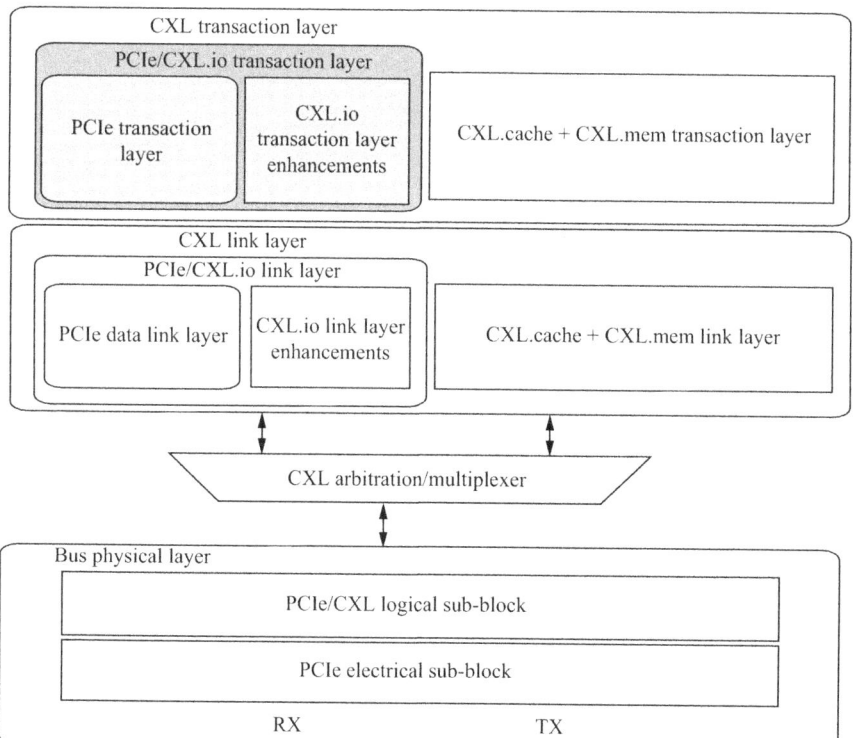

Fig. 4.32 CXL reusing the PCIe physical layer and bus [16]

The three types of CXL protocols can be flexibly combined to deliver the desired hardware features. CXL supports three types of devices:

Type 1: small-memory devices with computing power, such as programmable NICs. These devices have limited memory resources and must frequently access CPU memory to perform complex atomic operations and execute associated tasks. They can use CXL.io and CXL.cache to cache host memory to improve performance.

Type 2: large-memory devices with computing power, such as GPUs. These devices have abundant memory resources and interact with hosts in a more sophisticated manner. They can use CXL.io, CXL.mem, and CXL.cache to allow mutual memory access and caching with hosts, thereby improving performance under heterogeneous loads.

Type 3: large-memory devices without computing power, such as memory pools. These devices can use CXL.io and CXL.mem to expand memory size, improve memory access bandwidth, and add persistent memory without occupying local memory slots.

Asynchronous software drivers can simplify the programming interfaces in a CXL-based disaggregation architecture into shared memory interfaces with a unified memory space. This facilitates transparent architecture expansion, hardware maintenance, and upgrade of user programs while reducing the cost of pooling architecture.

4.5.2 CXL Type 1 Devices

Type 1 devices have special requirements for data access that cannot be satisfied by PCIe's Producer/Consumer ordering model. For example, a device cannot perform fetch or XOR atomic operations on host memory data based on PCIe. However, a Type 1 device can cache the host CPU memory using CXL.cache (illustrated by the green path in Fig. 4.33) and perform atomic operations with ownership of the cache

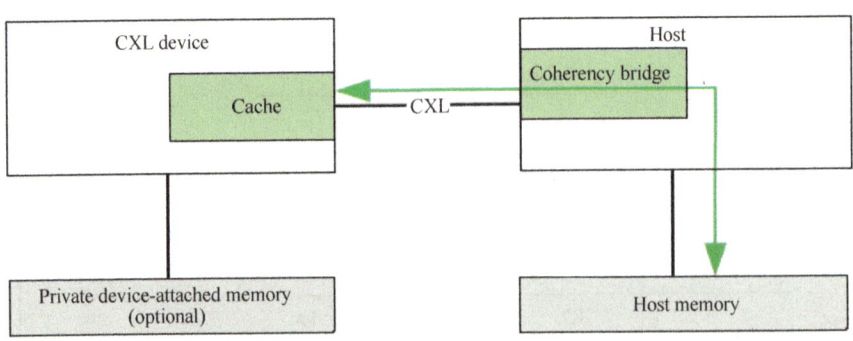

Fig. 4.33 Type 1 devices caching data in the host CPU memory [16]

4.5 Compute Express Link Protocol

line. In theory, this mechanism allows a device to implement any data access model and atomic command.

When a host needs to access the memory data cached by a device, the host temporarily loses permission to access the cache line. This triggers a sniffing process, which initiates a transaction to obtain cache line access permission and retrieve the latest data from the device. The cache line granularity, typically set at 64 bytes, is related to the granularity of internal cache management and the granularity at which sniffing can occur.

This type of device, such as an RDMA NIC, generally has a data input (inbound data) function. The process of writing host data to the device is as follows:

1. The device initiates a data write request and does not find the requested data in its cache.
2. The device's CXL bus controller sends a cache line request transaction to the host.
3. The host's CXL bus controller parses the transaction and reads the data from the host cache or memory. It also marks that the host does not have ownership of the cache line.
4. Finally, the host returns the requested data.

According to this process, when the host receives a cache line access request, the host finds that it does not have the cache line access permission. Therefore, it sends a cache line request transaction to the device to maintain cache coherency.

4.5.3 CXL Type 2 Devices

Type 2 devices are equipped with memory that is visible to hosts, as shown in Fig. 4.34. This type of device reads and writes data from its memory with extremely high bandwidth to boost performance. Type 2 devices use CXL.mem to enable hosts to access (read, write, and modify) the device-attached memory, as illustrated by the

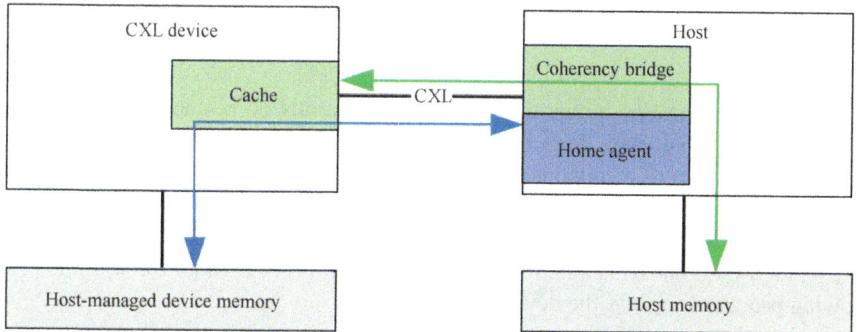

Fig. 4.34 Type 2 devices supporting access and caching of the host memory and mapping of the device-attached memory to the HDM for direct access by the host (load/store command) [16]

blue path in Fig. 4.34. This approach minimizes software and synchronization overheads to avoid impacting device acceleration. Device-attached memory that is mapped to the system coherent address space is called host-managed device memory (HDM).

Unlike with the HDM, the private device memory (PDM), for example, GPU memory, relies on conventional I/O interfaces. Since GPU memory is private, the host must use a software driver to replicate the PDM to host memory to read and write the PDM using commands. Similarly, devices must use the software driver for replication to read and write host memory. The HDM supported by CXL helps minimize memory replication between the host and the device.

During HDM access, sniffing is performed to ensure coherency. However, excessive sniffing may increase bus load. Although the HDM is attached to the device, the host may access a specific HDM more frequently than the device. When this happens, a Host Bias coherency model can be used to allow the host to treat the HDM as its own memory. When the device needs to access the HDM in Host Bias mode, it needs to send a request to the host to maintain coherency. By contrast, in a Device Bias coherency model, the host does not cache the HDM, so the device does not need to send a coherency request.

These two models help improve CXL performance in various access modes. Type 2 devices manage the coherency models at a specific granularity (using a bitmap to represent one page with 1 bit). A device needs to send a synchronization request to flush the host cache. The device-attached memory can be accessed after the synchronization request has been sent. Devices can employ various management mechanisms, such as software collaborative control or automatic hardware selection, to maintain a coherency model table. When a CXL device accesses the HDM, it checks the table first. If the HDM is set to Device Bias mode, the device can directly access the HDM and set the transaction packet to deny caching by the host when responding to host access requests. If the HDM is set to Host Bias mode, the host can directly access the HDM, and the device must first send a synchronization request before it can access the data.

4.5.4 CXL Type 3 Devices

Type 3 devices have no computing power and generally do not need to cache host memory. Therefore, they do not need to use CXL.cache. These devices primarily rely on CXL.mem to respond to memory access requests from the host (illustrated by the blue path in Fig. 4.35). Type 3 devices are typically byte-addressable data storage pools, such as memory pools or persistent memory pools.

When a CPU (host) is able to cache device data, it needs to go through the following process to access the device:

1. In the event of a memory access command (read or write), the host searches for the desired data in the CPU cache but cannot find the data.

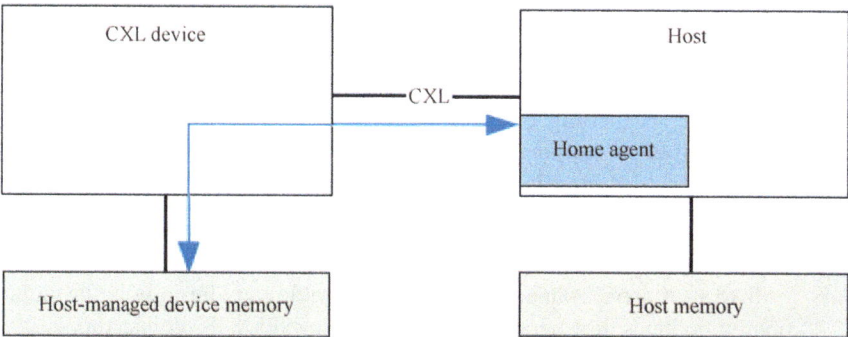

Fig. 4.35 Type 3 devices primarily relying on CXL.mem to provide memory expansion [16]

2. The access command, regardless of whether it is a read or write command, is translated into a read request aimed at retrieving the associated cache line to the CPU cache.
3. The memory management unit queries the read request and finds its destination address in the CXL bus address space. Then, the memory management unit translates the read request and forwards it to the CXL bus controller.
4. The CXL bus controller generates a CXL memory access transaction and sends it to the device's CXL bus controller via the CXL bus.
5. The device invokes the necessary modules, such as memory access transaction translation, address mapping, and the DRAM controller (or persistent memory controller if it is a persistent memory pool), to read the data.
6. Finally, the device's CXL bus controller returns the cache line.

A write request follows a similar process.

4.6 Summary

Storage protocols define how applications, hosts, and storage devices interact with each other. This chapter elaborates on the development of storage protocols, including the SCSI protocol, SCSI link bearer protocols, NVMe protocol, NVMe over Fabrics protocol, and the latest CXL protocol. These protocols have been established to meet users' data access and reliability requirements during the development of computer storage systems and to optimize the performance of new computer hardware. As we continue to develop new storage devices and technologies, storage protocols will also need to evolve to keep pace. It is important to seamlessly integrate traditional storage protocols with new storage architectures so that we can continue to drive rapid storage innovation.

4.7 Practice Questions

1. **What are the differences between Serial-Attached SCSI (SAS) and Fibre Channel (FC) regarding physical connections, data transmission modes, and the number of supported devices? When building a storage network, what factors determine which protocol should be used?**

   Answer: SAS vs. FC

   1. Physical connection:

      SAS uses serial connections and is applicable to point-to-point or multi-point topologies. It generally uses SFF-8088 or SFF-8482 connectors.

      FC uses optical fibers or high-speed copper cables and is applicable to point-to-point, point-to-multipoint, and multipoint-to-multipoint topologies. It generally uses LC or SC connectors.

   2. Data transmission mode:

      SAS enables high-speed serial data transmission and is well-suited for scenarios involving a small number of connected devices.

      FC also provides high-speed serial data transmission but applies to large-scale storage networks.

   3. Number of supported devices:

      SAS supports a small number of devices and is typically used for direct connections between servers and storage devices.

      FC supports a large number of devices and offers excellent scalability for establishing connections between enterprise-level storage systems.

   4. Performance and latency:

      SAS applies to latency-sensitive scenarios that require high performance, such as high-performance computing.

      FC provides low latency and high throughput, making it suitable for enterprise-class storage environments with high performance demands.

   5. Cost and deployment:

      SAS has a relatively low deployment cost, making it suitable for small- and medium-sized enterprises or small-scale storage environments.

      FC has relatively high deployment and maintenance costs and is applicable to large-scale enterprise-class storage.

      When selecting a protocol, it is important to consider factors such as storage requirements, budget, device scale, and performance requirements. SAS is most suited for small-scale storage environments with a limited budget, while FC is better suited for large-scale and high-performance enterprise-class storage systems.

2. **How does iSCSI enable the transmission of SCSI commands and data over a TCP/IP network?**

   Answer: The iSCSI protocol implements remote storage access by transmitting SCSI commands and data over a TCP/IP network. During this process, the iSCSI initiator triggers storage access by encapsulating the SCSI commands generated by applications into iSCSI protocol frames and sending them to the target device over a TCP/IP network. The iSCSI target, which is the target device, receives and parses the SCSI commands in the iSCSI frames, performs the corresponding read/write or control operation, encapsulates the operation result in the iSCSI frames, and then returns the frames to the iSCSI initiator. This establishes a transparent communication channel based on TCP/IP between the iSCSI initiator and target, enabling the host to access remote storage resources over the network just as it would access a local device. This flexible remote storage access method is widely used in modern data centers and network environments for convenient and efficient storage management.

3. **What is the queue model of the NVMe protocol, and why is it critical for improving storage performance?**

    Answer: The NVMe protocol has a Submission Queue (SQ) and a Completion Queue (CQ). The host submits NVMe commands in the SQ, which are then executed by the driver and their results placed in the CQ. This parallel processing and asynchronous communication model significantly enhance storage system efficiency.

4. **Why does the NVMe protocol provide higher performance than traditional SATA and SAS protocols on SSDs?**

    Answer: In comparison to traditional SATA and SAS protocols, the NVMe protocol maximizes the capabilities of SSD features. It facilitates multiple queues and parallel operations to leverage the enhanced parallel performance of SSDs, which greatly improves I/O concurrency and throughput. In addition, the NVMe protocol uses a low-latency design, simplifying the protocol layer and accelerating command processing speeds to reduce latency across the entire storage system effectively. NVMe SSDs using PCIe ports offer higher bandwidth and transmission rates compared to those using SATA ports. Furthermore, the NVMe protocol introduces features such as Command Queues, Doorbell registers, and advanced memory and cache management methods, further enhancing system concurrency and response speed. In summary, the NVMe protocol efficiently leverages SSD features like parallel performance, low latency, and high bandwidth, resulting in superior storage performance when compared to traditional protocols.

5. **How does the NVMe over Fabrics protocol overcome bottlenecks in traditional network storage and provide low latency and high bandwidth for remote storage access?**

    Answer: The NVMe over Fabrics (NVMe-oF) protocol effectively addresses bottlenecks found in traditional network storage, resulting in significant improvements in latency and bandwidth for remote storage access. NVMe-oF runs on high-performance network protocols like remote direct memory access (RDMA) and implements more efficient data transmission methods than traditional SCSI-based storage protocols by bypassing traditional protocol stacks. This streamlined design simplifies data processing along the transmission path, reduces CPU overhead, and greatly shortens the queuing and processing time of I/O requests. In addition, NVMe-oF facilitates direct memory access through protocols like RDMA, enabling a storage device to exchange data directly with a host without involving the host's CPU, thereby reducing storage access latency. This innovative architecture boosts performance for remote storage access, making NVMe-oF suitable for widespread use in large-scale data centers and storage networks.

6. **How does the NVMe over Fabrics protocol meet the low latency and high throughput requirements of high-performance computing (HPC) environments?**

    Answer: The NVMe over Fabrics (NVMe-oF) protocol offers substantial performance advantages for high-performance computing (HPC) applications. By transmitting NVMe protocol commands and data over networks, NVMe-oF meets the urgent requirements of HPC applications for low latency and high throughput. Its key advantages lie in dedicated network protocols, remote direct memory access (RDMA) technologies, and effective support for NVMe protocols.

    First, NVMe-oF uses dedicated network protocols and RDMA to enable data transmission over high-performance networks. RDMA facilitates direct memory access between hosts and storage devices, bypassing traditional protocol layers and minimizing communication latency. This capability enables storage devices to swiftly respond to latency-sensitive HPC applications.

    Second, NVMe-oF offers robust support for the NVMe protocol, renowned for its exceptional performance in local storage environments. In HPC scenarios, NVMe-oF fully leverages the advantages of the NVMe protocol, including the parallel queue model, asynchronous command processing, and low-latency design, to effectively fulfill the storage performance requirements of HPC applications.

7. **Describe the design objectives of the CXL protocol in detail, including its advantages for storage connections.**

    Answer: The CXL protocol is designed to deliver memory expansion, high performance, hardware accelerator support, and shared memory. The CXL protocol offers high bandwidth and low latency for storage connections, enabling efficient data transmission between computing devices and storage systems. Moreover, the protocol supports point-to-point, multipoint-to-point, and multipoint-to-multipoint topologies to address diverse hardware configurations and application needs. CXL also implements memory sharing through its memory expansion feature, enabling a computing device to remotely access the memory of another device. This promotes collaborative processing of shared data, ultimately enhancing system flexibility and performance. This comprehensive design makes CXL an innovative connection protocol that can play an important role in high-performance computing and storage.

8. **The CXL protocol defines three types of protocol interfaces: CXL.io, CXL.cache, and CXL.mem, which can be flexibly combined to deliver the desired hardware features. Explain how the three protocol interfaces can be combined to improve performance for small-memory devices with computing power (such as programmable NICs), large-memory devices with computing power (such as GPUs), and large-memory devices without computing power (such as memory pools).**

Answer:

Type 1: small-memory devices with computing power, such as programmable NICs. These devices have limited memory resources and frequently access CPU memory for complex atomic operations and associated tasks. They can leverage CXL.io and CXL.cache to cache host memory, thereby enhancing performance.

Type 2: large-memory devices with computing power, such as GPUs. These devices have abundant memory resources and engage with hosts in a more sophisticated manner. By utilizing CXL.io, CXL.mem, and CXL.cache, they enable mutual memory access and caching with hosts, enhancing performance under heterogeneous loads.

Type 3: large-memory devices without computing power, such as memory pools. These devices can use CXL.io and CXL.mem to expand memory size, improve memory access bandwidth, and add persistent memory without occupying local memory slots.

9. **What are the advantages and application scenarios of CXL.io, CXL.mem, and CXL.cache connections in high-performance computing (HPC) environments? How do they offer flexibility and performance benefits to facilitate collaborative operations among processors, storage, and accelerators?**

Answer: The CXL protocol defines three types of connections: CXL.io, CXL.cache, and CXL.mem. CXL.io provides high-bandwidth and low-latency I/O communication between devices, making it suitable for use with peripherals and network adapters. CXL.mem facilitates memory sharing and direct access, allowing processors and accelerators to access each other's physical memory directly. Its applications extend to large-scale data sets and high-performance computing. CXL.cache implements shared caches and improves data sharing and reuse, making it particularly suited for processing complex computing tasks. The three types of connections allow the CXL protocol to be flexibly used in high-performance computing environments, optimizing collaborative operations among processors, storage devices, and accelerators and addressing the requirements of diverse application scenarios, thereby enhancing overall system performance.

References

1. Worden DJ. Storage networks. Berkeley: Apress; 2004.
2. Khattar RK, Murphy MS, Tarella GJ, et al. Introduction to Storage Area Network, SAN[R/OL]. (1999–08) [2023-04-10].
3. Zhang F. High-speed serial buses in embedded systems. Singapore: Springer; 2020.
4. Hufferd JL. iSCSI: the universal storage connection. Boston: Addison-Wesley Professional; 2003.
5. NVMe Specifications Overview [EB/OL]. 2021. [2023-05-07].
6. NVMe over PCIe Transport Specification [EB/OL]. 2022. [2023-05-07].

7. Ellefson J. NVM express: Unlock your solid state drives potential [EB/OL]. (2013–08) [2023-04-11].
8. NVMe over Fabrics (oF) Specification [EB/OL]. 2021. [2023-05-07].
9. How Ethernet RDMA Protocols iWARP and RoCE Support NVMe over Fabrics [EB/OL]. 2016. [2023-05-07].
10. Rob Davis. Accelerating Flash Storage with Open Source RDMA [EB/OL]. 2019. [2023-05-07].
11. Shpiner A, Kim J, Spencer T, Renwick R. Understanding NVMe over Fabrics on TCP [EB/OL]. 2016. [2023-05-07].
12. Shan Y, Huang Y, Chen Y, et al. LegoOS: a disseminated, distributed OS for hardware resource disaggregation. In: USENIX, 13th USENIX symposium on operating systems design and implementation. Carlsbad: ACM SIGOPS; 2018. p. 69–87.
13. Lim K, Chang J, Mudge T, et al. Disaggregated memory for expansion and sharing in blade servers. ACM SIGARCH Comput Archit News. 2009;37(3):267–78.
14. Gu J, Lee Y, Zhang Y, et al. Efficient memory disaggregation with Infiniswap. In: USENIX, 14th USENIX symposium on networked systems design and implementation. Boston: ACM SIGOPS; 2017. p. 649–67.
15. Amaro E, Branner-Augmon C, Luo Z, et al. Can far memory improve job throughput? In: ACM. 15th European conference on computer systems. New York: ACM; 2020. p. 1–16.
16. Compute Express Link: The Breakthrough CPU-to-Device Interconnect [EB/OL]. 2023. [2023-05-24].

Chapter 5
Key-Value Stores

A key-value store, also known as a key-value database, is a storage system that stores, indexes, and looks up data using key values. Unlike in a conventional relational database, each record in a key-value store system has only two fields: a key and a value. A key is globally unique in a key-value database and is used to search for a record. A value is an unstructured binary string, and its semantics are generally opaque in the key-value database.

Key-value store systems have become one of the basic components of many computer systems and software. They can be seen as a functional component in storage servers, file systems, and large data warehouses. For example, a storage server can use a key-value store system to store mappings between LUN logical offsets and allocated physical spaces, a file system can use it to store directory structure and file attribute information, and a data warehouse can use it to store incoming data. Some new data services, such as distributed transaction databases, graph databases, and time-series databases, also use key-value databases as their storage components.

The key-value store system is an important part of the non-relational database (non-SQL, sometimes referred to as Not Only SQL, or NoSQL for short) family. Non-relational databases do not use relational models to describe data. Other common non-relational databases include document-oriented databases such as MongoDB and CouchDB, graph databases such as Neo4j, time-series databases such as Graphite, and columnar storage databases such as Cassandra and HBase. NoSQL databases provide weaker consistency assurance and weaker atomicity, consistency, isolation, and durability (ACID) assurance than relational databases. However, they can support distributed transaction processing and offer higher availability and scalability.

5.1 Basic Operations

The basic operations of the key-value store system include **put**, **get**, and **delete**. The **put** command is used to insert a new key-value pair into the store system, the **get** command is used to query the value corresponding to a key, and the **delete** command is used to delete a key and its associated value. Some key-value databases also subdivide the semantics of **put** into **insert** and **update**. **Insert** is used only when a key does not exist, and **update** is used only when a key exists.

In addition, some key-value store systems may support a series of extended interfaces. Common extended interfaces include batch interfaces, such as MultiGet and BatchWrite. Using a batch interface is generally more effective than invoking a basic interface multiple times. Batch interfaces can also guarantee that the returned sets come from a consistent version. Range search interfaces, such as NewIterator (lower, upper), return all data between the **lower** and **upper** keys. Persistence interfaces, such as Sync, allow users to explicitly specify when to ensure data is nonvolatile. There is high overhead associated with data persistence. Exposing the Sync interface enables a database to optimize by merging multiple records into a single write, which can greatly reduce the overhead. Snapshot interfaces can create snapshots of the current database with low overhead. The snapshot query result reflects the database status at the moment when the snapshots are created. Some key-value store systems further provide transaction interfaces, which vary in semantic richness and isolation level.

5.2 Key-Value Indexes

In terms of implementation, a key-value database may be divided into two parts: a storage module and an index module. The storage module determines a record's storage location and write format, and the index module provides the location information of the keys recorded on the storage device. The index module is an additional structure derived from the main data. It does not affect the data content, but it often determines the efficiency of data access, so it plays an important role. A key-value store database may also contain other modules, such as the page cache module, concurrency control module, and network module, which are not discussed in this section.

5.2.1 Hash Indexes

A hash table is a data structure that directly maps keys to storage locations. The hash table uses a hash function to map an input key to a hash value. The hash value is an integer that determines the storage location. Each storage location is called a bucket,

5.2 Key-Value Indexes

and it can store one or more records. The operation to determine the storage location is simple, resulting in a fairly fast lookup speed.

Generally, we rely on the hash function to evenly map keys to each bucket to increase space utilization and improve lookup performance. Therefore, we want the hash function to satisfy both uniformity and randomness. Uniformity means that any key can be mapped to any bucket. Randomness means that the hash function does not require a specific distribution of keys or depend on the relationship between input keys when mapping keys to buckets.

5.2.1.1 Hash Conflict

Each bucket has a limited amount of storage space. When newly inserted keys are mapped to a full bucket, the bucket overflows, causing what is known as a hash conflict. Different hash tables have different methods for handling bucket overflows.

For example, with typical chaining, each bucket maintains multiple records in a one-way linked list to store data. With probing, the system will try to insert records into other buckets related to the overflow bucket. Chaining is a closed addressing method, while probing is an open addressing method.

Open addressing methods are suitable for loads that mainly involve read and insert operations. They are not suitable for loads that involve delete operations. This is because keys may be stored in multiple buckets, making delete operations more troublesome. Open addressing methods are typically used to build a symbol table for a compiler. Key-value databases must properly support delete operations, so open addressing methods are generally not used.

5.2.1.2 Extendable Hashing Table

The dynamic hashing technique allows the hash table to dynamically scale out or in to accommodate larger or smaller data volumes. It does not require users to predict the maximum space required for the hash table in advance, which makes space utilization more efficient. It also avoids the performance degradation caused by record insertion.

The first dynamic hashing technique is extendable hashing, proposed by Ronald Fagin [1] in 1979. Extendable hashing adapts to changes in database size by dynamically splitting and merging buckets. Unlike the plain scale-out practice of locking the entire hash table and allocating twice the space, the scale-out of an extendable hashing table acts on only one bucket at a time, so it has less of an impact on concurrent read/write operations and lower performance jitters.

Extendable hashing can be viewed as an implementation that uses a trie (also called a prefix tree) to manage hash-valued bit strings. Structurally, an extendable hashing table consists of two parts: a directory and buckets. During queries, the hash value of a key is calculated first, and then the hash value is used as an index in the directory to obtain the directory entry. The bucket associated with the directory entry is the bucket corresponding to the key. A directory entry is associated with only one bucket, while a bucket may be associated with multiple directory entries at the same time. Whenever an inserted key is mapped to a full bucket, the system allocates a new bucket and migrates about half of the keys from the full bucket to the new bucket. The directory and directory entries then need to be modified accordingly.

When deleting a record, the system locates the corresponding bucket using a query and then deletes the matching record from the bucket. If the bucket is left relatively empty, the system merges the emptier buckets. This process is similar to the reverse process of scaling out.

5.2.1.3 Cuckoo Hashing

Another type of hash table that supports dynamic scale-out is the cuckoo hashing table, which was proposed by Rasmus Pagh and Flemming Friche Rodler [1] in 2001. Figure 5.1 shows its basic structure. When a hash conflict occurs, this hashing table pushes a record that already exists to another location in the table. This behavior is similar to the cuckoo bird's habit of pushing other birds' eggs out of the nest, hence the name.

Unlike other hash table implementations, cuckoo hashing uses two different hash functions, h_1 and h_2, to map a key to two locations, L1 and L2. When querying and deleting a key, the system checks both locations to determine whether the record exists. When inserting a key, the system will first try to insert it into L1, and if that fails, it will then try to insert it into L2. If attempts at both locations fail, the system will try to vacate L1 or L2 to make room for new keys. Vacating is accomplished by attempting to push a record into one of its alternative locations. The record's alternative locations may not be empty, so the vacating operation may involve moving a series of records in sequence. This cascading move will stop when the record

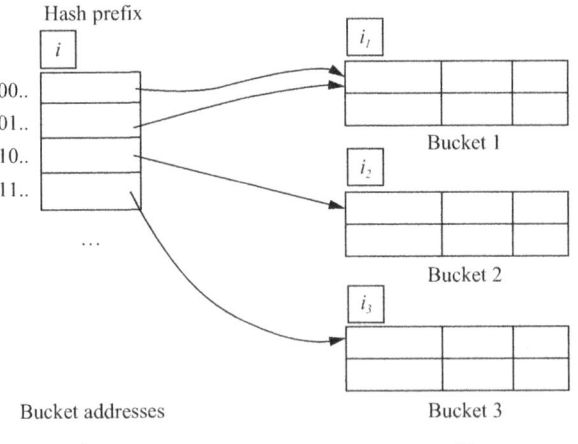

Fig. 5.1 Basic structure of the cuckoo hashing table

sequence becomes a loop or when the number of keys moved exceeds a preset value. If the vacating operation fails, a new hash table with twice the available space needs to be allocated, and all original records need to be moved to the new hash table.

There are many ways to implement cuckoo hashing. For example, three hash functions can be used instead of two to further increase the load factor of the hash table. Each bucket can be allowed to store multiple records to increase the load factor, or an additional overflow buffer can be maintained to store records that cannot be successfully inserted into the table.

In practice, cuckoo hashing is about 20–30% slower than hash tables using linear probing, mainly because checking one additional location may cause additional cache misses.

5.2.1.4 Using Hash Tables as Indexes

When using hash tables as indexes, you can set values to location identifiers on the storage device, such as the storage device number or block number.

The advantage of using hash tables as indexes is higher point query performance. However, there are three disadvantages. First, the keys in the hash table are not ordered, so the entire hash table needs to be scanned for a range query. Second, hash conflicts and dynamic scale-out cause severe jitters in access performance and long-tail delays. Third, it is often necessary to place the entire hash table in memory to achieve good performance, so sufficient memory is needed to store all keys. If there is not enough memory to store the entire hash table or to store all the keys, a large number of random small-granularity reads and writes will be generated during the query, and these random reads and writes can negatively affect the storage media.

5.2.2 B+ Tree Indexes

A B+ tree is a balanced *m*-ary tree, which consists of a root node, internal nodes, and leaf nodes. Figure 5.2 shows its structure. A B+ tree can be considered a B-tree variant optimized for block storage devices because it reduces the number of I/O operations required when searching the tree. To reduce the number of I/O operations, the depth of the tree needs to be as low as possible, and therefore the tree nodes need to have as high a fanout coefficient as possible. The B+ tree can be improved by removing values from the internal nodes and storing them in the leaf nodes instead. This change creates more space for storing keys and child node pointers. In addition, the B+ tree maintains sibling pointers in its leaf nodes, which

P_1	K_1	P_2	...	P_{n-1}	K_{n-1}	P_n

Fig. 5.2 B+ tree structure

makes range queries faster compared to the B-tree. The use of the B+ tree can be traced back to IBM's Virtual Storage Access Method (VSAM) in 1973.

An internal node in the B+ tree contains $n-1$ keys (K_1, ..., K_n-1) and n pointers (P_1, ..., P_n) to child nodes. These $n-1$ keys divide the key space managed by the current node into n parts. The keys are then assigned to n child nodes. The keys and pointers in the node are stored sequentially, meaning that, for $i < j$, we have $K_i < K_j$, and all keys in the P_i child node are smaller than all keys within the P_j child node. This reflects the orderliness of B+ tree records. For a leaf node, the first $n-1$ pointers point to the storage locations of $n-1$ records, and the last pointer P_n points to the next adjacent sibling node. Therefore, P_n links all leaf nodes together in the form of a linked list to facilitate range queries. Alternatively, the non-leaf nodes of the B+ tree can be viewed as a multi-level sparse index composed of leaf nodes.

When querying the key K in the B+ tree, the query will start from the root node and recursively go from the current node to a child node until it finally reaches the leaf node. Therefore, the query overhead associated with the B+ tree is proportional to the tree depth. If the key can be found in a leaf node, the query hits; otherwise, the record is missing.

The process of performing a range query within the B+ tree is similar to a point query. When querying records in the range of K_1–K_n, the first key that is not smaller than K_1 will be found first. After this, the next key will be found in size order until it reaches a key greater than K_n. Since sibling pointers are maintained in the leaf nodes of the B+ tree, the overhead associated with range queries tends to be low.

When inserting records into or updating records in the B+ tree, an attempt will be made to update the leaf nodes. This process may cause the size of the leaf nodes to exceed the permitted size (typically the block size of the storage device). For this reason, a node split needs to be performed on the B+ tree to split it into two half-full leaf nodes. After a leaf node is split, a pointer to the new leaf node needs to be inserted into the parent node to maintain the consistency of the tree structure. This may require a cascade of updates to the parent node, meaning that multiple nodes may be modified.

Theoretically, the depth of a B+ tree that stores N records does not exceed $\log_{\frac{n}{2}}(N)$. In practice, it only takes several I/O operations to complete a query on the B+ tree. The node size of the B+ tree is generally set to the block size of the storage device, which is usually 4 KB. Assuming that the size of a key is 12 bytes and the size of a pointer is 8 bytes, a node can store more than 200 keys and pointers. Even if a million records are stored in the B+ tree, a search only needs to access three nodes ($\lceil \log_{200} 1000000 \rceil = 3$). The root node is often stored in the memory for a long time, so the above example actually requires only two disk I/O operations.

There are several B+ tree variations. The first one is record inlining, where instead of storing pointers to records, the leaf nodes of the B+ tree store the records themselves. The benefit of this variation is that having direct access to a record (as opposed to indirect access through a leaf node pointer) requires one less random I/O operation. However, significantly fewer records can be stored inline on leaf nodes, so page-split events on leaf nodes will have to happen more frequently.

5.2 Key-Value Indexes

The second variation is prefix compression. With prefix compression, instead of storing the full key, the non-leaf nodes of the B+ tree store a prefix of the key long enough to distinguish the subtrees of the node. Since the keys are compressed, the nodes will have a higher fanout coefficient and the B+ tree will have a lower depth, and as a result, fewer storage I/O operations are required for access.

5.2.3 LSM Tree Indexes

Hash indexes and B+ tree indexes cannot deliver adequate performance in some scenarios. For hash indexes, when the memory cannot store the entire hash table, index-based access becomes much less efficient. Since the space required by the hash table is proportional to the sum of the sizes of all keys, it is impractical to store the entire hash table in the memory. For a B+ tree, it does not perform well under write-intensive loads and small-granularity key-value loads. Write-intensive loads cause random writes, which negatively affect the storage device. Since the B+ tree manages data at block granularity, small-granularity key-value loads can also cause significant write amplification.

The log-structured merge (LSM) tree is an index structure optimized for write-intensive loads. It also provides superior performance under small-granularity key-value loads. This index structure is commonly used in mainstream key-value databases, such as RocksDB [2] and LevelDB. It was first proposed by Patrick O'Neil in 1996 [3]. The version of the LSM tree that is currently used by storage databases such as RocksDB and LevelDB is more sophisticated than the prototype presented in O'Neil's original paper. It consists of three parts: memory table (MemTable), sorted string table (SSTable), and write-ahead log (WAL). A MemTable is an ordered associative array stored in the memory which is used to store the most recently written records. When the size of a MemTable exceeds the preset value, it will no longer accept new records and wait for serialization into an SSTable to be written to the storage device. The written records will simultaneously be added to the WAL to ensure crash consistency.

A MemTable can be implemented using any ordered associative array, such as red-black trees, Adelson-Velsky-Landis (AVL) trees, and heaps. In most instances, the MemTable is implemented using a heap. The main reason is that compared to the balanced tree structure, the heap offers better access performance in the case of multi-thread concurrency. Figure 5.3 shows the heap structure.

Figure 5.4 shows the LSM tree structure. When data is inserted into the LSM tree, records are first inserted into the MemTable, and the insert operation is submitted to the WAL. The WAL asynchronously persists the accumulated operations in the form of log. When a MemTable is full, it will be converted into an immutable MemTable and wait to be serialized into an SSTable and written to the storage device. MemTables have an ordered structure because they ultimately need to be serialized into ordered SSTable structures. SSTable structures need to be ordered to speed up queries.

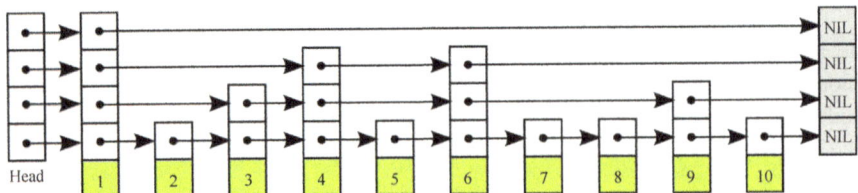

Fig. 5.3 Heap structure

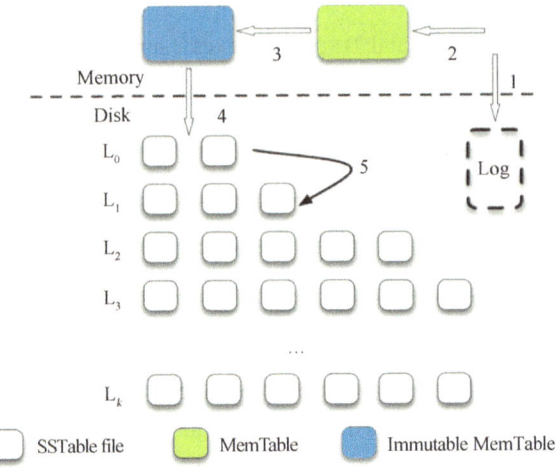

Fig. 5.4 LSM tree structure

As records continue to be inserted into the LSM tree, more and more SSTables will accumulate in the system. If these SSTables are not managed and merged, the system will run out of storage space. An even greater issue is the potential distribution of a key in every SSTable, which would significantly impact query performance. To solve these problems, the LSM tree manages SSTables hierarchically and periodically merges SSTables to optimize search performance.

Specifically, the LSM tree allocates the SSTables to $k + 1$ layers for management. The top layer is marked as L_0, and the bottom layer is marked as L_k. The $k + 1$ layers are shaped like a pyramid, with the top layer occupying a small space and storing newer records, and the bottom layer occupying a large space and storing older records. The size ratio of two adjacent layers meets a preset fanout coefficient, which is generally set to 10. Whenever the size of a layer exceeds the preset value, the SSTables in this layer will be automatically merged with the next layer. During the merge, only the most recent version of each key will be retained, and the deleted keys will be removed. The new SSTable will still keep the internal ordering. It is important to note that the SSTable of L_0 is not merged with any other layer but rather serialized from the full MemTable.

Two SSTables are said to be disjointed if the ranges of their stored keys do not overlap. We call a collection of disjointed SSTables in the same layer a sorted run.

5.2 Key-Value Indexes

This definition implies that all records within a sorted run are ordered, but ordering between any two sorted runs is not guaranteed as records may overlap. Each SSTable in L_0 is converted from the MemTable, so two SSTables may overlap, and each SSTable belongs to an independent sorted run. The number of sorted runs in other layers depends on the SSTable merging algorithm chosen for the LSM tree. There may be only one or multiple sorted runs in each layer.

The number of sorted runs is an important indicator for evaluating the orderliness of records in an LSM tree. The smaller the number of sorted runs, the more orderly the overall record arrangement and the lower the overhead associated with read requests. However, the overhead associated with the merge algorithm used to maintain SSTable orderliness will be higher as this is exacerbated by write-intensive loads. Conversely, the greater the number of sorted runs, the more disordered the record arrangement and the higher the overhead costs associated with read requests. However, the overhead costs associated with the merge algorithm will be lower.

There are two common merge algorithms: leveled compaction and tiered compaction. The two algorithms make different trade-offs when assessing the orderliness of the system. Leveled compaction produces more orderly systems than tiered compaction. Figure 5.5 shows leveled compaction.

Leveled compaction can ensure that each layer except L_0 contains only one sorted run. Since there are no duplicate keys in any of the layers (except the top layer), the space amplification coefficient is also better. This advantage is clearer when the load features write offsets. In this case, each merge operation will merge several SSTables in L_i and all intersecting SSTables in L_{i+1}. Once the merge is completed, the new SSTable will be written into L_{i+1}.

Tiered compaction does not involve SSTables in L_{i+1} when merging SSTables in L_i, so it does not and cannot guarantee that there is only one sorted run in each layer. This compromises the performance of read operations. However, since each merge operation involves a much smaller number of SSTables, the write amplification effect of the merge is not as palpable as with leveled compaction. More specifically, tiered compaction only merges several SSTables in L_i and stores the results directly in L_{i+1}.

Another special merge algorithm is first-in first-out (FIFO) compaction, which is different from leveled compaction and tiered compaction in that it does not guarantee that the written data is persistent. Old data is discarded during the merge. FIFO compaction has the lowest write amplification coefficient and the lowest merge

Fig. 5.5 Leveled compaction

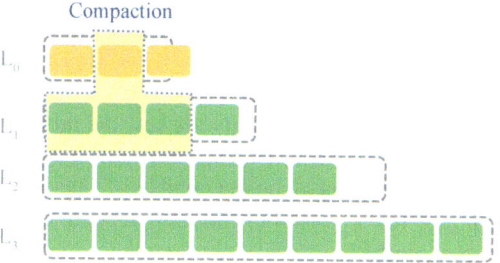

Table 5.1 Comparison of the three algorithms [2]

Algorithm	Write amplification	Maximum space usage	Average space usage	Average number of I/Os per read when the Bloom filter is used	Average number of I/Os per read when the Bloom filter is not used	Average number of I/Os per traversal
Leveled	16.07	9.8%	9.5%	0.99	1.7	1.84
Tiered	4.8	94.4%	45.5%	1.03	3.39	4.80
FIFO	2.14	N/A	N/A	1.16	528	967

overhead, so it works well for write-intensive loads. It is typically used to store server operation and maintenance logs. FIFO compaction maintains all SSTables in L_0. During a merge, the system will delete the oldest SSTable, which is similar to how other TTL-based management algorithms operate.

Table 5.1 summarizes data such as write amplification, space usage, and I/O overhead using the leveled, tiered, and FIFO compaction algorithms.

5.2.3.1 Bloom Filter

Finding a key in an LSM tree requires searching for the key in each layer and in each sorted run within each layer. This is a process that requires a large number of read I/O operations. To speed up this process, most LSM trees use a Bloom filter.

A Bloom filter is a probabilistic data structure with low space overhead that was first proposed by Burton Howard Bloom in 1970. A Bloom filter can be used to determine whether an element is in a set, but there is a risk of false positives, i.e., incorrectly reporting that an element is in a set when in fact it is not. There is no risk of false negatives, i.e., incorrectly reporting that an element is not in a set when in fact it is.

A Bloom filter generally consists of an array of m bits, all of which are initially set to 0. When adding a key to the Bloom filter, k different hash functions are used to map the key to k different hash values, and then the bits in these k positions are set to 1. When querying a key, the k hash values are calculated, and if all the bits in these k positions are 1, there is a high probability that the key exists in the set. If they are not all 1, the key does not exist in the set. A Bloom filter cannot be used to delete a key because it cannot determine which bits in the k positions should be zeroed out.

Bloom filters have a huge space advantage over maintaining an actual set. This is because the minimum requirement for maintaining a set is that it stores all keys, while maintaining a Bloom filter only requires constant space overhead. If each key pays an average overhead of 9.6 bits, the false-positive rate of a Bloom filter can be reduced to 1%. At 14.4 bits per key, the false-positive rate can be reduced to 0.1%. Bloom filters also have a huge time advantage over sets. The query time is $O(k)$ and this does not depend on the number of keys in the set. In addition, the Bloom filter's queries for k positions do not depend on each other and can be processed concurrently.

5.2 Key-Value Indexes

The LSM tree maintains Bloom filters for the SSTable to speed up read requests. Structurally, an SSTable contains several data blocks, and each data block contains several key-value pairs. The SSTable maintains a separate Bloom filter for each data block, and this is stored inside the SSTable. Before reading data blocks from the storage device, the Bloom filter is queried. Any blocks that the Bloom filter determines do not exist are skipped. Since most LSM trees maintain a cache of Bloom filters and since Bloom filters take up very little space, the process of querying Bloom filters often requires no additional I/O operations.

5.2.3.2 Comparison of the LSM Tree and B+ Tree

Figure 5.6 compares the read and write performance of the LSM tree and B+ tree. The LSM tree usually has greater write throughput and performs well under small key-value loads, while the B+ tree has greater read throughput. The LSM tree merges the inserted records and writes them into the WAL, and this reduces the write amplification. The LSM tree always writes data to the SSTable sequentially, which works well for storage media such as disks. By contrast, B+ tree writes are randomly distributed among disk blocks. When the LSM tree serves read requests, data searches need to be done layer by layer and sorted run by sorted run, and this has high overhead. When the B+ tree serves read requests, however, data searches only need to be done along the root node to the leaf nodes, so the read performance of the B+ tree is stronger.

Data compaction is an effective solution for the LSM tree, so it has lower space overhead than the B+ tree. The B+ tree is prone to leaving storage space unused due

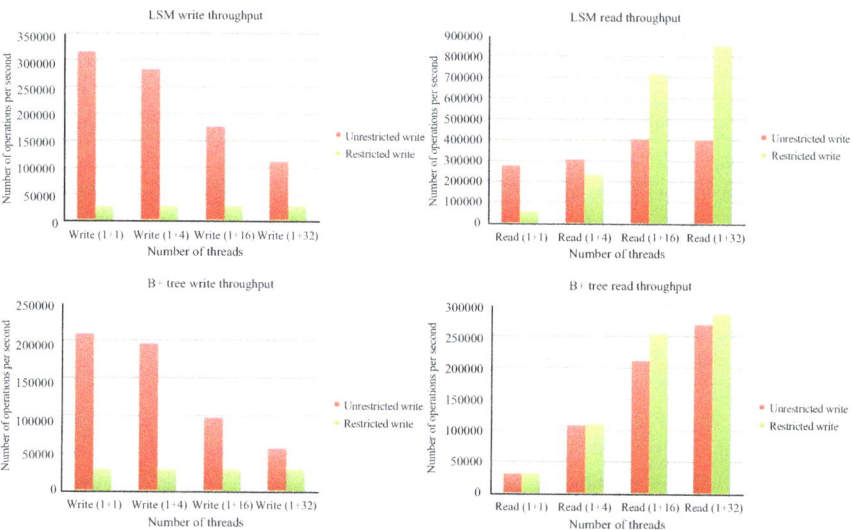

Fig. 5.6 Comparison of the read and write performance of the LSM-tree and B+ tree [4]

to fragmentation, especially during page splitting which tends to leave pages with free space. In contrast, the LSM tree is not page-oriented, and its SSTable merge algorithm periodically rewrites the SSTable to remove fragments.

Both the B+ tree and LSM tree experience write amplification but for different reasons. The write amplification of the B+ tree comes from record management in pages. Even if only a few bytes within a record have changed, the entire page needs to be rewritten. The write amplification of the LSM tree comes from multiple compactions of SSTables. The other difference is that the write amplification overhead of the LSM tree is produced in the background, while that of the B+ tree is produced in the foreground.

Merge operations that take place on the LSM tree often seriously affect the performance of foreground requests, causing throughput jitters and tail latency to increase. In comparison, the read and write latency of the B+ tree are not usually affected and remain more stable.

5.3 Data Layouts

To ensure persistence, key-value databases need to write records to storage media. There are generally two ways to update a record. One is to write the new value directly to the original storage location, which is called an in-place update. The other is to write the new value to a new location, which is called an out-of-place update. For out-of-place updates, this section describes the common log-structured data layout.

5.3.1 In-Place Updated Data Layout

Most storage devices support atomic write granularity of some size. An atomic write means that a write operation to a storage device either occurs completely or does not occur at all, which is to say that it will never be in a half-written intermediate state. This guarantee needs to hold even in the event of a system crash. We assume that a storage device has a block size of 4 KB. In general, a storage device is able to guarantee atomicity for writes at a granularity that corresponds to its block size. This guarantee makes it possible to update data in place.

Here is an example of when in-place updates can be used: When hash indexes are used, records are often stored on individually allocated blocks, and when B+ tree indexes are used, records are stored on block-sized leaf nodes. If an update to a record needs to be made and the block can still accommodate the updated data, then the record can be updated using an in-place update.

There are many benefits to using in-place updates. One benefit is that the storage location of the record does not change, so the system does not have to update the index. Another is that in-place updates do not lead to failed blocks, so it is possible to avoid the costs associated with this, such as the cost of garbage collection.

5.4 Crash Consistency

In-place updates also have limitations. First, if the records are too large and spread over multiple blocks, or if the blocks are not large enough to accommodate the updated data, then in-place updates are not possible, as crashes during the update will lead to inconsistent data. Second, in-place updates are also not possible if the update operation itself involves multiple blocks, as in the case of node splitting in the B+ tree. Additional mechanisms to support in-place updates involving multiple blocks will be described in Sect. 5.4.

5.3.2 Log-Structured Data Layout

When organizing data in a log-structured data layout, all insert, update, and delete operations are sequentially written to the storage medium in the form of log records. To serve read operations, the log-structured data layout will operate in conjunction with an index structure. The index structure identifies valid and up-to-date log entries, rendering the rest of the log entries invalid.

During database operations, the log entries corresponding to overwritten keys and deleted keys are invalidated, but they still occupy storage space. For this reason, the log-structured data layout periodically runs a garbage collection algorithm to reclaim the space occupied by invalid logs. A typical garbage collection algorithm traverses all indexes, copies all valid log entries, and consecutively writes them to new storage locations. As the log entries are copied to the new locations, the indexes are updated accordingly. Once everything has been copied over, the original log space can be fully reclaimed.

The advantage of organizing data into a log-structured layout is that all write operations are sequential, and this works well for most storage devices, especially disks. Due to the nature of out-of-place updates, a copy of the original data is retained when the data is modified to support concurrent data access and recovery in the event of a crash. The cost of running the garbage collection algorithm makes up a significant part of the overhead associated with a log-structured data layout.

5.4 Crash Consistency

A system is crash-consistent if it remains consistent in the event of a crash. Specifically, the system should ensure atomicity and persistence even after experiencing a crash, meaning that update operations in progress at the time of the crash either take full effect or do not take effect at all and that the update operations implemented before the crash are not lost. Updates involving multiple blocks are common in practice, such as possible node splits caused by updates to B+ tree indexes, tree rotations caused by updates to balanced tree indexes, and updates to records whose sizes exceed the block granularity. If there is a crash during an update operation that involves writes to multiple blocks, the writes may only take effect in some of the blocks. In this case, a mechanism would need to be used to bring the system back to a consistent state after the crash.

5.4.1 WALs

One way to ensure crash consistency is to use write-ahead logs (WALs). Essentially, WAL is a method to ensure crash consistency through two write operations. Specifically, changes are first written to another storage location before they are written to the database. This ensures that the system has at least one correct version at any given moment. WALs can be categorized into redo logs and undo logs.

Suppose we want to perform writes to n blocks. The redo log method is to write all the new values and addresses of these n blocks to the log before the actual writes occur. The actual writes to these blocks can be done only after the redo log has been completely written and checked for persistence:

- If the system crashes while the redo logs are incomplete, the system deletes these incomplete logs as if no update operation has been run.
- If the system crashes while the redo logs are complete, the system completely re-executes the writing of data blocks regardless of whether the data blocks have already been written. After this, the update is considered to have been completed successfully.

There are many ways to verify the integrity of redo logs. The first method is to reserve a checksum field for each log entry. The log is considered complete only if the checksum matches. The second method is to use a commit log entry to mark the log entry before it as complete.

The undo log method is the opposite of the redo log method. Assuming that writes to n blocks are to be performed, the undo log method does this by writing the old value and address of a block to an undo log before writing data to the block. Only after the writes to this undo log are finished can the write operation to the block be performed. When the actual writes to all n blocks have been completed, a commit log entry is written to the log, marking all operations as complete:

- If the system crashes when the commit record of the undo log has been written, the update operation is considered to have been completed successfully and nothing else needs to be done.
- If the system crashes when the commit record of the undo log has not yet been written, the system performs an undo operation and writes the old values to all blocks. After that, it is as if no update operation has been performed at all.

Redo logs and undo logs each have their own advantages and disadvantages. The redo log method requires that all redo log entries have been completely written before the actual update operation can be performed. This means that the system needs to know all the block addresses and new values involved in the update operation in advance. However, the system does not always have complete knowledge of this information. For example, during an interactive transaction, the next block to be updated depends on the user's next input into the system, and this cannot be known in advance. By contrast, the undo log method does not need to know any of this information in advance. It only requires that undo log entries are written before it writes data to a block. However, undo logs often do not perform as well as redo logs because redo log entries can be combined into one large I/O request and written at once, while undo log entries must be written sequentially in turn. In addition, an extra commit log entry must be written when using the undo log method, whereas such a commit log entry is not required for the redo log method.

5.4.2 Shadow Pages

Shadow pages are a type of copy-on-write (CoW) technique that avoids in-place updates to any block. When a block needs to be updated, the shadow page technique always allocates a new block and writes the new value to it. Subsequently, the system uses an atomic method to replace the old block with the new one, for example, by automatically updating pointers to bring the new block into effect:

- If the system crashes when the newly allocated block has taken effect, the update operation is considered to have been completed successfully and nothing else needs to be done.
- If the system crashes when the newly allocated block has not taken effect, the newly allocated block is discarded as if no update operation has been run at all.

The problem with shadow pages is that for data structures such as binary trees, the pointer in the parent node needs to be modified for a shadow page to take effect. Modifying the parent node also requires the use of the shadow page technique to ensure crash consistency. Therefore, modifications to one piece of data may cause a cascade of modifications to multiple pages.

5.5 Summary

Key-value store systems have simple interface semantics, deliver excellent performance, and support a large number of important applications, making them invaluable in the big data era. These systems cover all aspects of data storage. In terms of indexing, key-value store systems often use hash tables, B+ trees, or LSM trees. LSM trees are widely used because they support range searches and are suitable for storage devices. Regarding data layouts, key-value store systems use in-place updates or out-of-place updates. Out-of-place updates can be used with the log-structured data layout. For crash consistency, consistent versions are maintained using write-ahead logs (WALs) and shadow pages to tolerate exceptions such as server crashes. In the future, key-value store systems will utilize multiple storage media, such as persistent memory, to optimize cost-effectiveness. This will require re-examining existing mechanisms such as indexing, data layouts, and crash consistency.

5.6 Practice Questions

1. **What are the tradeoffs between space efficiency and cache performance in open and close addressing?**

   Open addressing does not require extra storage space to handle conflicts. Therefore, its data storage is more compact, the space efficiency is higher, and the cache performance is better.

   By contrast, closed addressing requires extra space for each element to store the pointer that points to the next element. Therefore, its space efficiency is lower. In addition, since elements are stored dispersedly, its cache efficiency is lower.

2. **What are the best and worst search lengths when N key-value pairs are inserted into an empty hash table? Consider open and close addressing methods.**

   Assume that the ith key-value pair is inserted. If open addressing is used, the best search length is 1, where the key is mapped to an empty bucket (the bucket corresponding to the key's hash value), and the worst search length is i, where the key and its previous keys are mapped to the same bucket.

   If close addressing is used, the best search length is 1, identical to open addressing. The worst search length is 1, where all keys are mapped to the chaining structure in the same bucket. The chaining structure allows new keys to be inserted at the head of the linked list.

3. **What are the benefits and drawbacks of using cuckoo hashing instead of chain hashing? What are the considerations in choosing between them for applications?**

   In contrast to chain hashing, which is affected by hash collisions, cuckoo hashing offers faster lookups, as it only needs to examine two positions determined by the two hash functions.

   Cuckoo hashing has unstable insertion latency due to emptying operations that cause cascaded records to move. This latency is more stable with chain hashing, which allows new records to be inserted at the head of the linked list.

   Cuckoo hashing is generally more space-efficient because its emptying mechanism allows for higher loading factors.

   Therefore, cuckoo hashing is suited to applications that are read- and space-intensive, whereas chain hashing is more suitable for applications that require low insertion latency.

4. **Describe how cuckoo hashing can be used with the Bloom filter.**

   The Bloom filter can be used as a cuckoo hashing table to store the fingerprints of all keys. The specific implementation is as follows: This hash table uses the *hash* function to determine the first storage location $L1 = hash(k)$ for key k. This location stores the key fingerprint *finger(k)*, which can be taken as the lower 8 bits of $hash(k)$. Storing only fingerprints ensures the required storage space is much smaller. When k needs to be moved [in this case, only *finger(k)* and $L1 = hash(k)$ are saved among the data related to k], the second storage location can be determined as $L2 = L1 \wedge hash(finger(k)) = hash(k) \wedge hash(finger(k))$, where $\wedge$ is the bitwise XOR. $L2$ is determined in this way to exploit the properties of the XOR operation. That is, if $A \wedge B = C$, then the XOR of any two values of the three values of ABC can yield the third value. With this feature, $L1$ and $L2$ can be calculated using each other to implement cascaded movement operations in cuckoo hashing.

5. **Assume the data set size is N, and the node size of the B+ tree is B. Calculate the read/write amplification of the B+ tree**.

    Each internal node contains $O(B)$ child nodes and the leaf node contains $O(B)$ records. Therefore, the height of the B+ tree is $O((\log N)/(\log B))$.

    Write amplification: Each insertion writes data to the leaf node. Regardless of the actual size of the data, B needs to be written every time. Therefore, the write amplification is B.

    Read amplification: A query in the B+ tree needs to search from the root node to the leaf node. Therefore, the read amplification is determined by the tree height $O((\log N)/(\log B))$.

6. **Assume the data set size is N, the fanout coefficient between LSM tree layers is k, and the size of a single SSTable at each layer is B. Calculate the read/write amplification of the LSM tree**.

    Similar to the B+ tree, the number of LSM tree layers is $L = O((\log N)/(\log B))$.

    Write amplification: Each record will be written k times at this layer before being merged into the next layer. Therefore, the write amplification is $k \times L = O(k \times (\log N)/(\log B))$.

    Read amplification: The size of the ith layer from the end is $N/(k^i)$, and there are $N/(B k^i)$ blocks in total. The complexity of performing a binary search is $\log N/(B k^i)$. By summing up all L layers, you obtain the read amplification $O(\log2(N/B) /\log k)$.

7. **For which applications are the B+ and LSM trees suitable? Give some examples**.

    LSM trees are suitable for write-intensive applications, such as log systems and time series databases. In practice, LSM trees generally store stock market, IoT device, and meteorological observation data.

    B+ trees are suitable for read-intensive applications, like transaction processing that supports high-frequency queries in traditional relational databases. In practice, B+ trees are often used to serve historical record queries; document and archive data queries, population data, and other public record queries. They are also used to build education and scientific research databases, store and retrieve academic articles, etc.

8. **LSM trees support two compaction methods suitable for different scenarios. What workloads are leveled compaction and tiered compaction suitable for?**

   Leveled compaction works well for read-intensive workloads, while tiered compaction works well for write-intensive workloads.

   When tiered compaction is in use, fewer copies occur during merging. Therefore, the write overhead is lower. For read operations, keys in the same layer may intersect, so key searches within a layer often require reading SSTables multiple times. Therefore, the read latency is higher and the latency is more unstable.

   When leveled compaction is in use, more copies occur during merging. Therefore, the write overhead is higher. For read operations, it ensures keys in the same layer do not intersect, so the key search within the layer is very efficient (only one SSTable needs to be read in most cases). Therefore, the read latency is low and stable.

References

1. Pagh R, Rodler FF. Cuckoo Hashing. In: der Heide AFM, editor. Algorithms—ESA, vol. 2161. Heidelberg: Springer; 2001.
2. Dong S, Kryczka A, Jin Y, et al. Evolution of development priorities in key-value stores serving large-scale applications: the {RocksDB} experience. In: 19th USENIX conference on File and Storage Technologies (FAST 21). Berkeley: USENIX Association; 2021. p. 33–49.
3. O'neil P, Cheng E, Gawlick D, et al. The log-structured merge-tree (LSM-tree). Acta Informatica. 1996;33(4):351–85.
4. Gorrod A. Btree vs LSM [EB/OL]. (2017-08-24) [2023-04-25].

Chapter 6
File Systems

Users can directly operate and manage data on secondary storage devices, but this approach is complex, error-prone, and unreliable. Multiprogramming and time-sharing systems require easy and reliable sharing of file data on high-capacity secondary storage devices. File systems are designed to meet these requirements. This chapter describes the basic concepts and implementation of file systems and provides an example to illustrate the implementation of a standalone file system. For details about distributed file systems, see Chap. 8.

6.1 Basic Operations

A file system is a module in an operating system for accessing and managing information. File systems are used to store, search for, update, share, and protect user and system information in a unified manner. They provide users with a full set of effective methods for using and operating files.

Major file system functionalities are categorized as follows:

Namespace: A global namespace is provided for the creation, maintenance, searching of files and directories. It enables file access by name and provides access interfaces for users.

Storage management: Users can allocate and reclaim file storage space and manage the mapping between logical and physical addresses.

Advanced features: Users can share and protect files and ensure data reliability and consistency.

A file is a dataset stored on a secondary storage device and bound to a file name string. A file can be much larger than a computer's memory. A file's lifespan is usually much longer than the duration of a computing task and sometimes longer than the service life of the computer that stores it. In a file, data is typically organized into a one-dimensional array of bytes or blocks. Applications can build complex data storage patterns on top of this one-dimensional array. Generally, a file needs to allow multiple processes or even multiple computers to access file data simultaneously. In addition to data, each file has metadata that describes different attributes of the file data. The metadata of a file is typically stored in a data structure called an index node (inode). Each inode has a globally unique identifier called an inode number. A file's metadata usually includes the file's inode number, name, type, size, data block size, access control information, owning user identifier, creation and modification time, and data block storage location.

A directory is a special type of file. Its data records information about a group of files and directories. When directory B is located within directory A, B is a subdirectory of A, and A is the parent directory of B. Each subdirectory and file in a directory is called a directory entry. Each directory entry records the name and inode number of the subdirectory or file. Figure 6.1 shows an example of a file system directory tree. The inode number is used to locate the metadata and data of the subdirectory or file. Generally, associated files are stored in the same directory, and associated directories are stored in the same parent directory. In this way, a directory tree structure is established to effectively manage data. Each directory has two special directory entries, which are "." and "..", where "." indicates the current directory and ".." indicates its parent directory.

The basic file operations are as follows:

Create: creates a file, where an amount of storage space and an inode number are allocated to the file, and a corresponding directory entry is created in the directory.

Delete: deletes a file, where the storage space and inode number used by the file are released, and the corresponding directory entry is deleted from the directory.

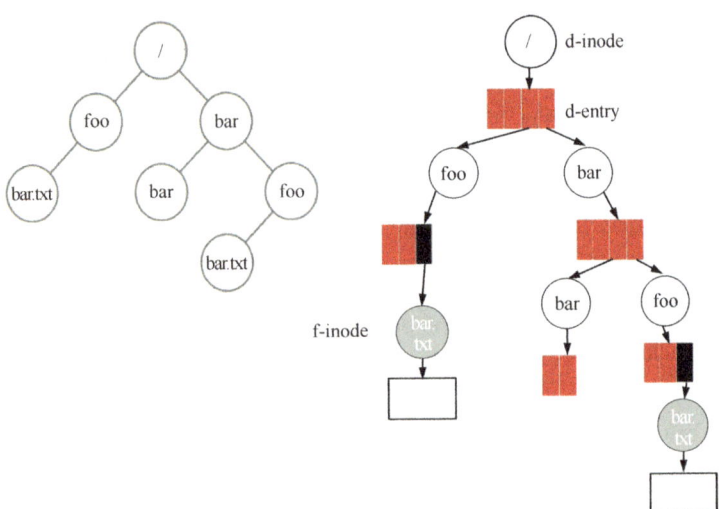

Fig. 6.1 Example of a file system directory tree [1]

Open: opens a file, where the file's and all its parent directories' metadata are loaded into memory, and the file's metadata is associated with a file handle.
Close: closes a file, where the association between the file handle and file metadata is canceled, and all modifications are persisted to secondary storage.
Read: reads a file's data from secondary storage to memory, starting from the location indicated by the file pointer.
Write: writes an application's data from memory to secondary storage, starting from the location indicated by the file pointer.
Append: appends data to a file.
Seek: changes the location indicated by the current file pointer.
Rename: renames a file and can relocate it.
Getattr: obtains a file's metadata.
Setattr: modifies a file's metadata.
The basic directory operations are as follows:
Mkdir: creates a directory.
Rmdir: deletes a directory. (Generally, the directory must be empty.)
Link: creates a directory entry in a directory and links it to an existing file.
Unlink: deletes a directory entry linked to a file from the directory entry. The file is also deleted if no other links point to it.
Readdir: reads a directory entry in a directory.
Rename: renames a directory and can relocate it.

6.2 File System Implementation

Section 6.1 introduced two concepts about file systems from a user perspective—file and directory—as well as the interfaces provided by file systems. In this section, we will discuss the issues to be considered when constructing file systems from the perspective of their inner workings.

6.2.1 A Simple File System

File systems can be categorized based on the media they use. Some file systems are built on disks, such as ext4 and New Technology File System (NTFS). Others are built to operate over networks, such as network file system (NFS). In Linux, there are also virtual file systems, such as procfs. Each file system is designed differently depending on the specific media it uses. In this section, we will start with a disk-based simple file system.

Assume that data of any byte size can be read from and written to at any location on a disk, in the same way memory is accessed. When creating the first file, a question arises: What information needs to be recorded if we only store one file?

The answer is the file size. If recording a file's size requires 8 bytes, we can store the file in the way shown in Fig. 6.2.

We only need to read the first 8 bytes to know the file length, and we can read the subsequent n bytes to obtain the file data. This means that $8 + n$ bytes are required to store a file. Reading a file requires two disk read operations. If we want to append data to the file, we need to first obtain the file length (the first 8 bytes) and then append content to the file (after $8 + n$ bytes). This changes the file length. We need to change the initially recorded file length (the first 8 bytes). So, there are two write operations and one read operation.

Now let us assume that multiple files are stored and the second file is placed right behind the first one. In this case, the first file's length can be obtained from the first 8 bytes, and its file content can be obtained by reading the part after the first 8 bytes. But what should we do about the second file? The answer is we first find the location where the second file's length is stored and then we obtain the file content. See Fig. 6.3.

If there are a large number of files, we need to find the first k-1 files before locating the kth file. This is a very slow process, so it is better to record the starting location of each file. The data that records a file's information, such as the starting location and file length, is called metadata. Currently, each file's metadata only includes the file length and starting location, occupying the same amount of space. Therefore, it is best to place all the metadata at the beginning so that any file can be quickly found. Figure 6.4 shows the inode [2] structure in a UNIX file system.

The inode size is fixed. Let us assume it is 16 bytes. When the third file needs to be accessed, the starting location of the third file inode is calculated as follows: 16 bytes $\times$ 2 = 32 bytes. We can read the inode to obtain the starting location and length of the third file and read the content of the third file from the starting location

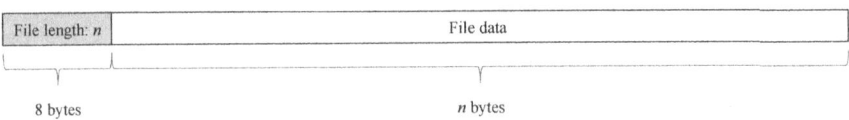

Fig. 6.2 Storage of a single file

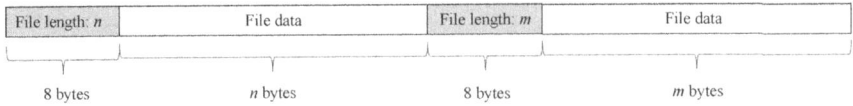

Fig. 6.3 Storage of multiple files

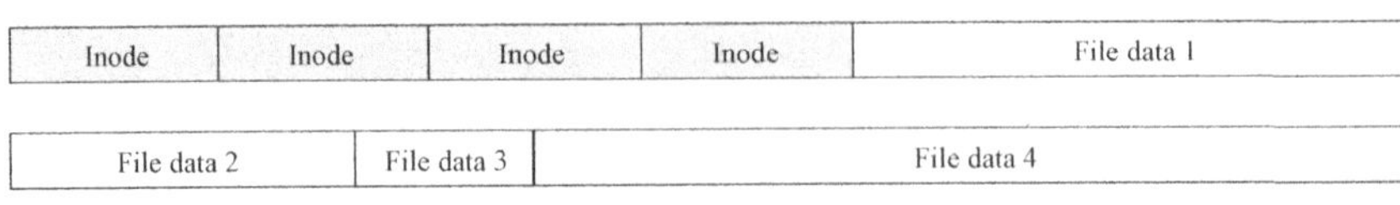

| Inode | Inode | Inode | Inode | File data 1 |
| File data 2 | File data 3 | | File data 4 | |

Fig. 6.4 Inode structure

6.2 File System Implementation

recorded in the inode. In this case, reading a file still requires two disk read operations.

By abstracting the inode structure, we can store more file metadata. In a real file system, an inode also stores additional information such as access time, permissions, and creation time. Figure 6.5 illustrates an example of a **hello.txt** file in a Linux operating system. The first character "-" in the first column indicates that this is a file. If the first character is **d**, it means that this is a directory. The subsequent characters in this column indicate the read, write, and execute permissions of the file. The second column indicates the number of file links. The third and fourth columns indicate the user and group to which the file belongs. The fifth column indicates the file size. The sixth, seventh, and eighth columns indicate the file modification time. The last column indicates the file name.

There is still a problem with the preceding example. In the inode structure shown in Fig. 6.4, if more files need to be added, there is no space for storing inodes. Therefore, more inodes should be planned, and space must be pre-allocated to them. However, this will limit the number of files supported by the file system, but this is not a serious problem. For example, the number of files supported by mainstream file systems like NTFS and ext3 is also determined during initialization.

Therefore, disk space is specially divided into two regions. One stores inodes and the other stores data, as shown in Fig. 6.6.

As mentioned above, the number of inodes can be specified during file system initialization. The specified quantity needs to be recorded as well. A file system also has other information such as the total size, type, and free space. This is the entire file system's global information, which is independent of files. This information is different from file data and does not belong to any inode. Therefore, we need to find another place to store it. In a file system, this global information is typically stored in a superblock. Based on the preceding disk division, an extra region is added, as shown in Fig. 6.7.

```
root$ ls -alF
total 1504
drwxr-xr-x  19 root root   4096 Jul 14 14:29 ./
drwxr-xr-x  19 root root   4096 Jul 14 14:29 ../
drwxr-xr-x   2 root root   4096 Aug 20  2021 boot/
drwxr-xr-x   9 root root   2820 Jul 14 14:29 dev/
drwxr-xr-x 114 root root   4096 Jul 14 14:29 etc/
-rw-r--r--   1 root root   4096 Jul 14 15:44 hello.txt
```

Fig. 6.5 Example of file metadata in Linux

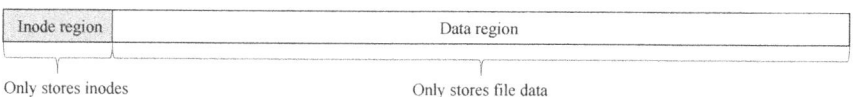

Fig. 6.6 File system structure with more inodes

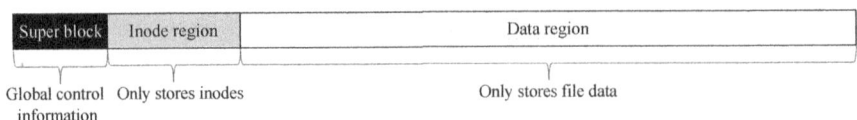

Fig. 6.7 File system with a global block

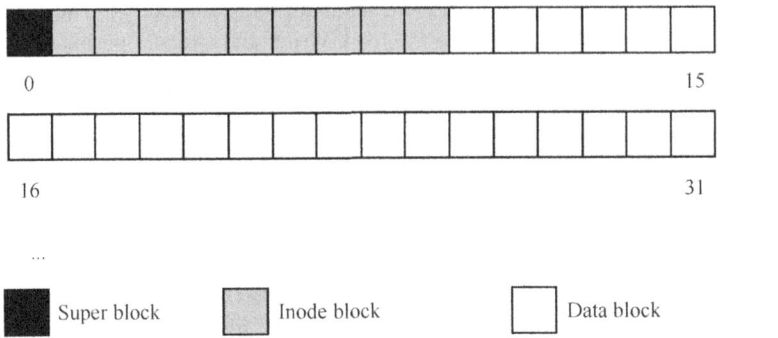

Fig. 6.8 File system with reads/writes at the block granularity

On a disk, a sector (typically 512 bytes) is the minimum storage unit. The read/write granularity cannot be smaller than the sector size. For the file system mentioned earlier, we made the assumption that disks could be read and written at any random granularity. Now, we need to modify the file system to support data reads and writes at the sector granularity.

From the perspective of a file system, a disk is a set of sectors. If a file system accesses data based on the sector size or its integer multiple (referred to as a block), disk reads and writes are much easier.

So, we modify our file system based on a block size of 1 KB. There is only one superblock. Its occupation of one or two blocks does not waste much space. An inode does not contain much information, and its size is typically 128 bytes. Therefore, one block can house eight inodes. Files need to meet alignment requirements. The beginning of each file must be aligned based on the block size. Each file wastes at most the last [1024—(size % 1024)] bytes (where size indicates the file size). With these modifications, the file system changes to the format shown in Fig. 6.8.

Now, a number of blocks exist in the file system. Only the corresponding inodes know the usage of these blocks. If new space needs to be quickly allocated to a new file, what should be done? Is it necessary to access all files? If it is, the overhead is too high. Therefore, the allocation of each block should be recorded. In addition, a dedicated space is needed to record the allocation of inodes.

The preceding file system can create, modify, and read files, but it lacks an effective way to organize files. Having all files in one place would lead to chaos. Therefore, we need an important function to manage files—directory.

To implement the directory function, we do not need to significantly modify the file system. A directory is also a type of file, but its content is the mapping between

6.2 File System Implementation

the names and inode numbers of the files in it. Note that these files can also be directories. When it comes to reading a directory, file names and their corresponding inode numbers are read from the directory. In this way, we can obtain information about all the files in it. When a file is created in a directory, an entry is added to the directory to record the mapping between the new file's name and inode number. This approach enables files to be created in directories.

When no directory is created, where is a file placed? A file system needs an entrance to find files in sequence. Therefore, a root directory is required. The root directory inode is typically fixed, and its number is stored in the superblock. The root directory enables us to manage files. In this way, all files and directories can be found recursively from the root directory.

Once a directory is added, file operations become different. For example, assume that a file exists in the root directory. Opening the file involves reading the root directory inode, root directory data, file inode, and then file data. This means that when a directory exists, opening a file requires first opening its directory and then reading file data from the directory. The process of opening a directory is essentially a process of opening all the related files until the root directory is reached. The modification of a file also involves the modification of the directory content. Creating a file will also cause inode information and inode allocation information to be modified.

A single file operation modifies not only a single file, but also the inode, inode allocation, data block, data block allocation, and upper-layer directory information.

So far, we have looked at how to design a simple but versatile file system. Figure 6.9 shows the complete structure of a simple file system.

For this type of file system, we need to consider two issues: how to allocate disk space to files and how to index files through inodes.

In this section, we have only divided disk space into several parts and established a relationship between them. However, when creating or modifying a file, file data is dynamically added or deleted. Therefore, a dynamic allocation mechanism is

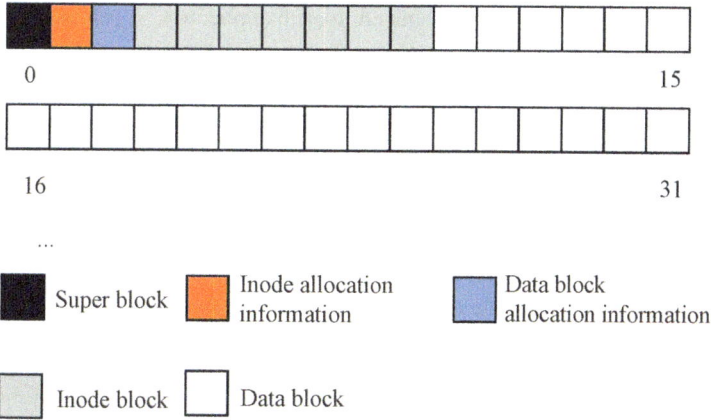

Fig. 6.9 Complete structure of a simple file system

required for the data region. Moreover, files vary greatly in size. How should inodes record data block locations? We will discuss this in Sect. 6.2.3.

6.2.2 Namespace Management

Storing data as files on secondary storage is just the beginning of constructing a file system. When it comes to namespace management, the main concern is how to enable applications to locate their desired files. File systems provide a namespace for naming and finding files. The efficiency of namespace management determines how efficiently applications can access their requested data from among terabytes of data or tens of millions of files.

Typically, a file system uses a directory tree to manage its namespace. Unlike flat namespace management that uses key-value indexes, the hierarchical structure of a directory tree can process more complex data organization. For a medium-sized dataset, the depth of the tree structure increases gradually as the data size grows. Therefore, file indexing based on a directory tree delivers high performance. When a dataset becomes excessively large, a file system can divide the directory tree to implement more fine-grained namespace management, thereby supporting larger-sized data.

A file system's directory tree has only one root directory, which is the root of the directory tree. A non-leaf node in a directory tree corresponds to a directory in the file system, and a leaf node corresponds to a file in the file system. In a directory tree, each file has a corresponding path name. A path name is a character string that includes the names of all the intermediate directories in the path from the root directory to the target file, as well as the name of the target file, all connected using separators. A path name starting from the root directory is called an absolute path name, for example, **/home/alice/db/data**. If a directory tree has many levels, always searching for a file from the root directory can cause high latency. Therefore, the concept of current directory is introduced to each application. A relative path name includes the names of all intermediate directories in the path from the current directory of an application to the target file, as well as the name of the target file, all connected using separators, for example, **db/data**. Using a relative path name can shorten the path length for file queries.

On a computer, we can deploy different types of file systems on different secondary storage devices, but the operating system has only one global namespace. The file system to which the global namespace belongs is called the root file system, which is initialized during the operating system startup. Before accessing the file system on a secondary storage device, we must incorporate the file system into the global namespace. This operation is known as mounting the file system. Once a file system is mounted to a directory in the global namespace, any subsequent access to the directory is handled by that file system. We can perform the unmount operation to detach a mounted file system from the global namespace.

6.2 File System Implementation

On the same secondary storage device, we can also use multiple file systems to manage data. A secondary storage device can be divided into multiple partitions. Each partition is an independent logical device and can be formatted using a file system. We use the first or second data block on the secondary storage device to store a partition table. This table is used to index information about partitions. Each item in the table records a partition's starting logical block number (LBN) and length. Generally, a partition table is processed by the operating system's device driver. Requests initiated by a file system are associated with the partition where the file system resides. The device driver layer automatically converts these requests based on the starting location of the partition.

However, there are still a number of challenges in managing a file system's namespace through a directory tree. First, a directory tree is difficult to scale to support ultra-large files. Although we can divide a directory tree for finer-grained management, the load between directory subtrees may not be balanced. Moreover, each directory subtree still faces the same scalability problem. Second, when a directory contains an excessively large number of directory entries, such as in an image processing scenario where each frame of a video is stored as a file in the same directory, searching for a file in that directory will be very slow. In this case, we can speed up file search by dividing a large directory into multiple subdirectories. Third, we usually store files in specific directories based on their features. A file can be stored only in one directory. Therefore, a file cannot be indexed based on its multiple features. A possible solution is to index the same file through links in multiple directories.

6.2.3 Storage Management

In Sect. 6.2.1, we categorized blocks in a file system into data blocks and inode blocks. Files come in different sizes and may occupy different numbers of data blocks. Therefore, we should consider how to allocate data blocks to files and how to manage free space.

6.2.3.1 Space Allocation

Contiguous Allocation

The simplest practice is to store all the file data contiguously. By doing this, we only need to locate the first data block to find the entire file.

In Fig. 6.10, each file is stored in contiguous blocks on a disk. In this case, an inode only needs to record the location of a file's starting data block and the file length to find any data block of that file. Contiguous allocation requires that all blocks of a file be contiguous on disk.

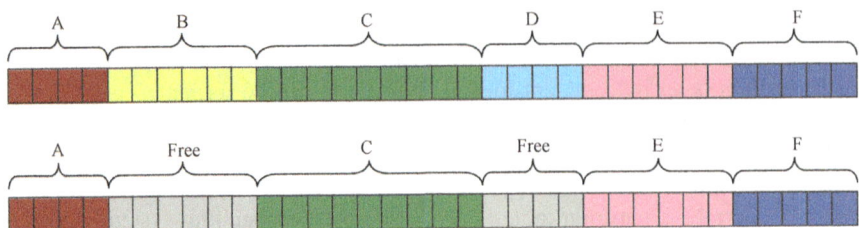

Fig. 6.10 Schematic diagram of contiguous allocation

Contiguous allocation has the following advantages: First, it enables simple and fast indexing. A file can be found easily using the location of its first file block and the file length. Second, it delivers high performance in random access. A file's blocks are stored contiguously, enabling easy index calculation. Therefore, a single file block does not require complex search operations. Third, it enables simple mapping from logical to physical addresses. Assume that we need to find the data presented by the *Nth* byte of a file, where the location of the file's starting block is *start*, and the size of a single block is *BSIZE*. Then the data should be located in the block whose number is *(N/BSIZE) + start*, and the offset in the block is *N % BSIZE*.

Contiguous allocation has the following disadvantages: First, deleting a file can cause fragmentation, which is similar to that caused by memory allocation. In Fig. 6.10, for example, if we delete file B and then create a larger file, the space previously occupied by file B cannot be reused. If a newly created file is smaller than file B, there will be a smaller amount of remaining free space. Defragmentation takes a long time. Second, creating a file or increasing a file's length can cause a high overhead. Upon creating a file, all the needed space must be allocated in advance. To increase a file's length, if the blocks located after the current file have already been allocated, an amount of contiguously available space on a disk must be found to store the entire new file, like file E shown in Fig. 6.10.

NTFS uses the contiguous allocation method to store large files.

Linked Allocation

Contiguous allocation enables easy data reading but makes it difficult to add, delete, or modify data. This is similar to an array, which enables quick indexing but has difficulties in adding and deleting elements. A linked list is a data structure different from an array. Can storage space be allocated in a way similar to a linked list?

As shown in Fig. 6.11, a file is scattered across different disk blocks, and these blocks are connected using pointers, just like a linked list. From the first block, a linked list can find subsequent blocks in sequence. Therefore, an inode only needs to record a file's starting data block, length, and ending data block.

6.2 File System Implementation

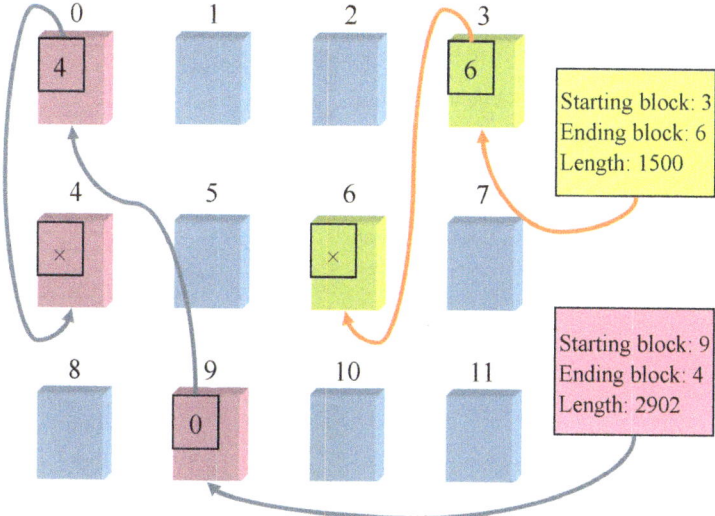

Fig. 6.11 Schematic diagram of linked allocation

Linked allocation has the following advantages: First, it enables high disk space utilization because its non-contiguous storage approach allows any free block to be selected for space allocation. Second, adding and deleting file content is straightforward. The system simply needs to connect new blocks to the linked list when adding file content.

Linked allocation has the following disadvantages: First, its random-access efficiency is low because access to any subsequent block needs a traversal from the first block. Second, its reliability is poor because the loss of a single data block in the link can make subsequent data blocks inaccessible.

The Xerox Alto file system uses linked allocation.

Indexed Allocation

Contiguous allocation is prone to disk fragmentation, which hinders file length appending. Linked allocation reduces the efficiency of random file access. Indexed allocation is a balanced method for ensuring file access efficiency and reducing file modification costs.

In indexed allocation, the location of each data block is recorded using a pointer, and an inode is used to store data indexes, as shown in Fig. 6.12. Indexed allocation needs to store more data block pointers than the other two allocation methods.

Indexed allocation has the following advantages: First, the disk space utilization is high. Similar to linked allocation, indexed allocation does not require file data blocks to be contiguous. Instead, any free data block can be allocated to a file. Second, the random access efficiency is high. The location of each file data block can be found quickly through indexing.

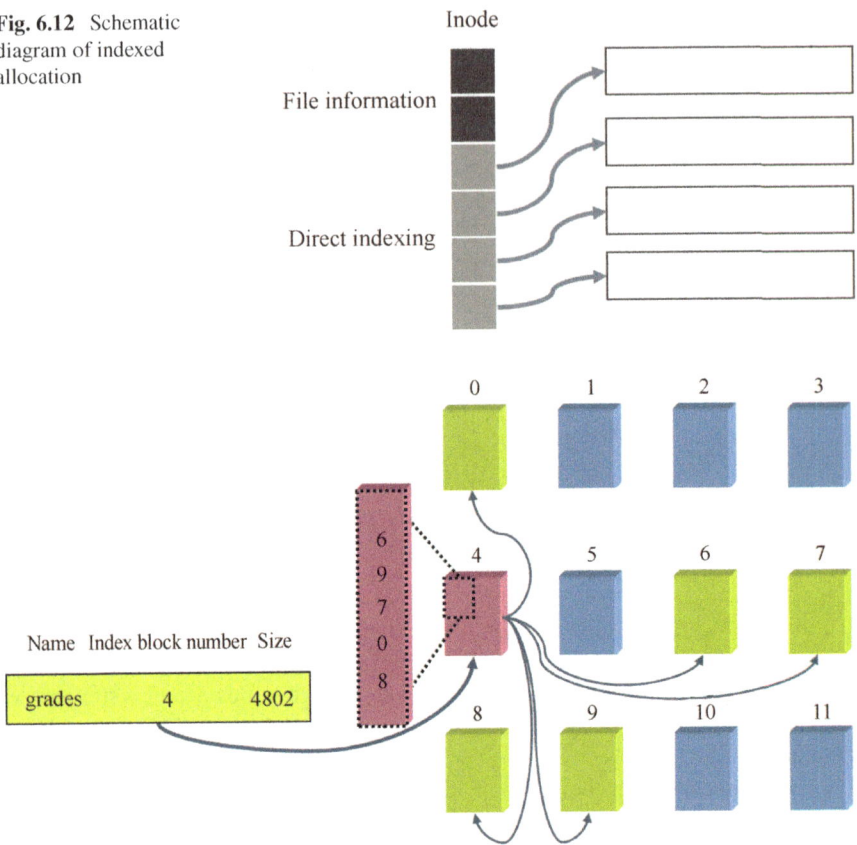

Fig. 6.12 Schematic diagram of indexed allocation

Fig. 6.13 Schematic diagram of indirect indexing

The disadvantage of indexed allocation is the additional space overhead. Regardless of the file size, one or more index blocks are needed to record data block locations. This disadvantage is more noticeable with smaller files.

The size of a single index block is limited, as is the number of pointers that can be stored in an index block. The upper limit of the file size depends on the index block size. We can use indirect indexing to support larger files. See Fig. 6.13 for its schematic diagram.

Some data blocks are directly indexed. When these blocks are insufficient for storing file data, indirect indexing is implemented. The indirect index pointer points to a data block that only stores indexes. The pointers in this block point to the actual data blocks. The pointer overhead is low. Therefore, one block can store a large number of pointers, enabling a multifold increase in the supported file size. If a large file cannot be stored using doubly indirect indexing, triply indirect indexing can be implemented.

6.2 File System Implementation

The greater the number of indirect indexing levels, the larger the supported file size. However, the overhead caused by random file access also increases. If direct indexing is used, the location of a data block can be found by accessing the inode only once. If indirect indexing is used, the inode may need to be accessed multiple times, which causes an additional overhead.

The ext2 file system uses multi-level indexing.

6.2.3.2 Free Space Management

Disks typically have free blocks. When allocating data blocks to a file, scanning all blocks in sequence can cause a huge overhead. Therefore, we need to record information about free blocks to enable quick allocation of data blocks to files.

Management Using a Bitmap

A block is either in the used or unused state. A bitmap efficiently records, queries, and modifies the status information of a block while occupying only a small amount of space.

One bit is enough to indicate the status of a block. The value 0 indicates that a block is not used, and 1 indicates that a block is used. The space occupied by a bitmap varies depending on the disk size and block size. Assume that a file system uses 4 KB as the block size and the disk size is 80 GB. Then, the required bitmap space is 80 GB/4 KB × 1 bit = 2.5 MB.

Management Using a Linked List

A free block can also be used to store information about other free blocks, thereby saving disk space. All free blocks can be connected in the form of a linked list, similar to linked management used for file block allocation. When using a linked list for management, there is no need to record additional information. However, a linked list performs poorly in random access, making it difficult to quickly determine whether a block is free. This hinders the full utilization of data locality during data storage.

Issues that May Occur During Free Block Management

Free block information needs to be maintained on both the operating system and disks, which causes consistency issues. Changed information must be promptly written to disks. Otherwise, crashes can cause storage space leakage or data loss. Writing data back to disks causes a high latency overhead. Free space management is an important factor that slows down a file system.

6.3 File System Instance: ext2

ext2 is a classic file system [1] in Linux. This section uses ext2 as an example to describe the implementation of a real file system.

6.3.1 Disk Data Structure

When ext2 is created on a disk partition, the partition is divided into multiple blocks. The first block is the boot block occupied by the operating system and is not managed by the file system. Other blocks are organized into different block groups. For details, see Fig. 6.14.

The data bitmap in a block group has a fixed block size and represents a limited quantity of available data blocks. Therefore, one partition may need to be divided into multiple block groups. Each block group consists of a superblock, a block group descriptor table, a data bitmap, an inode bitmap, an inode table, and data blocks. Both the super block and block group descriptor are copies of those in block group 0, and they describe the file system status. The functions of the other parts are the same as those described in Sect. 6.2.

6.3.2 Directory Tree Structure

In ext2, a directory is a special file that stores directory entries. A directory entry is the mapping between a file name and an inode. Unlike an inode, a directory entry does not have a one-to-one mapping. The Linux system has a link mechanism. Therefore, the same file can be accessed via different names. That is, multiple directory entries may point to the same inode. Fig. 6.15 shows the ext2 directory tree structure.

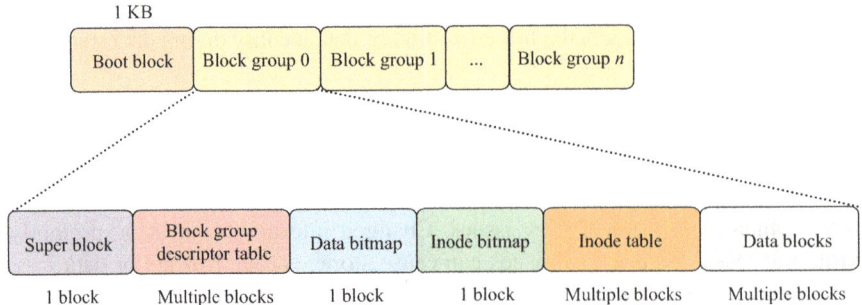

Fig. 6.14 Structure of the ext2 file system

6.3 File System Instance: ext2

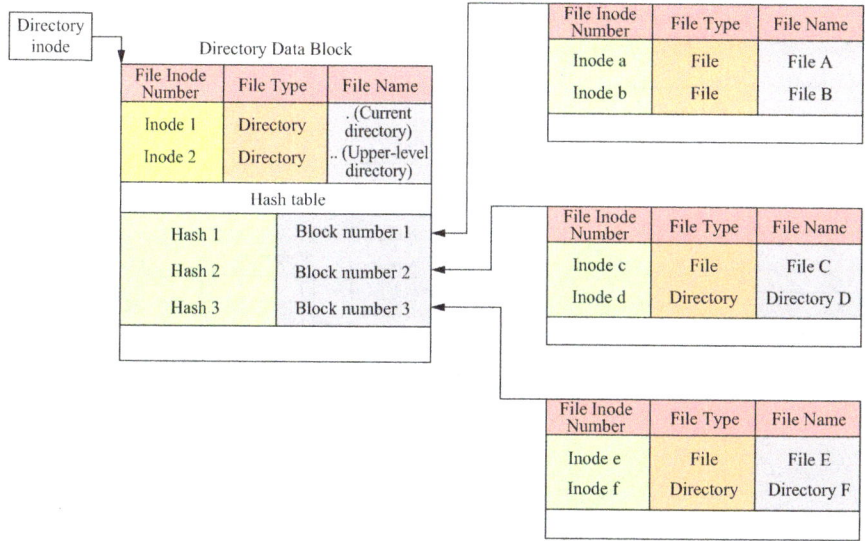

Fig. 6.15 ext2 directory tree structure

A directory can contain a large number of files. In a search for a target file, traversing all files in sequence can be extremely inefficient. Therefore, an ext2 directory uses a hash table to match file names, allowing for quick file addition and deletion with only a slight increase in space overhead.

ext2 organizes all files into a tree structure based on directories.

6.3.3 Data Block Addressing

ext2 uses multi-level indexing to store data and supports a maximum of three-level indirect indexing. A disk inode typically has 15 index pointers. Twelve of them are direct pointers, and the remaining three are singly, doubly, and triply indirect pointers. Figure 6.16 shows the multi-level indirect indexing structure.

Storing a file can fall into one of the following four cases: If the file size does not exceed 12 data blocks, direct indexing is used. If the file size exceeds 12 data blocks, singly indirect indexing is used. If singly indirect indexing is insufficient for storing the entire file, doubly indirect indexing is used. If doubly indirect indexing is insufficient for storing the entire file, triply indirect indexing is used.

If the block size is 1 KB, one data block can store 256 pointers (1 KB/4 bytes = 256). If direct indexing is used, 12 data blocks are required. If singly indirect indexing is used, 256 data blocks are required. If doubly indirect indexing is used, 256^2 data blocks are required. If triply indirect indexing is used, 256^3 data blocks are required. Therefore, the maximum size of a single file is $(12 + 256 + 256^2 + 256^3) \times 1 \text{ KB} \approx 16 \text{ GB}$.

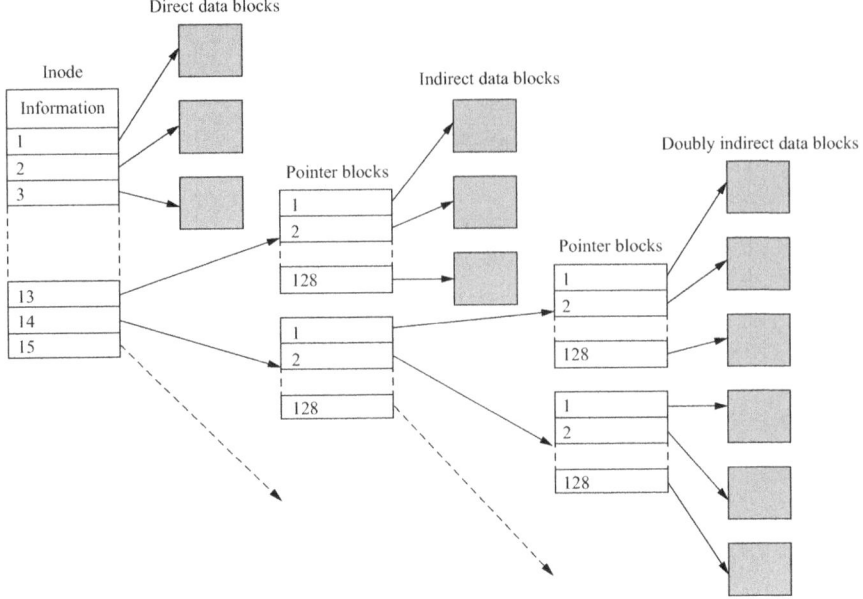

Fig. 6.16 Multi-level indirect indexing structure

6.4 Summary

By abstracting the hierarchical directory tree, a file system enables upper-layer applications to share data easily and reliably. The internal implementation of a file system is more complex than a key-value store system, including how to establish, maintain, and search the namespace and how to allocate, reclaim, and map file storage space. It is predicted that file systems will be optimized for different applications. For example, for internet and AI dataset management applications, file system capabilities for processing massive amounts of small files will likely be optimized. For high-performance computing applications, file system bandwidth for reading and writing large data blocks will likely be optimized.

6.5 Practice Questions

1. **What function does a superblock serve in a file system?**

    A superblock stores a file system's metadata, such as the file system version and data block size. It manages space allocation by tracking the number of free blocks and maintains a file system's integrity information, including the number of mounts, mounting times, and checksum. Additionally, it provides layout information, such as the location of the inode table and the starting data block, and records the features supported by the file system.

6.5 Practice Questions

2. **Can multiple file systems coexist on one storage device? How?**

   Yes, multiple file systems can coexist on one storage device through partitioning. Partitioning divides the storage device into independent areas, each known as a partition. Once partitioned, each partition can be formatted with a specific type of file system. The operating system can then mount each partition to a location in the directory tree, enabling multiple file systems to operate on the same storage device.

3. **What are soft and hard links in a file system? How does a file system implement them?**

   A hard link is a direct pointer to file data in a file system. When creating a hard link, a new entry is added to the directory, pointing to the inode of the linked file. Inside the inode, a link counter is maintained. When the link count drops to zero, both the inode and its associated resources are released.

   A soft link, also known as a symbolic link, is a special type of file containing a text string indicating the path to another file. When creating a soft link, a new file containing the original file's path is generated.

4. **Applications such as data center services and large model training can create either deep directories (containing numerous levels of subdirectories) or fat directories (containing a large number of files). Both types of directories can degrade a file system's performance. Analyze the cause of performance deterioration and propose solutions.**

   Deep directories increase path parsing time because the file system must parse directories level by level, potentially involving mutiple I/O operations. To mitigate deep directories, a flat directory structure can be used to distribute files across a preset number of subdirectories, avoiding the creation of new subdirectory levels.

   Fat directories make directory item management inefficient as searches for specific files require traversing numerous directory entries. To address fat directories, they can be divided into multiple subdirectories, reducing the number of files in each subdirectory and improving management efficiency.

5. **A distributed file system stores file paths in key-value pairs, whereas a local file system (e.g., ext4) manages paths via a directory tree. Why?**

   A distributed file system needs horizontal expansion to handle large amounts of data and high access concurrency. Key-value pairs help evenly distribute data to prevent node overload.

   In contrast, a local file system primarily serves users. Managing paths via a directory tree is more user-friendly. In addition, a local file system stores data on a single local device, resulting in lower overhead for maintaining the tree structure compared to a distributed environment.

6. **Many file systems, such as Apple File System (APFS), provide a snapshot function that preserves the state of a file system at a certain point in time. Briefly describe how this function is implemented in file systems.**

   The copy-on-write (COW) technique is used. With this technique, when a file or block in a file system is modified, the original data is not overwritten. Instead, the modification is written to another location on the disk, and the metadata of the file system is updated to point to the new block.

7. **F2FS is a file system designed for flash storage devices. Summarize its optimizations tailored for flash storage devices.**

   F2FS is based on a Log-structured File System (LFS), utilizing a log structure to maximize the high sequential write performance of SSDs and mitigate flash wear caused by frequent writes and erases. It implements flash-friendly garbage collection and wear-leveling policies to monitor block wear and dynamically adjusts data placement for balanced wear across flash storage units. F2FS also incorporates adaptive zone management to enhance storage efficiency.

8. **ext4 is one of the most widely used file systems. Summarize the major differences between ext2, ext3, and ext4.**

   ext2, an early Linux file system, provides basic file system functions but lacks journaling, potentially leading to long recovery times after system crashes.

   The most significant update to ext2 found in ext3 is the addition of journaling, which significantly reduces the time required for system checks (fsck) after a system crash, thereby improving reliability.

   ext4 supports larger single file and file system sizes, delayed allocation, and faster file system checks (fsck).

References

1. Bovet DP, Cesati M. In: Lijun C, Qiongsheng Z, Hongwei Z, editors. Understanding the Linux kernel. 3rd ed. Beijing: China Electric Power Press; 2007.
2. Arpaci-Dusseau R, Arpaci-Dusseau A. Operating systems: Three easy pieces [EB/OL]. (2020-08-08) [2023-04-10].

Chapter 7
Network Storage Architectures

The development of computer-related technologies and the increasing service loads of computers pose higher requirements for storage capacity and speed. Although storage arrays can increase capacity and speed by stacking more storage units in a single device, they address only single-disk limitations. The requirements for the entire storage system, including scalability, stability, and shared access, are not fully satisfied. The development of storage protocols such as the Small Computer System Interface (SCSI) and Fibre Channel (FC) has made network storage a mainstay. There are different types of storage architecture, including direct-attached storage (DAS), centralized network storage such as network-attached storage (NAS) and storage area network (SAN), and distributed storage such as parallel storage, peer-to-peer (P2P) storage, and cloud storage. This chapter will introduce and discuss these storage architectures.

7.1 DAS

The concept of DAS was proposed only after network storage emerged. Unlike network storage, DAS does not involve a network or network device. Any storage architecture with hard disk drives (HDDs) or solid-state drives (SSDs) directly connected to a computer can be referred to as DAS.

Based on the location of a storage device relative to the computer, DAS can be classified into internal DAS and external DAS.

Internal DAS: The storage device is located inside the computer, as shown in Fig. 7.1. The physical storage space of internal DAS is limited and is generally used for system startup.

External DAS: The storage device is located outside the computer, and it is not connected to the computer via a network. Such a storage device can be shared by

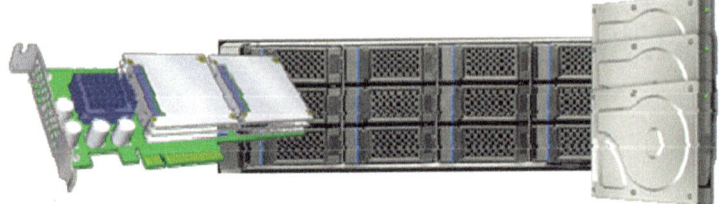

Fig. 7.1 Internal DAS

Fig. 7.2 External DAS

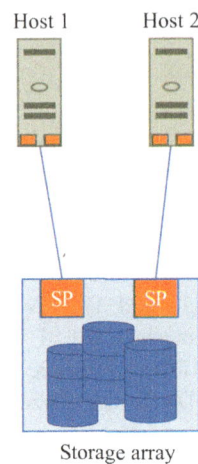

multiple computers, as shown in Fig. 7.2. External DAS offers more storage space than internal DAS.

DAS connects to the computer through various interfaces rather than over the network or via network devices. For internal DAS, a storage device is usually connected to the computer through a host bus adapter (HBA), which provides high-speed connections and reduces the load on processors for data storage tasks. Common interface protocols of external DAS include SCSI and Serial Advanced Technology Attachment (SATA). The SCSI protocol features high performance and high reliability and is widely used in servers and workstations, but it is complex and expensive. By contrast, the SATA protocol is cost-effective and widely used in personal computers. External DAS allows connection to multiple disks, and redundant array of independent disks (RAID) technology can be used to form disk arrays for capacity expansion and fault tolerance.

Internal and external DAS can be used together to achieve certain capacity and fault tolerance targets. However, DAS is limited by physical performance and features poor scalability and restricted transmission distance. Therefore, DAS should be used for scenarios in which storage systems must be directly connected to servers and only a small storage capacity is required, or for scenarios where servers are scattered and establishing a network connection is not possible.

7.2 NAS

NAS is a file system sharing device connected to an IP local LAN. Though users may need to share files, it is not easy to share files between multiple computers. For example, if one computer is powered off, the files on that computer become inaccessible. A dedicated storage server can solve this problem. Users can access the dedicated storage server over the network to obtain corresponding data. This was the prototype of early NAS. The main idea behind it is to separate storage devices from servers and manage data separately.

7.2.1 Architecture Characteristics

Figure 7.3 shows the NAS architecture, which contains a dedicated storage server and network adapter. The storage server usually contains a storage array, such as a RAID subsystem, and a lightweight operating system for the NAS system. This operating system runs a file system optimized for NAS.

A client can use the IP-based network file protocol to send multiple file-level I/O requests to the NAS system to operate files stored in the NAS system. Communication protocols between the NAS system and clients generally include the Network File System (NFS) and Common Internet File System (CIFS).

NAS offloads file I/O operations to dedicated storage servers, reducing the data management load and releasing processor and memory resources on clients. In addition, standardized customization of network file protocols allows NAS to be compatible with multiple platforms and feature strong flexibility.

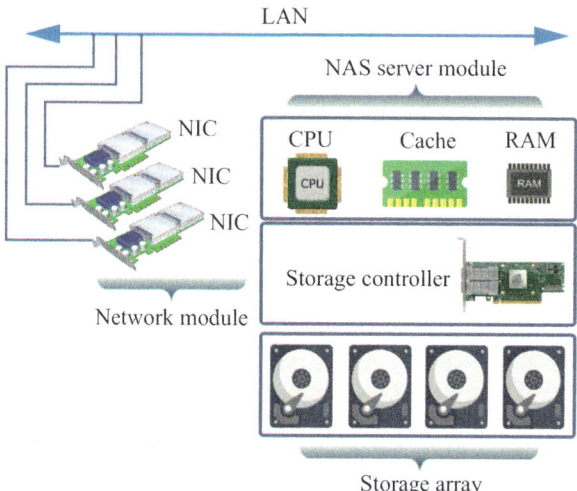

Fig. 7.3 The NAS architecture

7.2.2 Network-Based File Access Protocols

7.2.2.1 Protocol Type

Clients can operate files on a NAS system using various network-based file access protocols, such as NFS and CIFS. These are the best protocols for Linux-based and Windows-based network architectures, respectively.

The NFS protocol was proposed by SUN in 1984. It allows users to access files on a remote server similarly to how they would access local files. The NFS protocol adopts the client/server model based on the remote procedure call (RPC) mechanism. NFS is the most popular network-based file access protocol for UNIX and Linux systems because it delivers high performance and is very flexible.

CIFS was proposed by Microsoft. It is commonly used for network file sharing between Windows hosts. Unlike the NFS protocol, the CIFS protocol is network connection-oriented and requires high network reliability. In addition, CIFS is a stateful protocol and is highly sensitive to faults.

7.2.2.2 I/O Operation Process

Figure 7.4 shows the process of an I/O operation in NAS:

1. An I/O request sent by a client is encapsulated into TCP/IP packets and sent to the NAS system over the network.
2. The NAS system converts the I/O request into block-level I/O operations to execute the corresponding request in the physical storage area.
3. The NAS system encapsulates the I/O request execution result using a file protocol and sends it to the client via TCP/IP packets.

Unlike on common servers, the aforementioned I/O operation is processed by the dedicated operating system and file system within the NAS system. The dedicated operating system and file system have optimized multiuser connections and

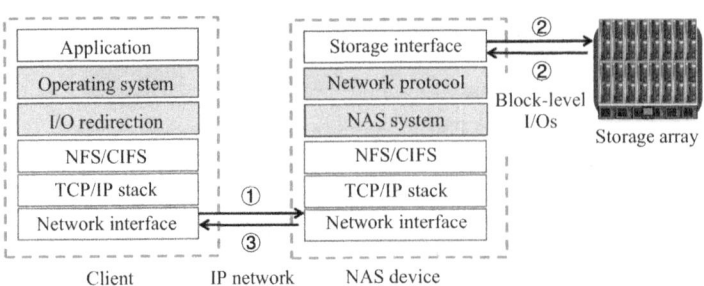

Fig. 7.4 The NAS I/O operation process

concurrent I/O operations. In addition, these systems are based on open standard protocols and can be used by different vendors.

7.2.3 Application Scenarios

NAS systems can be used as centralized storage systems for small and medium-sized enterprises to reduce their data storage and management costs. In addition, enterprises can use NAS systems to improve functions such as data sharing, backup, and fault tolerance. NAS systems can provide both high-performance access and file-level security assurance, making them suitable for scenarios that require frequent concurrent file-sharing access.

NAS systems are also commonly used by individuals looking to store multimedia data. A consumer can easily set up a LAN-based NAS system at home. NAS devices are smaller, cheaper, and more flexible than traditional rack-mounted storage devices. As the price of NAS devices continues to decrease, they are becoming more and more popular for home use.

7.3 SAN

A storage area network (SAN) interconnects computers and storage devices through dedicated links to centralize data storage and management within a relatively independent SAN [1]. The SAN architecture consists of storage devices, networks, and servers, as shown in Fig. 7.5. In a broad sense, any storage form that interconnects computers, dedicated storage networks, and storage devices can be called SAN. However, due to the commercial success of the Fibre Channel (FC) protocol, the term SAN has become synonymous with FC-based network storage architecture.

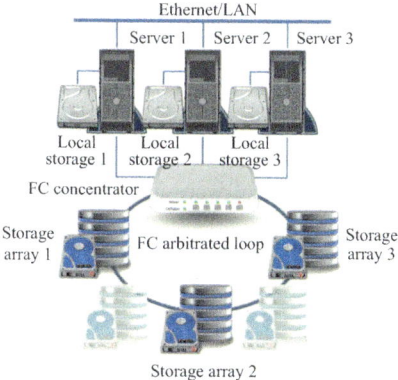
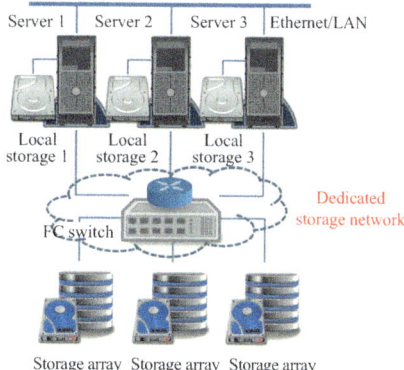

Fig. 7.5 The SAN architecture

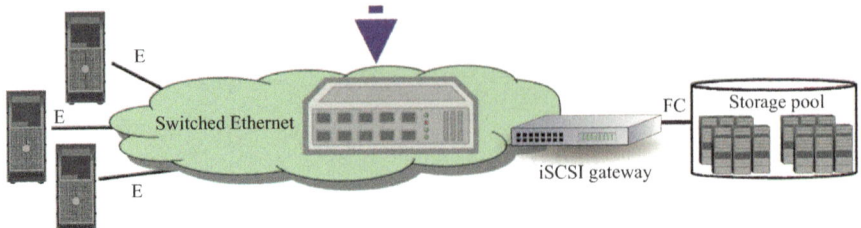

Fig. 7.6 The IP SAN architecture

7.3.1 Architecture Characteristics

A SAN can form a storage array of different storage devices, such as HDDs and SSDs, and manage them using a RAID system, thereby increasing both storage capacity and fault tolerance. An FC-based SAN requires a dedicated network, which is expensive. The emergence of Internet SCSI (iSCSI) also brought TCP-based SAN (IP SAN), as shown in Fig. 7.6. iSCSI allows the SCSI protocol to directly run on IP networks, improving the performance of data operations over Ethernet. In addition to delivering exceptional performance, IP SAN reduces the cost of network architecture construction. In a SAN, servers are responsible for communicating with storage devices. Operating systems and file systems running on these servers are used to operate data. Additionally, SAN management software is used to organize and manage the servers, storage devices, and network.

7.3.2 Core Components

7.3.2.1 Management Software

The servers, storage devices, and networks in a SAN are managed at the software management layer. Management software can be installed on one or more servers. It can be classified as in-band management software or out-of-band management software [2] based on whether the data path and control path are separated, as shown in Fig. 7.7. In-band management software transmits data and control information over the same network, while out-of-band management software uses a dedicated network to manage the SAN.

Although in-band management software is easy to implement, it may cause a system bottleneck on the I/O path. In addition, it has a low fault tolerance. If the storage network is faulty, management information cannot be sent to storage devices, rendering storage device management unavailable.

Out-of-band management software does not depend on the storage network. When there is a fault on the storage network, it can still be used to manage storage

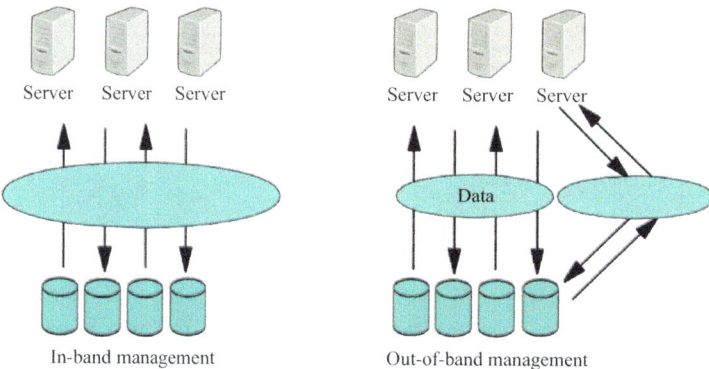

Fig. 7.7 The different types of management software

devices and rectify the fault. However, out-of-band management software cannot automatically detect the topology of a SAN. In addition, agent software needs to be installed on the server to communicate with the management software.

7.3.2.2 File System

A SAN does not provide file abstraction; it only transmits data blocks. Therefore, a server must use its local file system to manage the network storage devices in the SAN. If multiple file systems access storage devices in a SAN at the same time and share logical units (LUNs), data errors are likely to occur. Therefore, a dedicated file system is required to manage the sharing of physical storage space between multiple machines. The shared-disk file system (SDFS) enables multiple computers to directly access disks. It provides concurrency control and isolation mechanisms to ensure that data is not damaged or lost when multiple clients access disks simultaneously.

7.3.3 Application Scenarios

Block-level data access makes SANs suitable for scenarios requiring large storage capacity and frequent data access. SANs are highly scalable, have high fault tolerance, and can support disaster recovery. Therefore, they can also be used in scenarios requiring high data security, such as data centers, and can be applied in the financial industry. To improve the utilization of existing storage devices, data centers use SANs to connect storage resources, allowing shared access to expensive storage resources. Cloud storage also generally uses SANs. SANs can reduce the cost of managing storage resources more effectively than DAS and NAS systems.

Table 7.1 Comparison of NAS and SAN

Item	NAS	SAN
Access interface	File	Block
Cost	Low (Ethernet)	High (dedicated FC network)
CPU usage	Low	High (hosts need to run file systems)
Deployment complexity	Easy	Complex

7.3.4 Comparing NAS and SAN

NAS and SAN are typical network storage architectures. They are based on redundant storage arrays and have similar architectures. Table 7.1 compares NAS and SAN from different angles to offer a more comprehensive analysis of the two architectures.

7.4 Object Storage

Object storage abstracts data blocks at a higher level and provides a set of new standard access interfaces. It integrates the platform-agnostic access advantage of NAS and the scalability advantage of SAN. As a result, it is widely used in large-scale cluster systems.

7.4.1 Architecture Characteristics

The object storage architecture consists of user components and storage components, as shown in Fig. 7.8. The user components provide logical data structures for user applications, such as files, paths, and interfaces for accessing these data structures. The storage components are located on intelligent storage devices and organize specific data blocks on physical disks. Users access storage devices through object interfaces. Object storage does not put storage on hosts or provide block interfaces for users in the same way that traditional file systems do.

Object storage offers several advantages:

Data sharing: Storage objects combines the advantages of files and blocks since they can be directly accessed like data blocks and shared on different operating systems like files.

Self-management: An object storage device manages and allocates its own storage space, instead of being managed by the file system on a host. The most direct benefit of this is that space management is separated from storage applications. Storage devices can independently reorganize data to improve performance and adjust backup and failure recovery policies.

7.4 Object Storage

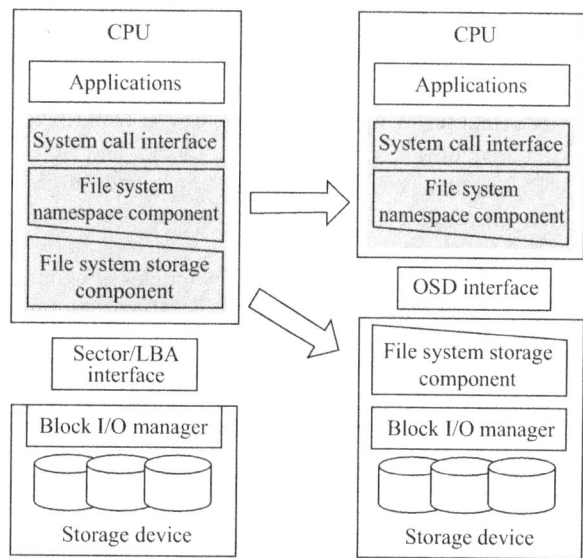

Fig. 7.8 The object storage architecture [3]

High security: An object is a collection of logical bytes that contain the storage method, data attributes, and storage security policies. Therefore, object storage greatly improves the file-based data layout, service quality, and security.

7.4.2 Core Components

Object storage re-divides the basic storage functions of data storage into access, control, and management. The data layout (logical-to-physical mapping) of legacy file systems is implemented at the object layer. This involves the following core concepts:

Object: This is the most basic unit for storing data in an object storage system. An object stores data from one or more files and contains data attributes that define file-based access methods and security policies. Each object has a unique ID. A client accesses an object using parameters such as the object ID, starting address, and data length. There are multiple types of objects. For example, a root object is used to define the basic attributes of a storage device, a group object provides directory abstraction for an object set, and a user object stores actual data.

Object-based storage device (OSD): An OSD is sufficiently intelligent and is equipped with an independent CPU, memory, network system, and disk system. The core functions of OSDs are as follows: First, OSDs are used to store data, manage object-related data, and place data on disks. OSDs do not provide block interfaces. When a client initiates a data request, data is accessed based on the unique object ID and offset information. Second, OSDs maintain an

intelligent data layout. The dedicated CPU and memory resources configured in OSDs work together to optimize the data distribution and improve disk performance. Third, OSDs are responsible for managing object metadata. The metadata has a similar structure to a conventional inode and usually includes detailed information about data blocks and object length. Unlike traditional NAS systems where metadata is managed on file servers, object storage systems centralize the management function, which significantly reduces the overhead for clients.
- Metadata service (MDS): An MDS guides the interaction between clients and objects and provides three functions. The first is object storage access. The MDS constructs and maintains the file directory tree so that clients can directly interact with objects. The MDS grants access permission to clients during the interaction, and each access request is verified by the OSD before permission is obtained. The second is file and directory management. The MDS creates a hierarchical structure for files in the storage system and supports the creation, deletion, and access control of directories and files. The third is cache consistency maintenance. The object storage file system provides a client cache scheme to improve client performance. The MDS notifies clients of file changes so that clients can refresh their cache and avoid problems caused by cache data inconsistency.
- Client of the object storage file system: As with traditional file systems, a client must be installed on a compute node to access the OSD. The client allows applications to perform read and write operations on the OSD in the same way as accessing a standard Portable Operating System Interface (POSIX) file system.

7.5 Parallel Storage

Parallel storage is mainly used in the high-performance computing (HPC) field. The HPC cluster uses the interconnection technology to connect a large number of computer systems and uses parallel computing technology to organize the connected systems in a way that enables them to process large-scale computing tasks quickly. For example, China's supercomputer Sunway TaihuLight has a peak computing speed of 125 petaFLOPS. HPC has high requirements for processors, memory, and storage. Storage technologies have long been a weakness in IT development, especially in the HPC field. Although thousands of processor units work together to perform large-scale computing tasks, the limited I/O concurrency of the storage system cannot meet the data access requirements of the computing cluster, severely impacting overall system performance. As a result, parallel file systems are being widely studied and applied. Technologies such as data caching and sharing, fine-grained concurrency control, and native support for parallel computing programming models all play important roles in improving the I/O parallel capabilities of storage systems. This section describes the architecture characteristics and key technologies of the parallel storage system.

7.5 Parallel Storage

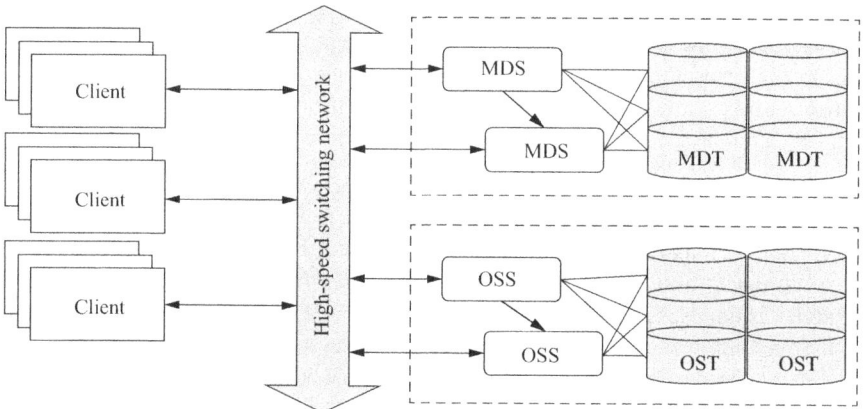

Fig. 7.9 The Lustre storage system architecture used in the Sunway TaihuLight supercomputer [4]

7.5.1 Architecture Characteristics

The storage structure of a supercomputing system evolves alongside its computing capability. In the early stages, when the system is relatively small, the I/O requirements can be met by directly mounting the file system client to the compute node. As the computing scale increases, I/O nodes are added to the supercomputing system. The compute nodes forward I/O requests to the I/O nodes for execution. SSDs have become more popular, and so has configuring SSDs on I/O nodes to provide aggregated read/write bandwidth. New network and storage hardware technologies are being rapidly developed, and they are transforming the structure of the supercomputing storage system.

For example, China's Sunway TaihuLight supercomputer uses the open-source file system Lustre as the back-end storage system. Lustre uses the MDS and object storage server (OSS) to store metadata and data, respectively. Two MDSs work in primary/secondary mode to manage metadata targets (MDTs) on storage nodes, while two OSSs work in primary/secondary mode to manage object storage targets (OSTs) on storage nodes, as shown in Fig. 7.9.

7.5.2 Key Technologies

Large-scale parallel file systems (such as Lustre and GPFS) are the most widely used storage systems for high-performance computers. High-performance computers pose increasing requirements on the throughput and concurrency capabilities of file systems because massive data needs to be processed and analyzed, and intermediate results need to be periodically stored during long-term computing. Although computing tasks can be allocated to tens of millions of cores for concurrent execution, the concurrent I/O capability of the file system cannot increase as it is constrained by the software stack design and CPU computing capability. As a result, the storage operation efficiency of HPC applications is low, and I/O loads interfere with each other, causing a bottleneck that restricts the computing task performance. The key technologies of the parallel file system are as follows:

Parallel I/O: In high-performance computing scenarios, a large number of processor cores send parallel I/O requests to the parallel file system when executing tasks. Therefore, it is essential to improve the parallel I/O capability of the storage system. The basic idea of parallel I/O is to distribute data to multiple storage nodes as much as possible and obtain an aggregated I/O bandwidth by leveraging the parallel access to and channels of multiple storage devices, thereby fully utilizing hardware resources. Common file distribution modes include the cyclic placement, hash scheme, and user-defined modes. Using a proper file distribution mode will significantly improve I/O node load balancing and concurrent I/O access.

Metadata management: Before accessing the file data, a working process must access the metadata server to obtain information about the location of the related file data. Since metadata servers need to manage a global view of the entire file system, mainstream file systems adopt centralized architectures and use one metadata server to manage the metadata of the entire file system. This architecture avoids distributed transactions in metadata processing and has advantages like simple architecture, stability, and reliability. However, the metadata server can easily become a single-point performance bottleneck as all metadata requests are sent to it, hindering scalability. By contrast, in a distributed metadata architecture, the metadata is distributed to multiple metadata servers. Existing metadata distribution schemes include subdirectory division (e.g., CephFS [5]) and hash-based division (e.g., IndexFS [6]). In a distributed architecture, multiple metadata servers can concurrently process the metadata requests initiated by a client. However, the adjacent layers of a file system directory tree structure depend on each other. For example, some metadata operations (e.g., creating, deleting, or renaming a file) modify multiple metadata items at the same time. If these metadata items are distributed to different servers, a distributed transaction scheme is required to coordinate the modification. Therefore, a distributed architecture greatly increases the complexity of metadata management. To sum up, the merits of each architecture should be comprehensively analyzed and evaluated based on the actual load characteristics of HPC services.

Cache technology: HPC clusters usually use the storage-compute decoupled architecture. Data access requests initiated by compute nodes need to be transmitted to storage nodes over the network, resulting in a large number of network I/O operations. As a result, the data access bandwidth is limited and the latency increases. The emergence of new storage media such as the Non-Volatile Memory Express (NVMe) high-end SSDs made equipping compute nodes with high-speed storage for read/write caching more popular. For example, in a read-intensive service scenario, the compute node cache may read data to the compute node in advance using prefetch technology, thereby reducing latency and bandwidth consumption caused by cross-network data access. In this scenario, the file access mode determines the prefetch efficiency. Write-intensive service scenarios are prone to write performance bottlenecks. For example, scientific computing applications in fields such as meteorology and geoscience periodically output a large number of intermediate results to the parallel file system. In addition, checkpoint files are periodically written to prevent system faults. These write modes are characterized by unpredictability and high concurrency. Therefore, storage devices on the compute nodes can also be used as a burst buffer to cope with sudden write requests for intermediate results or checkpoint files. Once the computing is complete, data in the burst buffer is flushed to the back-end storage nodes, and the burst buffer is cleared.

7.6 P2P Storage

The peer-to-peer (P2P) technology has developed rapidly in the fields of instant messaging, file sharing, and streaming media and has become one of the important technologies for constructing large-scale Internet applications. This section describes the architectural characteristics, technological developments, and typical applications of P2P storage systems.

7.6.1 Architecture Characteristics

The P2P storage system adopts a decentralized architecture and organizes storage nodes into a storage network in a peer-to-peer layout. In a traditional C/S architecture, a client initiates a request, and a server receives and processes the request. In a P2P network architecture, each node is both a request initiator and a request processor, as shown in Fig. 7.10.

The P2P storage system architecture has the following features:

Decentralization: The P2P storage system distributes storage resources to all nodes in a peer-to-peer manner. Unlike conventional storage systems, the P2P storage system does not need a centralized metadata server.

Scalability: In a P2P storage system, more users not only mean higher service requirements but also more system resources and stronger service capabilities. As a result, the P2P storage system can always meet user requirements, and its scalability is unlimited.

Availability: Storage services are distributed on user nodes. Faults on some nodes or networks have little impact on other nodes. Therefore, storage services have high availability.

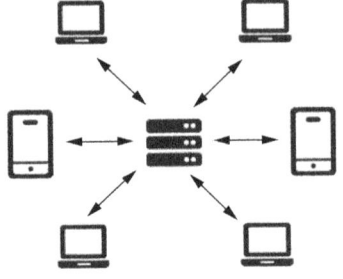 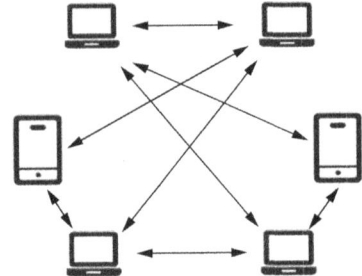

Traditional C/S centralized architecture   Decentralized P2P architecture

Fig. 7.10 A comparison of the architectures in a P2P storage system and a traditional storage system

Privacy: The P2P storage system does not require a node to process requests in a centralized manner. Therefore, the risks of eavesdropping and personal data leakage are greatly reduced.

Load balancing: In a decentralized P2P architecture, data requests do not need to be sent to one server for centralized processing. Each node functions as both a client and a server, and storage resources are evenly distributed on each node. This means loads are balanced across the entire system.

7.6.2 Key Technologies

The key technologies of P2P storage systems include the structured overlay network, data distribution technology, and data redundancy technology.

7.6.2.1 Structured Overlay Network

The structured overlay network interconnects and organizes all nodes at the application layer. This ensures that any two nodes can communicate with each other and that any node can dynamically enter or exit the system. To ensure network connectivity between nodes, a global namespace is used to manage nodes and provide a globally unique name for each node. On a P2P network, a node name is generally a unique value in the namespace. For example, the Chord routing algorithm [7] uses a ring space as the namespace, as shown in Fig. 7.11. When a new node is added to the system, it selects a location from the ring and uses the corresponding value as its unique identifier. When the node obtains its unique identifier, its relationships with adjacent nodes are also determined. It can then obtain other nodes' IP addresses based on these adjacency relationships. For example, nodes 13 and 3 are located on either side of node 0, which has just entered the system. Based on the adjacency relationships of

7.6 P2P Storage

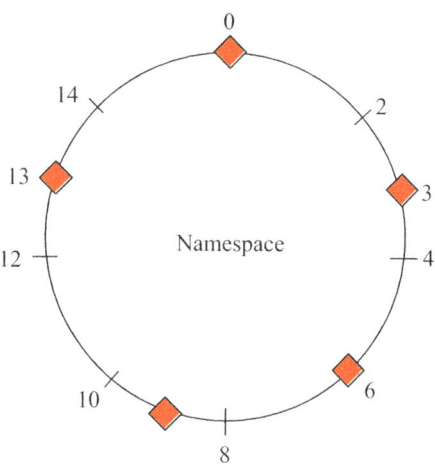

Fig. 7.11 Namespace of a structured overlay network

nodes 13 and 3, node 0 can get information about each node in the entire system. The routing table of each node in the system is updated at the same time. The process for a node to exit the system is much more complex and is not described here.

7.6.2.2 Data Distribution Technology

To store data in a structured overlay network, you need to name the data to determine its specific storage location. The data naming conventions are similar to the node naming conventions. For example, if the name of file A is 7, file A should be stored on the node named 7. If there is no node 7, a range-based responsibility mechanism can be defined. In this case, the next node after the file location can be used to store the file. In the example below, the node responsible for file 7 is node 9. However, when new nodes enter or exit the system, they may change the mapping relationship between file 7 and node 9, causing data access failures. Data distribution schemes can be used to prevent this. There are two major data distribution schemes: direct distribution based on the distributed hash table (DHT) and directory-based distribution.

Direct data distribution based on the DHT means that data is directly stored on a node that is responsible for the data location. When a node exits the system, its successor node is responsible for storing the data. This method is easy to implement but has an obvious limitation: the entry and exit of nodes result in the movement of a large amount of data, which is a waste of network bandwidth. In directory-based distribution, data is randomly distributed to nodes in the system, and the location mappings of these nodes are regarded as directory information and stored in the node determined by the DHT. When data is read, a mapping relationship of the data is obtained first, and then the node that is storing the data is accessed. The indirect data distribution mode, together with the data redundancy technology, ensure that only a small amount of directory information needs to be moved when node members change. However, the indirect distribution mode increases the number of network hops and the response latency.

7.6.2.3 Data Redundancy Technology

On a P2P overlay network, massive numbers of storage nodes can freely enter or exit the system. Data redundancy technologies are required to ensure data availability during node changes.

Currently, the most widely used data redundancy technologies are multi-copy redundancy and erasure code redundancy. Multi-copy redundancy simultaneously stores multiple copies of the data to secure nodes. Erasure code redundancy divides the data into k parts, generates r copies of parity data using an encoding algorithm, and distributes the $(k + r)$ pieces of data to different secure nodes. Any t ($t \geq k$) parts of the data can be used to restore the original data.

7.6.3 Application Scenarios

P2P storage systems are mainly used to share files, where a user uploads a file to one node and other nodes download the file. The more nodes on the network, the faster the download speed. The first P2P storage application was Napster, developed by several college students in the 1990s to share MP3 music files. This was not a regular P2P storage system because MP3 files were stored on personal computers, but their location information was centralized. Users had to access the centralized server for file locations and then read the MP3 file in P2P mode. This was the prototype of modern P2P storage systems. Gnutella, developed in 2000, uses a real P2P network where each node is a peer. The survival time of a data packet decreases by one after each routing, preventing infinite loop transmissions due to incorrect routing tables.

7.7 Cloud Storage

As globalization accelerates, people's demands for data increase. More data are generated as the Internet expands. Efficiently and securely storing and managing this data has become a new challenge for storage systems. Cloud storage is one way of coping with these challenges.

The purpose of cloud storage is to store and manage large amounts of data. It combines distributed storage, redundant storage, load balancing, and virtualization technologies to construct a highly available, cost-effective, customizable, and scalable storage platform. The platform provides personalized services for different users through web interfaces.

Cloud storage offers users secure and available online storage services. They do not need to manage the underlying storage architecture themselves. Instead, they

7.7 Cloud Storage

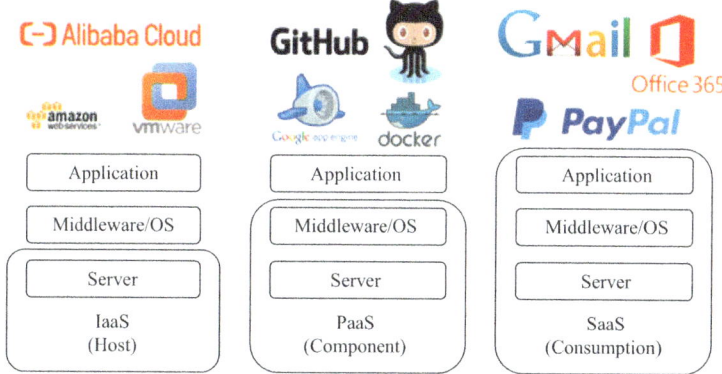

Fig. 7.12 The basic architecture of cloud computing

can use the storage services provided by service providers, who manage and operate the storage platforms.

Cloud storage evolves alongside cloud computing, which provides three service modes depicted in Fig. 7.12: infrastructure as a service (IaaS), platform as a service (PaaS), and software as a service (SaaS). The cloud storage offered by cloud platforms falls under IaaS and SaaS.

IaaS provides basic computer functions as a service for users. Users need not build a data storage center but can instead rent infrastructure from service providers and access these functions via the network, akin to renting them from a third-party provider. For example, Amazon's Simple Storage Service (S3) is a typical IaaS. It is an object storage service that uses buckets to store objects, allowing users to store an unlimited number of objects.

SaaS service providers develop application software on their servers, allowing users to customize the software as needed. Once customization is complete, users need not set up their own software systems. Instead, they can access the software functions provided by the service provider through the network. Currently, many service providers provide SaaS-level storage services. Examples include Google Docs and Tencent Docs. Google Docs, developed by Google, offers collaborative office functions and allows users to store locally edited documents on the cloud.

For users, the benefits of cloud storage include avoiding the need to purchase their own servers or hire manpower or source materials to construct their storage platform. Cloud storage has revolutionized storage and will continue to play a key role in future informatization.

7.7.1 Architecture Characteristics

The architecture of cloud storage differs from that of traditional storage systems. Figure 7.13 illustrates the cloud storage architecture.

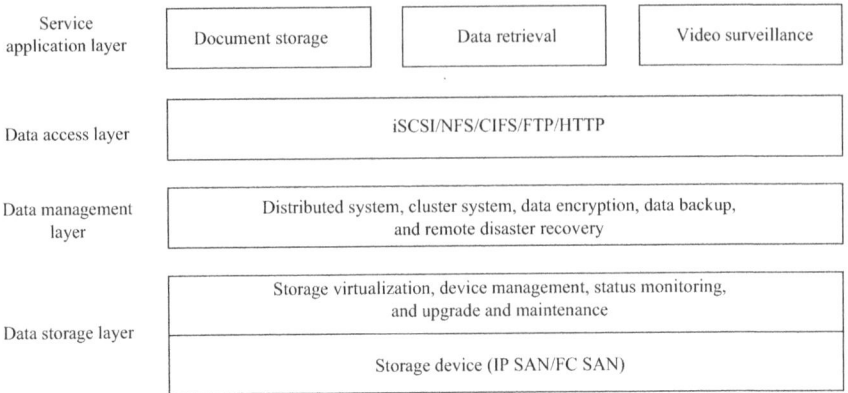

Fig. 7.13 Cloud storage architecture [8]

First, unlike traditional storage, cloud storage offers storage functions as a service. The underlying layer of cloud storage typically comprises a distributed cluster. When providing services, cloud storage employs hardware virtualization technology to conceal real computer hardware from users, creating a simulated computer environment on real hardware devices to meet the storage requirements of different users. In contrast, traditional storage architecture is tailored to specific application fields and relies on dedicated hardware.

Second, cloud storage can store massive amounts of data and has high scalability. Users can add machines to the cluster to increase storage capacity. Conversely, traditional storage architecture has poor scalability and is expensive to maintain and manage.

Third, the data management layer is a key part of cloud storage. A cloud storage platform uses distributed storage technology to facilitate collaboration among multiple devices. Through these technologies, cloud storage vendors can provide users with secure and reliable storage services. By encrypting, backing up, and implementing remote disaster recovery for stored data, cloud storage vendors can ensure the availability and integrity of data stored on the cloud.

Lastly, diverse hardware devices with varying levels of performance can be used for cloud storage. These devices are employed to set up the data storage layer and are interconnected via the network.

7.7.2 Application Scenarios

7.7.2.1 Public Cloud

A public cloud generally refers to a cloud computing service owned and operated by a third-party service provider, offering services to multiple users through the Internet. It is also the most common cloud deployment mode. All software and hardware resources and other supporting infrastructure in the public cloud are generally deployed outside the enterprise and are owned, managed, and controlled by the cloud service provider. The public cloud is usually used to deliver services such as web-based emails, online office applications, storage, and development environments. Currently, mainstream public cloud storage services in the market include Amazon S3, Alibaba Cloud, Huawei Cloud, and Baidu Wangpan.

The core attribute of a public cloud is its resource-sharing service. These resources are provided to users either free of charge or on a pay-per-use basis. Users can use services in a pay-per-use mode without investing in infrastructure construction, thereby significantly reducing the threshold for migrating data to the cloud and investment costs.

A public cloud offers several advantages:

Cost-effective: Instead of purchasing software and hardware resources, users only pay for the services they use, eliminating resource operation and maintenance costs.
Reliable: With massive server resources and professional service assurance teams to continuously maintain resources, cloud vendors can effectively prevent service disruptions caused by faults.
Scalable: Public cloud services can easily provide resources on demand to meet users' evolving service development requirements.

Therefore, public cloud services are the best choice for most startups and small- to medium-sized enterprises. They free enterprises from infrastructure deployment and O&M, allowing them to focus on their service development and innovation.

7.7.2.2 Private Cloud

Different from the public cloud, a private cloud is built for the exclusive use of enterprise customers. Enterprises own the cloud infrastructure and use cloud technology to centrally manage various storage resources through private clouds. Storage resources are efficiently used to meet specific requirements of different services, ensuring effective data security and service quality control. Generally, the private cloud is deployed within the firewall of the enterprise data center or in a secure hosting environment. Operating and maintained on a private network, the private cloud provides exclusive access rights to enterprise users.

The core attribute of a private cloud is dedicated resources. Enterprises that own a private cloud can easily virtualize and customize storage resources for cross-department sharing.

A private cloud offers the following advantages:

Flexible: Enterprises can customize cloud environments to meet specific service requirements.
Secure: Resources are exclusive to enterprises, ensuring higher levels of control and privacy. Therefore, private clouds are usually used by government agencies,

financial institutions, and other medium- and large-sized organizations that require greater control over the cloud environment. Private clouds can be customized based on actual service requirements to achieve higher flexibility and privacy.

7.7.2.3 Hybrid Cloud

Hybrid cloud storage connects private and public cloud storage to local resources, enabling data sharing between private and public clouds. It combines both public and on-premises or private clouds, integrating software, hardware, and network connections to manage data resources akin to local storage. Hybrid cloud storage exists in various forms, including CPFS storage, storage arrays, and distributed storage. For example, the hybrid cloud storage array brings together the best features of public cloud storage and traditional storage arrays. It offers ease of use, flexibility, efficiency, and reliability, along with functions like cloud tiering, synchronization, and caching. Customers need not modify their existing IT architecture or worry about protocol compatibility between local devices and cloud storage. The hybrid cloud storage array also features cross-region multi-copy protection to ensure data reliability.

A hybrid cloud offers the following advantages:

Flexible: The architecture can be customized to accommodate various programs and workloads, allowing for seamless data migration between the public cloud and traditional infrastructure.

Cost-effective: Workloads can be run in either the public or private cloud environment as needed, leading to cost reductions and long-term budget savings.

Scalable: Cloud resources are scalable, making hybrid cloud an ideal solution for growing services requiring additional storage resources and enhanced performance.

The hybrid cloud storage architecture consists of the public cloud, private cloud, or local servers, integrators, and data architecture. Each hybrid architecture includes at least one public cloud component that delivers hardware infrastructure functions. Private clouds or local servers cater to organizations requiring infrastructure, albeit at a higher cost than public clouds. Effective communication between each component of the data center architecture is crucial, facilitated by an integrators such as wide area network (WAN), virtual private network (VPN), or application program interface (API). The data architecture standardizes the data layer and connection process, offering a unified approach to managing the data lifecycle.

7.8 Storage Virtualization

Nowadays, storage requirements have changed dramatically due to storage system expansion. Traditional siloed IT infrastructure falls short in meeting service demands. When purchasing storage infrastructure, customers must consider long-term business expansion and anticipate the need for ramping up storage resources ahead of peak seasons. In this case, the storage scale in the initial procurement is usually much larger than needed, wasting a great amount of resources. In the traditional mode, different services operate on dedicated IT systems, leading to an inability to directly share hardware resources and resulting in low resource utilization. In addition, sudden increases in business scale pose significant challenges to the scalability of the storage system. In response to these challenges, storage virtualization technology has emerged as a solution. With the advancement of new infrastructure such as cloud computing and data centers, storage virtualization has evolved from conventional storage virtualization to modern software-defined storage (SDS), hyperconverged infrastructure (HCI), etc.

7.8.1 Basic Concepts

In a broader context, storage virtualization can be traced back to the inception of storage systems. Initially, users accessed disks through their linear block addresses (LBAs), which enabled the virtualization of the logical address space into a three-dimensional structure comprising cylinders, tracks, and sectors. Currently, the definition of storage virtualization is still evolving. Simply put, a logical abstraction layer is added between physical storage systems and servers so that servers do not directly interact with storage hardware. The addition, deletion, splitting, and combination of storage hardware are independent of servers. Storage virtualization technology decouples the storage system from the physical hardware, allowing administrators to manage storage resources as a pool that can be allocated and reallocated as needed. This leads to greater flexibility and efficiency in storage utilization. Currently, storage virtualization has been applied to different storage layers, breaking through the initial definition and extension of storage virtualization.

The Storage Networking Industry Association (SNIA) classifies storage virtualization into virtualization object, location, and implementation mode, as shown in Fig. 7.14. Virtualized objects, or objects created during virtualization, include disks, tape systems, data blocks, file systems, files, or records. Disk virtualization enables users to access disks through block addresses without having to know the internal hardware details of the disks. Block virtualization employs RAID groupings to consolidate multiple hard disks into logical volumes, each emulating a standalone physical hard disk. Users can read and write to logical volumes just like regular hard disks but with the benefit of RAID-based data protection. A typical example of file system virtualization is the virtual filesystem switch (VFS) based on the Linux kernel, which was introduced by Sun Microsystems when defining NFS. VFS allows operating systems to use different file systems to manage storage devices. It defines a set of standard interface layers between physical file systems and services and abstracts file system details so that different file systems are the same in the Linux kernel and other processes running in the system.

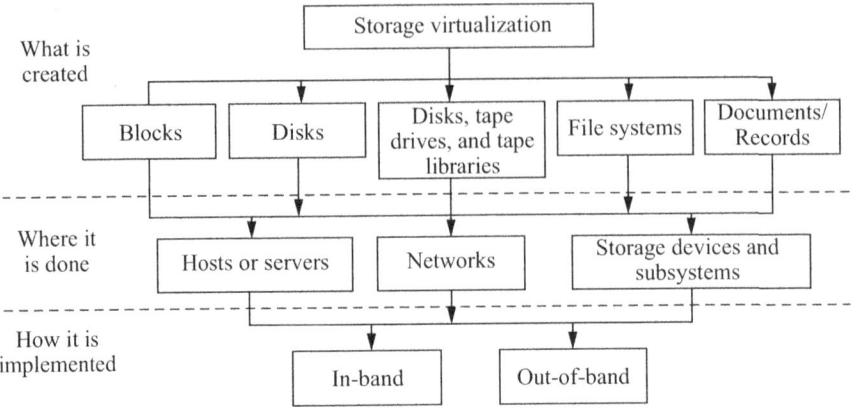

Fig. 7.14 Implementation of storage virtualization [3, 9]

Based on where virtualization is done, storage virtualization is classified into host-based, storage device-based, and network-based. Host-based virtualization enables the storage space of a server to span multiple heterogeneous disk spaces and is generally used for mirror protection between different arrays. Despite its benefits, host-based virtualization has several drawbacks, including high resource consumption, performance degradation, and compatibility issues with operating systems and applications. Storage device-based virtualization protects and migrates data by adding virtualization functions to storage controllers. The process is host-transparent, occupying no host resources and providing rich data management functions. However, the virtualization can be implemented only inside devices, lacking interoperability. Network-based virtualization is implemented by adding a virtualization engine to the SAN. It can be used for heterogeneous storage system integration and unified data management. Despite similar advantages with storage device-based virtualization, the disadvantages are that some vendors have weak data management functions and low maturity.

In terms of implementation methods, virtualization may be classified into in-band and out-of-band virtualization, according to whether the control path for transmitting metadata in storage virtualization is the same as the data path for transmitting data. In-band virtualization technology combines data access and virtualization management functions (including replication, mirroring, and continuous data protection). That is, storage virtualization is implemented in the network path from the host to the storage device. The advantage of in-band virtualization technology lies in the high compatibility between servers and storage devices. In addition, virtualization and data management functions are implemented using dedicated hardware, thereby avoiding the occupation of host computing resources. However, a fault inside the virtualization device may cause system breakdown. On the contrary, the out-of-band virtualization technology completes the virtualization process before data read or write. The virtualization engine is placed outside the access path from the host to the storage device. Therefore, compared with in-band virtualization,

7.8 Storage Virtualization

out-of-band virtualization can only be implemented based on the storage network. The advantage of out-of-band virtualization technology is its high availability when a virtualization device is faulty, ensuring uninterrupted system services. However, this method causes the consumption of system resources. In addition, most products that adopt this approach usually lack powerful data management functions.

SNIA's classification of storage virtualization demonstrates that storage virtualization is adapting to various user environments in different forms and reflects the complexity of storage virtualization technologies. Various types of storage virtualization also have basic commonality: hiding the complex structure of underlying storage components with a set of methods and providing some advanced storage services. Storage virtualization technology frees the capabilities of logical devices from being restricted by a single physical device. The capacity and performance of logical devices can be dynamically expanded. In addition, multiple methods can be used to optimize the storage resources of the entire system, reducing the total cost of ownership (TCO) and improving the service quality of the storage system.

7.8.2 Key Technologies

Storage virtualization has spawned many new technologies, products, and companies. As early as 1987, scientists at the University of California, Berkeley, proposed redundant arrays of independent disks (RAID) technology. The initial goal of RAID was to enhance storage performance and provide data recoverability after a disk fails. RAID technology is a milestone in the development of storage virtualization and is widely used. It has attracted a large number of researchers and there are three main research categories.

- Data layout optimization: Data layout optimization techniques primarily fall into two kinds: distributing data blocks and parity blocks uniformly across all disks (e.g., RAID 4 and RAID 5) and leveraging locality properties to scatter data blocks across different disks for better access efficiency (e.g., left-symmetric RAID 5). The parity declustering layout proposed by researchers uses as few disks as possible for data recovery so that other disks can process application I/O requests. This layout is further extended and optimized, for example, to be used in the Panasas PanFS file system. There are also some optimizations that are sensitive to load characteristics or application scenarios. For example, the disk cache disk (DCD) [10] uses an extra disk as the cache to convert small random writes into large log writes. The HP AutoRAID [11] hierarchical storage system divides storage space into RAID 1 and RAID 5. It differentiates read and write operations to improve storage bandwidth and reduce data redundancy. ALIS [12] and BORG [13] reorganize frequently accessed blocks (and block sequences) so that they are placed sequentially in a dedicated disk area.
- Storage reconstruction optimization: Many researchers recognized the importance of shortening data reconstruction time and proposed various methods to improve

reconstruction performance. Some research focuses on designing better data layouts within disk groups. For example, Kahn et al. proposed the cyclic Reed–Solomon codes [14] to minimize the I/O operations required for data recovery and degraded reads. Menon and Mattson proposed a technology called distributed sparing in disk arrays [15], which uses parallel free disks to enhance storage performance. Other methods optimize the workflow of storage reconstruction, such as disk-oriented reconstruction (DOR) [16] and pipelined reconstruction (PR) [17]. Additionally, some task scheduling techniques [18] can be used to optimize rate control for storage reconstruction. Chinese scholars have also proposed storage reconstruction optimization methods, such as priority-based multi-thread reconstruction optimization [19], S2-RAID (skewed sub-arrays in the RAID structure [20]), and fast reconstruction method for degraded RAID sets [21].

Storage expansion optimization: Researchers have been exploring new methods to improve expansion efficiency of storage systems. The addition of a new disk to the storage system triggers data migration to achieve balanced data distribution and maximize storage resource utilization. One type of storage expansion method is random RAID [22], which significantly reduces the amount of data to be migrated. However, multiple expansions cause unbalanced data distribution. More storage systems use deterministic layouts to organize data. Gonzalez et al. proposed a gradual assimilation algorithm to increase the capacity of RAID 5 [23]. However, the use of an incremental, serial data migration strategy with immediate metadata updates impedes the migration process, resulting in lower performance. The storage team from Tsinghua University came up with the reordering window characteristic for scaling a RAID volume [24, 25] and then proposed a series of methods that significantly improve the scaling efficiency.

7.9 Software-Defined Storage (SDS)

7.9.1 Basic Concepts

Before digging into SDS, you need to understand the concept of software-defined. Software-defined means that software is used to define hardware functions and provide various virtualized, flexible, and customized functions for hardware, maximizing system running efficiency and resource utilization. For example, in the network field, software-defined networking (SDN) is an IT infrastructure method that abstracts network resources into virtual systems. It separates network forwarding from network control to create physical and programmable networks that can be managed in a centralized manner. SDN allows O&M teams to control network traffic in complex network topologies through a centralized panel, eliminating the need to manually configure each network device.

Similarly, SDS is a storage architecture that separates storage software from hardware. With storage virtualization as the core, SDS defines server, storage, and network resources in a data center using software and automatically allocates and

7.9 Software-Defined Storage (SDS)

manages these resources. At its core, SDS embodies two key principles: First, it eliminates the software dependence on specialized hardware (e.g., NAS or SAN), enabling them to run on industry-standard servers or x86 platforms. Second, SDS emphasizes the centrality of software in delivering the core functionalities of storage virtualization technologies, including volume management, RAID, data protection, snapshots, cloning, and replication. As shown in Fig. 7.15, SDS can be divided into application layer, control layer, and infrastructure layer. The infrastructure layer abstracts and combines various storage devices to provide abstract storage resources for the control layer. The control layer classifies and divides storage resources and allocates storage resources based on data center requirements. The application layer provides different storage interfaces for various applications to meet their storage requirements.

The existing SDS has the following characteristics.

Scale-out architecture: Storage space can be dynamically increased or decreased.
Commodity hardware: Different storage resources, such as local storage devices and network storage devices, can be added.
Resource pooling: All storage resources must be integrated into a unified logical pool, which can be dynamically allocated based on resource requirements.
Abstraction: All storage devices can be collected and abstracted in a unified pool, which is compatible with various storage interfaces.
Automation: Different types of storage can be automatically defined and dynamically expanded based on user requirements.

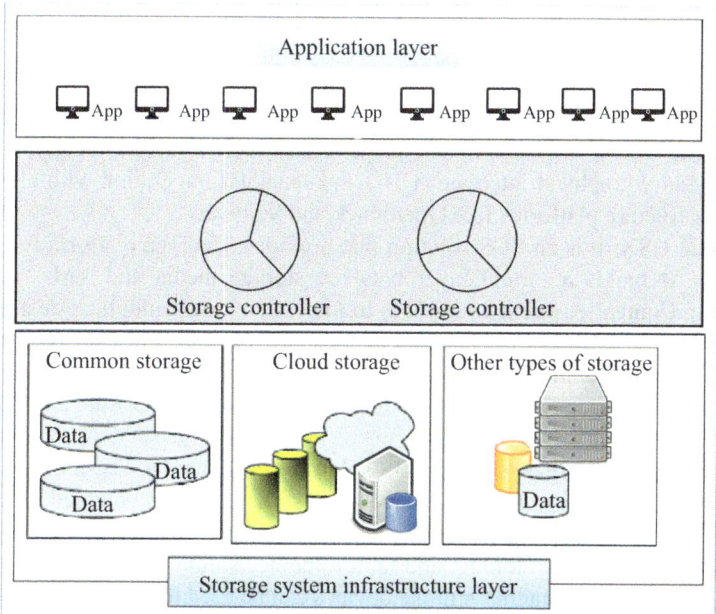

Fig. 7.15 SDS layers [26]

Programmability: Many APIs are available to provide visible control for the resources, integrating several system components to enable system automation.

Policy-driven: Different security, reliability, and quality of service (QoS) are provided for different users.

7.9.2 Representative Systems

Due to the diverse offerings of storage vendors or manufacturers, there are multiple solution approaches available, each with its own strengths in terms of performance, capacity, and expandability.

IBM Storwize: The main feature of IBM Storwize is its numerous APIs supporting virtualization environments, ensuring efficient processing of massive and growing data. The storage layer provides file and block storage access interfaces and scalable storage management for cloud storage.

EMC STaaS: EMC proposes the concept of storage as a service (STaaS), simplifying storage management; providing direct access to files, objects, and blocks in the storage resource pool; and dynamically adding new storage arrays. Different VMs provide abstract storage access for different users. In addition, EMC provides open access to its storage virtualization software platform ViPR. Unlike IBM's upper-layer interfaces, EMC also provides lower-layer interfaces for different storage vendors or enterprises to adapt to their hardware storage products. In terms of abstract storage access interfaces, new big data capabilities are also integrated into ViPR with a Hadoop Distributed File System (HDFS) Data Service.

Nexenta SMARTS: Nexenta's SMARTS solution delivers on six key pillars: security, manageability, availability, reliability, lower TCO, and scalability. This solution offers a range of security and reliability features, including clone, snapshot, backup, end-to-end verification, and adaptive data recovery. Additionally, it boasts a user-friendly graphical interface (GUI) and seamless integration with cloud service infrastructure platforms like OpenStack and VMware.

Atlantic USX: It is an SDS solution that provides a storage platform for virtual machines. It builds a virtual layer between storage media and VMs, uses the HyperDup Content-Aware data services to implement data deduplication and compression, and optimizes I/O channels between virtual and physical machines for efficient virtual machine I/O access. In addition, USX integrates memory and external storage to provide efficient access for remote desktops, XenApp, and VMware Horizon.

Ceph: Ceph is an open-source SDS solution that uses the Crush algorithm to achieve high scalability. A virtual object storage layer is implemented at the bottom layer to provide file, object, and block storage for the upper layer. Ceph supports data compression, cloning, and fault tolerance solutions. In terms of storage media, various heterogeneous storage devices can be used to form a unified virtual storage space. It is widely used in OpenStack to provide efficient storage access for VMs.

7.10 Hyperconverged Infrastructure   175

Gluster: Gluster is also an open-source SDS solution that supports various modules and can be combined randomly. It is a distributed scale-out file system that does not need metadata servers. Using the directory-based dynamic hash algorithm, Gluster expand the number of nodes to tens of thousands. Gluster is widely used in ultra-large supercomputing centers and data storage centers.

In addition to the previously mentioned SDS solutions, there are other options such as Maxta, DataCore, and CloudBytes. However, these solutions fall short of fully addressing SDS requirements. For instance, not all SDS solutions provide seamless access to files, objects, and blocks or offer a unified storage access space.

In brief, while existing SDS solutions strive to deliver improvements in scalability, security, reliability, and cost-effectiveness, they still face certain challenges.

7.9.3 Key Challenges

So far, SDS has not been clearly defined. Building an SDS presents significant challenges in integrating and coordinating resources, including dynamically allocating interfaces for block, file, and object storage, supporting data migration, data reliability, data compression, copy, QoS, providing advanced user APIs, metadata access, fault tolerance, and monitoring the system. In the future, with the development of storage hardware and the Internet, SDS will face new problems. For example, as new nonvolatile storage devices and compute express link (CXL) are widely used in storage systems, the storage hardware delay continuously decreases, but software overheads caused by storage increase. While integrating with high-speed network devices, SDS also needs to adapt to the development of some high-speed network devices for new adjustments. For example, when adopting SDS in Internet of Things (IoT) services, traditional data storage systems cannot meet high concurrency and high-performance persistency requirements.

7.10 Hyperconverged Infrastructure

Data centers typically require both a virtual computing platform and storage platform, and these components are often independent of one another. The most typical solution is the virtualization and storage consolidation solutions of Ceph and OpenStack. Their biggest feature is that storage and computing are separated. In this architecture, storage and computing resources can be scaled separately. However, two systems need to be deployed independently, increasing the management burden and cost. To build a data center, you need to consider the coordination of each software stack, which consumes huge management costs. To address this issue, the industry proposes the concept of hyperconverged infrastructure (HCI). The essence of "hyper" is virtualization. VMs provide computing resources, SDN networks the VMs, and SDS provides storage for the VMs. Therefore, HCI integrates both the philosophy of SDS and unified software management of computing and network resources [21].

7.10.1 Basic Concepts

In traditional virtual machine cloud platform solutions, computing virtualization and network virtualization are integrated. For example, the OpenStack cloud platform has implemented the fusion of SDN and virtual machine computing. Therefore, the most essential aspect of HCI is the addition of SDS, which is mainly represented by distributed file systems, distributed block storage, NAS clusters, and so on. However, HCI does not directly combine traditional SDS with virtual machine computing. Instead, it integrates computing, network, and storage into a unified platform to reduce management complexity. Typically, an x86-based physical device can provide a complete set of virtualization solutions, scaling out computing and network resources together for fast deployment of the data center, as shown in Fig. 7.16.

According to IDC, HCI is a system that integrates core storage, computing, and network functions into a unified software solution or device. The existing HCI system aims to achieve on-demand expansion, quick deployment, easy management, and elastic scaling. With on-demand scaling, the data center can continuously update hardware based on service requirements to achieve efficient utilization of physical resources. Software and hardware are integrated into an independent machine that can be directly used after interaction, achieving fast deployment. A unified management interface is provided to manage computing, storage, and network resources in a unified manner, eliminating the need for separate O&M. By supporting the distributed architecture, HCI implements linear expansion of performance and capacity. It eliminates node restrictions, avoids single points of failure, and supports operations such as backup, fault tolerance, and deletion.

The concept of HCI was initially proposed by Nutanix. Its integration solution is to provide distributed storage, virtual networks, and virtual machines on a single

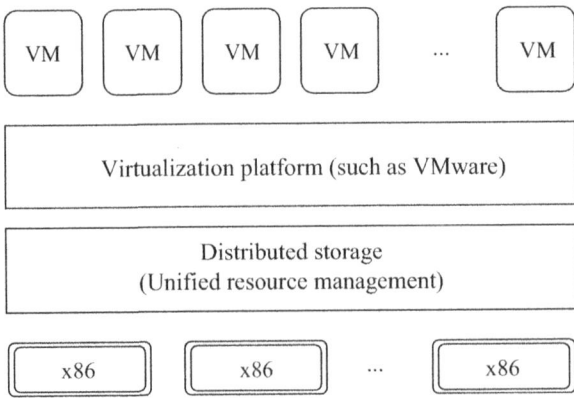

Fig. 7.16 Hyperconverged infrastructure

physical machine. Additionally, multiple physical machines can be horizontally scaled out to support data sharing, replication, fault tolerance, and compression across multiple machines. In the HCI solution, users only need to add physical machines to expand the capacity of a data center, eliminating the need for complex software configuration and overhead. It provides a fast and easy channel for building small computing centers, creating large business value. In addition, the efficient data access mechanism for HCI is also a hot research topic in the industry.

7.10.2 Key Technologies

Currently, there are two types of HCI solutions. One is software-only solutions, which support storage resource consolidation on existing hardware. During deployment, you only need to configure software on the existing hardware. Typical examples are SANsymphony of DataCore Software and ScaleIO of Dell EMC. The other is software–hardware integrated solutions, which integrate software and hardware into the same server. Typical examples are vSAN of VMware and Nutanix. In the existing Nutanix solution, controller VMs (CVMs) are proposed to replace the controllers in SAN storage. CVMs have two main advantages. First, CVMs of multiple nodes can form a distributed storage management module to provide data backup, deduplication, and copy functions. Second, storage controllers can directly access the CVMs by using data in the VMs, thereby avoiding the overheads of conventional complex software stack. In the CVM, a flash memory is usually added as a local cache to support low-latency and high-concurrency data access of VMs.

The storage HCI architecture also has inherent defects:

Limited scale-out flexibility: HCI features flexible and on-demand scaling and deployment using a minimum number of nodes. In HCI, the computing nodes, storage performance, and memory must be scaled at the same time, not meeting the requirement of single-aspect scaling. In addition, some vendors have requirements on the minimum scaling unit, which limits the expansion flexibility. When the cluster reaches a certain scale, the system architecture complexity increases nonlinearly, cluster management becomes more difficult, and the probability of hardware faults and self-recovery increases greatly.

Single storage mode: Only block storage access interfaces are provided. These interfaces are applicable to VM access in most cases. However, a data center must provide data persistence services for files and databases, which cannot be deployed on the existing HCI architecture and require additional hardware and systems.

I/O and compute resource scheduling difficulty: As network device and storage device performance increases, competition for computing resources between these components becomes the primary cause of unstable VM operations. For high-speed network and storage, the existing interrupt mechanisms cannot meet the latency requirement. As a result, performance of existing storage network hardware cannot be fully unleashed.

7.10.3 Representative Systems

Currently, mainstream HCI vendors include Nutanix, Huawei, VMware, SmartX, and Sangfor. Each has its own competitive product features and distributed storage implementation solutions. Table 7.2 compares some typical HCI products.

In terms of metadata management, the distributed block storage architecture of most HCI products adopts the following two solutions:

1. An independent metadata service module that manages the mapping between data blocks and storage devices, queries and update indexes of data requests, data placement, and load balancing policies. In the HCI scenario, computing and storage resources reuse the same server hardware. Therefore, the metadata service module may optimize the data placement policy according to the relative location relationship between computing and storage so that most I/O requests can be completed inside the current server, thereby reducing the I/O path and greatly reducing the overhead of network data transmission.
2. Using the consistent hashing algorithm. This method determines the specific storage location by calculating the hash value of the full path where the current data is located. The main advantage of this method is that it eliminates the performance bottleneck caused by the centralized metadata management module, making the data access path lightweight and efficient. The disadvantage is that the opportunity of local I/O optimization is lost and directory operations are

Table 7.2 Comparison of representative HCI products

Comparison item	Nutanix XC series	EMC VxRail series	Huawei product series	SmartX
SDS	NDFS	vSAN	FusionStorage	ZBS
Metadata architecture	Independent metadata service	–	Consistent hashing	Independent metadata service
Virtualized computing support	Almost all virtualization platforms	Only vSphere supported	Hyper-V not supported	Hyper-V not supported
Storage virtualization and integration mode	The storage service runs on an independent VM on the hypervisor	The storage service runs as a kernel module inside the hypervisor	The storage service runs on an independent VM on the hypervisor	The storage service runs outside the hypervisor
Resource consumption	4 cores 24 GB memory	10% CPU 32 GB memory	≥ 4 cores ≥ 64 GB memory	3–4 cores 16 GB memory
Management and O&M platform	Prism	vCenter	FusionCube Center	Fisheye
Delivery mode	Appliance software (for small- and medium-sized enterprises)	Appliance	Appliance	Appliance software

inefficient. For example, when a folder is renamed, the hash values of all subfiles need to be recalculated and migrated to a new server node.

Storage and virtualization consolidation solutions are classified into the following types:

1. Storage modules are embedded in the hypervisor as kernel modules. In this architecture, storage modules can directly access storage devices without passing through the hypervisor, significantly reducing performance loss. Typical products include EMC VxRail.
2. The storage module runs on a VM, and the VM functions as a Virtual Storage Appliance (VSA). In normal cases, the VSA needs to forward requests to access storage devices through the hypervisor. To improve performance, VSA uses the I/O passthrough technology to directly access storage hardware, decoupling virtualization from storage. Typical products of this solution include Nutanix.
3. The storage module runs outside the hypervisor but belongs to the same software stack as the hypervisor. This architecture is applicable to the Kernel-based Virtual Machine (KVM) platform. Typical products include SmartX, which features efficient I/O paths and high performance.

7.10.4 Concept Comparison

This section compares several emerging storage concepts that are easy to confuse. Storage virtualization is the technical basis for various storage architectures. It decouples physical details from the logical abstraction of storage devices and provides various storage functions and semantics. SDS and HCI are often used for data centers. Generally, storage virtualization is used on dedicated hardware devices, while SDS has no restrictions on devices and is based on storage virtualization. SDS provides data management functions as services. HCI is an integrated solution that combines computing and storage, providing near-disk computing as much as possible. A core component of HCI is SDS. The relationship between containers and applications drives application awareness of storage and HCI storage, the core of which is the new development of storage virtualization.

7.11 Summary

Various storage architectures and concepts emerge one after another. These new storage architectures and concepts have both common features and new differences. Typical storage architectures, such as DAS, SAN, NAS, and OSD, define differentiated interfaces between storage and computing. They have advantages and disadvantages in performance, maintenance costs, and scalability. Parallel, P2P, and cloud storage systems are designed to meet different storage requirements.

7.12 Practice Questions

1. **Please describe the differences between DAS, NAS, SAN, and OSD.**

    Answer: There are many differences between them in terms of usage, architecture, and working mode. DAS is used for local storage on a single node via SATA or SAS. It does not allow data to be shared with other nodes. NAS is used for file sharing and data backup. NAS devices run file systems and provide file access interfaces which can be accessed through network protocols (such as CIFS and NFS). SAN is used for mission-critical services such as virtualization and databases. SAN devices are interconnected using Fibre Channel or iSCSI and provide block access interfaces. OSD is used to store and retrieve a large amount of unstructured data and provides object access interfaces which can be accessed through HTTP.

2. **What are the functions of the SAN system management software? What are the differences between in-band management and out-of-band management?**

    Answer: The SAN system management software provides functions such as storage device management (adding, configuring, and deleting devices), route management, storage virtualization, security management, performance monitoring, backup and restoration, and bandwidth management. In-band management relies on the storage network and therefore may affect network performance. However, it can provide real-time management and monitoring of storage devices and networks. Out-of-band management is an independent management channel that does not interfere with the transmission of storage data. It is usually used to perform key management tasks to ensure system availability and stability.

3. **Compared with traditional block or file access interfaces, what are the advantages of object access interfaces of object storage?**

    Answer: Object access interfaces integrate the advantages of file and block access interfaces. In file storage, the tree structure and path access mode allow people to understand, memorize, and access data. However, to find a file, a computer needs to break down the path and search for it level by level, which hinders the storage performance. In block storage, once a logical block is mounted to a client, other clients cannot access the data in that logical block. In addition, the block can only be accessed by programs on the client after it has been partitioned and a file system has been installed. In contrast, in object storage systems, data is stored in a single-plane structure, allowing for higher scalability and better performance.

4. **What are the main features of parallel storage systems? What are the most important aspects to focus on during the design process?**

    Answer: Parallel storage systems improve performance and throughput by leveraging parallel processing and data distribution. They are used for large-scale data storage and processing in scenarios that require high performance, low latency, and scalability. These characteristics make them an important storage solution for high-performance computing, big data analysis, and scientific

7.12 Practice Questions

research. Designing a parallel storage system is a complex and challenging task. The following aspects must be considered:

1. Data distribution and balancing: Data should be evenly distributed across storage nodes to avoid hotspot problems and unbalanced loads.
2. Concurrency and consistency: Data consistency and integrity must be ensured when multiple concurrent read/write operations are supported.
3. Data access performance: This includes optimizing data access paths, reducing the time of disk addressing, and making data caching more efficient.
4. Scalability: Effective expansion policies and load balancing mechanisms are critical.

5. **Please list some common storage products at the IaaS, PaaS, and SaaS levels, respectively.**
   Answer:
   - IaaS: Amazon EBS, Microsoft Azure, Google Cloud, Alibaba Cloud, etc.
   - PaaS: Amazon RDS, Microsoft Azure SQL Database, Alibaba OceanBase, etc.
   - SaaS: DropBox, Tencent Docs, etc.

6. **Please briefly explain the difference between storage virtualization and software-defined storage (SDS).**
   Answer: Storage virtualization is a technology that abstracts multiple physical storage devices (such as hard drives, storage arrays, etc.) into a single virtual storage pool and provides a unified interface to applications and operating systems through a virtualization layer. The goal of storage virtualization is to simplify storage resource management and improve storage utilization. SDS is a broader concept that separates storage functionality from hardware and uses software to implement data management and storage operations. SDS includes various technologies. Besides storage virtualization, it also includes different types of storage solutions such as object storage, distributed storage, block storage, and file storage.

7. **What is the background of the proposal of hyperconverged infrastructure (HCI)? What are the typical application scenarios?**
   Answer: Conventional data center architectures typically require separate servers, storage devices, network devices, and management on hardware from multiple vendors, resulting in complex hardware configurations and management. Modern applications and businesses demand flexible IT infrastructure to rapidly scale resources up or down as needed, which conventional architectures may struggle to meet. In response to these challenges and demands, HCI is proposed as a new data center architecture model. Its goal is to integrate computing, storage, network, and management functions into a unified, highly integrated hardware and software platform.

8. **What are the technical methods for accessing storage devices in virtual machines? How is the specific performance?**
   Answer:
   1. Virtual disks, a common type of virtual storage device, are typically used for hard disk storage in virtual machines. Various underlying virtualization

approaches, such as hypervisor-based pure software virtualization and VirtIO, can be used.
2. Physical disk pass-through typically provides better performance because of the lower overhead of the virtualization layer. However, it may not be suitable for all virtualization platforms and use cases.

References

1. Shu J, Li B, Zheng W. Design and implementation of a SAN system based on the fiber channel protocol. IEEE Trans Comput. 2005;54(4):439–48.
2. Zhang G, Shu J, Xue W, et al. Design and implementation of an out-of-band virtualization system for large SANs. IEEE Trans Comput. 2007;56(12):1654–65.
3. Peglar R. Storage virtualization I: what, why, where and how; 2007 [2024-04-15].
4. Palanivel K, Li B. Anatomy of software defined storage challenges and new solutions to handle metadata, report. Minneapolis: University of Minnesota; 2013.
5. Weil S, Brandt S, Miller E, et al. Ceph: a scalable, high-performance distributed file system. In: USENIX Association. Proceedings of the 7th symposium on operating systems design and implementation (OSDI'06). Berkeley: USENIX Association; 2006. p. 307–20.
6. Ren K, Zheng Q, Patil S, et al. IndexFS: scaling file system metadata performance with stateless caching and bulk insertion. In: IEEE. Proceedings of the international conference for high performance computing, networking, storage and analysis (SC'14). Piscataway: IEEE Press; 2014. p. 237–48.
7. Ganesan P, Manku GS. Optimal routing in chord. In: Society for industrial and applied mathematics. Proceedings of the fifteenth annual ACM-SIAM symposium on discrete algorithms (SODA'04). Philadelphia: Society for Industrial and Applied Mathematics; 2004. p. 176–85.
8. Zhou K, Wang H, Li C. Cloud storage technology and its applications. ZTE Commun. 2010;8(4):27–30.
9. Shu J, Li S, Zhang G. cunchu xunihua yanjiu zongshu [Overview of storage virtualization research]. Commun CCF. 2017;13(6):14–24.
10. Yang Q, Hu Y. DCD—disk caching disk: a new approach for boosting I/O performance. In: ACM. Proceedings of the 23rd annual international symposium on computer architecture. New York: ACM Press; 1996. p. 169–78.
11. Wilkes J, Golding RA, Staelin C, et al. The HP AutoRAID hierarchical storage system. ACM Trans Comput Syst. 1996;14(1):108–36.
12. Hsu WW, Smith AJ, Young HC, et al. The automatic improvement of locality in storage systems. ACM Trans Comput Syst. 2005;23(4):424–73.
13. Bhadkamkar M, Guerra J, Useche L, et al. BORG: block-reorganization and self-optimization in storage systems (2007-07-01) [2023-04-10].
14. Khan O, Burns R, Plank J, et al. Rethinking erasure codes for cloud file systems: minimizing I/O for recovery and degraded reads. In: USENIX Association. Proceedings of the 10th USENIX conference on file and storage technologies. Berkeley: USENIX Association; 2012. p. 20.
15. Menon J, Mattson D. Distributed sparing in disk arrays. In: IEEE. Proceedings of the thirty-seventh international conference on COMPCON. San Francisco: IEEE CS Press; 1992. p. 410–21.
16. Holland MC. On-line data reconstruction in redundant disk arrays. Pittsburgh: Carnegie Mellon University; 1994.
17. Lee JYB, Lui JCS. Automatic recovery from disk failure in continuous-media servers. IEEE Trans Parallel Distrib Syst. 2002;13(5):499–515.

References

18. Lumb CR, Schindler J, Ganger GR, et al. Towards higher disk head utilization: extracting "free" bandwidth from busy disk drives. In: USENIX Association. 4th symposium on operating system design and implementation. Berkeley: USENIX Association; 2000. p. 87–102.
19. Tian L, Feng D, Jiang H, et al. A popularity-based multi-threaded reconstruction optimization for RAID structured storage systems. In: USENIX Association. 5th USENIX conference on file and storage technologies, FAST 2007. San Jose: USENIX Association; 2007. p. 277–90.
20. Wan J, Wang J, Xie C, et al. S2-RAID: parallel RAID architecture for fast data recovery. IEEE Trans Parallel Distrib Syst. 2014;25(6):1638–47.
21. Wu S, Jiang H, Feng D, et al. Workout: I/O workload outsourcing for boosting RAID reconstruction performance. In: USENIX conference. 7th USENIX conference on file and storage technologies. Berkeley: USENIX Association; 2009. p. 239–52.
22. Goel A, Shahabi C, Y YD, et al. Scaddar: an efficient randomized technique to reorganize continuous media blocks. In: IEEE. Proceedings of the 18th international conference on data engineering (ICDE). San Jose: IEEE; 2002. p. 473–82.
23. Gonzalez J, Cortes T. Increasing the capacity of RAID5 by online gradual assimilation. In: SNAPI. Proceedings of the international workshop on storage network architecture and parallel I/Os. New York: Association for Computing Machinery; 2004. p. 17–24.
24. Zhang G, Shu J, Xue W, et al. SLAS: an efficient approach to scaling round-robin striped volumes. Trans Storage. 2007;3(1):1–39.
25. Zhang G, Zheng W, Shu J. ALV: a new data redistribution approach to RAID-5 scaling. IEEE Trans Comput. 2010;59(3):345–57.
26. Palanivel K, Li B. Anatomy of software defined storage challenges and new solutions to handle metadata, report; 2013 [2024-04-15].

Chapter 8
Distributed Storage Systems

As data volume continues to increase, a single server fails to meet the storage requirements of applications. Therefore, distributed storage systems have come into being. These systems connect a large number of servers over a network, distribute data across these servers, and provide specific storage interfaces for upper-layer applications. In addition to standard server hardware, many companies have developed dedicated hardware for distributed storage systems to achieve higher reliability and performance, lower costs, and energy saving. Based on storage interface classification, common distributed storage systems include distributed key-value store systems, distributed object storage systems, distributed block storage systems, and distributed file systems.

8.1 Typical Architecture of Distributed Storage Systems

Figure 8.1 shows the typical architecture of a distributed storage system, including a coordination server, storage servers, and application servers.

The coordination server is a centralized component of the distributed storage system. It records the system-wide cluster configurations, including a set of live storage servers, and data distribution information, recording the mapping from data partitions to storage servers, in a highly reliable manner. In a distributed key-value store system, a data partition generally consists of key-value pair data included in a segment of continuous key space. In directory tree management of a distributed file system, a data partition may be a subtree of a directory tree. The coordination server monitors the survival status of storage servers, usually by sending heartbeat messages periodically to storage servers. It also performs load balancing based on the load status of storage servers. For example, when finding that a storage server bears too many requests, the coordination server generates a data migration task and migrates some data partitions in the storage server to another storage server. To prevent the coordination server from becoming a performance bottleneck, storage

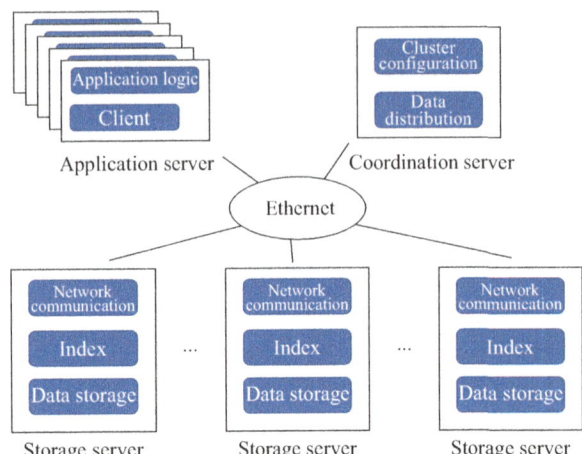

Fig. 8.1 Typical architecture of a distributed storage system

servers and application servers usually cache cluster configurations and data distribution information locally

Storage servers store data of data partitions according to the data distribution information and mainly include three modules: the network communication module, index module, and data storage module. Generally, the network communication module uses a mature remote procedure call (RPC) library to receive and send network requests; the index module queries final locations of data, such as a hash index or a B+ tree index; and the data storage module stores data persistently on nonvolatile media. In addition, these storage servers run distributed protocols, mainly including the distributed replication and the distributed commit protocols.

The application servers run upper-layer application logic, and upper-layer applications invoke storage interfaces of the distributed storage system through clients to store and query data. To reduce network access, the clients sometimes cache data of the storage servers.

8.2 Key Measurement Indicators of Distributed Storage Systems

In terms of key measurement indicators, distributed storage systems focus on performance, scalability, consistency, and availability.

8.2.1 Performance

Performance indicators mainly include throughput, bandwidth, and latency.

Throughput is the number of operations the entire distributed storage system can process per unit time.

8.2 Key Measurement Indicators of Distributed Storage Systems

Bandwidth is the number of bytes read and written in a distributed storage system per unit time.

Latency refers to the duration from the time when a client sends a request to the time when the client receives a response for each operation. Common latency indicators include median latency and tail latency (such as P99 latency). Modern distributed applications often have a high request fan-out coefficient. For example, rendering a web page generates dozens or even hundreds of independent requests and waits for the responses to all requests. Therefore, the tail latency of distributed storage systems is critical to the visible latency of applications.

8.2.2 Scalability

Scalability refers to the ability to efficiently increase the throughput, bandwidth, and capacity of the system when more storage servers are added.

From the perspective of project practice, the key to achieving high scalability of distributed storage systems lies in two points. The first is to avoid using centralized components in a data access path, for example, a coordination server. Most existing distributed storage systems cache information in the coordination server to clients to avoid the bottlenecks associated with centralization. The second is to avoid distributed transactions, which are common when a request needs to be completed by multiple storage servers in collaboration. This ensures atomicity but causes an excessively long critical section and a large number of serialized conflicting requests. To this end, some distributed storage systems relax the atomic semantics of operations.

8.2.3 Consistency

Consistency refers to a degree of consistency of an operation sequence observed by different clients. Strong consistency, causal consistency, and eventual consistency are common consistency levels, but their degree of consistency (listed here in descending order) is inverse to their performance (listed in ascending order). See Fig. 8.2.

Strong consistency, also referred to as linearizability [1], means that the operation sequence of a distributed storage system observed by all clients is the same, and the sequence is consistent with the physical time. Under this consistency, the semantics of the entire distributed storage system are the same as a sequence in which a server sequentially performs operations according to a request arrival sequence. Figure 8.2a shows the operation sequence of four clients in strong consistency mode. Clients 1 and 2 write data, and clients 3 and 4 read data. The four clients have the same observation sequence: [client 1 write ($a = 5$)], [client 2 write ($b = 1$)], [client 3 read ($b = 1$)], [client 4 read ($a = 5$)], [client 4 read ($b = 1$)], [client 1 write ($c = 2$)], [client 3 read ($c = 2$)], and [client 3 read ($a = 5$)]. The sequence is consistent with the actual physical time sequence of these operations.

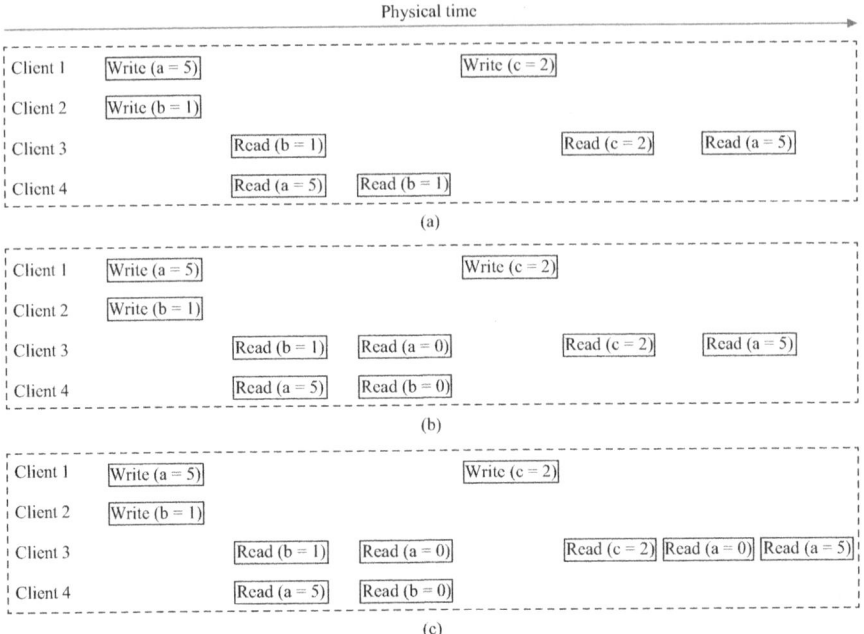

Fig. 8.2 Consistency level example. (**a**) Strong consistency. (**b**) Causal consistency. (**c**) Eventual consistency. Note: The initial values of a, b, and c are 0

In the case of causal consistency, the operation sequences of the distributed storage system observed by different clients may be different, but each sequence complies with the causal relationship. The causal relationship "→" is defined by the following three rules: First, if operation A is performed first and then operation B is performed on the same client, A → B; second, if operation B reads data written by operation A, then A → B; third, there is transitivity, and when A → B and B → C, then A → C. Figure 8.2b shows the sequences of operations observed by four clients under causal consistency. It can be seen that the write sequence of a and b observed by client 3 is [client 2 write ($b = 1$)] and [client 1 write ($a = 5$)] and that observed by client 4 is [client 1 write ($a = 5$)] and [client 2 write ($b = 1$)]. In a real distributed storage system, this sequence inconsistency may be caused by the asynchronous replication protocol: A write operation is completed and acknowledged by a client when it reaches the primary copy. Then, some clients read the data of the latest version from the primary copy, whereas other clients read the data of an earlier version from secondary copies. In this example, all operations have a causal relationship, for example, [client 1 write ($a = 5$)] → [client 1 write ($c = 2$)] and [client 1 write ($c = 2$)] → [client 3 read ($c = 2$)]. According to transitivity, [client 1 write ($a = 5$)] → [client 3 read ($c = 2$)]. Therefore, 5 can be observed when client 3 reads a, which is consistent with the operation [client 3 read ($a = 5$)].

Eventual consistency is the weakest consistency level. It only ensures that all clients can view the latest updated data when no write operation is performed in the system. Figure 8.2c shows the sequence of operations of four clients under eventual consistency. Compared to Fig. 8.2b, client 3 cannot see the update to a after seeing the update to c, which violates causal relationship.

8.2.4 Availability

Availability refers to the capability of a system to provide normal services. There are two specific expressions:

Time-based: Normal service time/(normal service time + service unavailable time).
Based on the number of requests: Number of successfully executed requests/total number of requests.

A distributed storage system runs on a large number of servers, causing a high likelihood of breakdown, be it a storage device or server. High-availability distributed storage systems must be able to quickly recover from breakdown. Specifically, systems must have the following capabilities:

Redundant data: The distributed replication protocol, which repeatedly stores each piece of data on different servers, or erasure coding (EC), is used for this purpose. EC is used for multiple data blocks stored on different servers, generates parity blocks, and stores the parity blocks on other servers. In many cases, breakdowns can occur between servers. For example, servers in a rack may break down together due to a rack power failure. Natural disasters such as typhoons and earthquakes may cause servers in an entire data center to fail to provide services. Therefore, for distributed storage systems that require high availability, data copies are distributed to different availability zones (AZs), a practice that isolates faults because AZs can be located in different data centers or even across continents.

Quick monitoring of breakdown: Distributed storage systems usually use the timeout mechanism to monitor breakdowns. Assume a server (A) periodically sends a heartbeat network packet to another server or a coordination server (B). If B fails to receive the heartbeat network packet of A on time, a timeout event occurs, and server A is determined to already have broken down. This mechanism makes a trade-off between accuracy and the speed at which breakdowns are detected. When the timeout interval is set to a small value, the system can quickly detect breakdowns but at the cost of accuracy. For example, if A experiences software latency (such as garbage collection on a Java virtual machine), heartbeat network packets cannot be sent on time, and A will be incorrectly considered to be in the breakdown state. By comparison, if the interval is set to a large value, the system can accurately detect breakdowns but not in a timely manner. For example, A may have broken down but is detected after the timeout period expires, resulting in the failure of some data in the system to be provided for a long time.

Quick data recovery: When detecting that a server breaks down, distributed storage systems need to recover the data and distribute the original data access requests of the server to other normal servers. If a system uses the distributed replication protocol for data redundancy, the recovery time is related to the system's consistency. When the system ensures low consistency (e.g., eventual consistency), instant recovery can be implemented because any copy can be read or written. When the system ensures strong consistency, the recovery protocol should be executed to ensure that clients can read the latest successfully submitted data after the recovery. If the system uses EC for data redundancy, lost data needs to be recalculated based on parity blocks during recovery.

8.3 Distributed Key-Value Store Systems

A distributed key-value store system is a distributed extension of single-node key-value store. The system stores key-value pairs in a large number of storage servers and provides a simple key-value access interface for applications. A unique primary key is used to get, put, or delete the corresponding key-value pair. With no dependency between key-value pairs, a distributed key-value store system has excellent scalability, with its operations requiring no distributed coordination. With a relatively small size (mostly ≤ 1 MB), a key-value pair is generally stored in a single server, and since data does not need to be divided into blocks, the system ensures low access latency because one key-value operation involves only one server.

Distributed key-value store systems are widely used in the back-end storage of large-scale Internet applications (such as online shopping and social media). For example, Amazon uses Dynamo [2] to store the data of shopping websites (including offering content and shopping cart content). Distributed key-value store systems can also further support other distributed storage systems. For example, some distributed file systems use distributed key-value store systems to store the file metadata. Distributed key-value store systems can be classified into two types based on the scale: distributed key-value store systems across data centers or distributed key-value store systems in a single data center. Distributed key-value store systems across data centers (such as Dynamo [2]) are often used for global services. With these systems, data is replicated in multiple data centers to achieve redundancy with low levels of consistency. However, the following benefits are provided: high availability, tolerance of the breakdown of an entire data center, and low latency (requests from a user can be sent to the data center closest to the user for processing). Distributed key-value store systems in a single data center (such as RAMCloud [3]) store key-value data in a single data center, guaranteeing high levels of consistency. In addition, special hardware (such as RDMA NICs) in the data center can be used to optimize performance.

In special scenarios, some distributed key-value store systems provide dedicated interfaces to extend semantics.

Range search: All key-value pairs in the specified key range are returned. For example, when the primary key of a distributed key-value store system is the user ID, and the value is the user data, all user data whose user IDs are in the [10, 100] range can be obtained through range search. If distributed key-value store systems support range search, the underlying index used by each data partition is generally an ordered index, for example, an LSM tree and a B+ tree.

Transaction operation: A transaction operation ensures atomic access to multiple key-value pairs. Atomicity herein consists of two parts: atomicity of persistence, that is, all modifications made to multiple key values in a transaction must be persisted in storage media, and no partial modification exists; and atomicity of concurrency, that is, the result of concurrent execution of multiple transaction operations is the same as the result of serial execution of these transactions. Because key-value pairs in one transaction operation may be stored on different servers, distributed key-value store systems ensure the atomicity of the transaction operation through lightweight distributed concurrency control protocols (e.g., optimistic concurrency control) and transaction logs.

Secondary indexes: Distributed key-value store systems that support secondary indexes allow users to use secondary keys to access key-value pairs. For example, if the primary key of a distributed key-value store system is a user ID, the secondary key is a user's birth date and the value is user data, and then data corresponding to a user of a specific birth date can be accessed. In contrast to the uniqueness of the primary key, multiple pieces of data can be accessed through a secondary key.

8.3.1 Typical Distributed Key-Value Store Systems

This section describes two typical distributed key-value store systems, including the Dynamo system [2] proposed by the industry and the RAMCloud system [3] proposed by academia. The former focuses on high availability and scalability, while the latter centers on low latency and high throughput of the system.

8.3.1.1 Dynamo

Dynamo is one of the earliest commercial distributed key-value store systems. It was proposed by Amazon and designed for high availability and scalability. It is used for many of Amazon's core services, such as online shopping, which require high availability. Therefore, Dynamo sacrifices consistency to ensure successful write operations of applications even if a server breaks down. In addition, to cope with the growing amount of data, Dynamo supports dynamic addition of servers and replication of key-value data across multiple data centers.

System Interfaces and Semantics

Dynamo provides only semantic assurance of eventual consistency. It has the following two interfaces:

get (key): returns the key-value pair corresponding to the primary key and the corresponding context. Because Dynamo uses a replication protocol with eventual consistency assurance, multiple key-value pairs of different versions may be returned. In this case, it is necessary to resolve the conflict and select a unique key-value pair. The context contains the metadata (such as version information) of key-value pairs.

put (key, context, value): writes the key-value pair <key, value> to the Dynamo system. Among them, "context" is the context returned from the previous "get" operation and is used to check for conflicts between the key-value pairs of different versions in the Dynamo system.

Replication Mechanism and Read/Write Process

Dynamo uses a quorum-based replication mechanism to achieve high reliability for data storage. Each key-value pair is copied to N servers. In addition, each key-value pair has a preference list, which contains M servers ($M > N$). The preference list is ordered, where the first server in the normal state serves as the coordinator of the key-value pair and is responsible for initiating a data replication operation.

When a client initiates a put/get operation, the request is routed to the corresponding coordinator. The coordinator then broadcasts the request to the first N servers in the normal state in the preference list (including the coordinator itself). Regarding the put and get operations, the coordinator must wait for W and R replies, respectively, in order to return the operation result to the client. Generally, $W + R$ should be greater than N, which ensures a high likelihood that the latest written data will be returned for the get operation. However, in the case of a server breakdown or unstable network connection, the coordinator and the first N servers in a normal state will change. As a result, key-value pair data corresponding to a primary key in different servers may be inconsistent. Dynamo uses vector clocks to trace the causal relationship between key-value pairs of different versions and thus check version conflicts.

Vector clocks are classical methods for tracing causal relationships between events in distributed storage systems. Assume that there are P servers in a cluster, which are respectively marked as server 1, server 2, ..., and server P. Any server i maintains an array whose length is P, which is referred to as a vector clock and denoted as V_i. When the system is initialized, all elements in V_i are 0. V_i has the following three update rules: When a local event occurs on server i (e.g., data storage), $V_i[i]$ is increased by 1. When server i sends a network packet (a sending event), $V_i[i]$ is increased by 1. The vector clock is carried in the sent network packet. When server i receives the network packet of server j (a receiving event), $V_i[i]$ is increased

8.3 Distributed Key-Value Store Systems

by 1 first, and $V_i[k]$ is updated to the larger value in $V_i[k]$ and $V_j[k]$ for all k ($1 \leq k \leq P$) (where V_j is the vector clock contained in the network packet of server j).

Vector clocks can accurately capture the causal relationship between events. If each element in the vector clock of event A is greater than or equal to the corresponding element in the vector clock of event B, a causal relationship exists between A and B, that is, B → A. If two vectors cannot be compared, there is no causal relationship between the two events.

In Dynamo, each version of key-value pairs generates a corresponding vector clock. When the client initiates a put request, the request carries the context returned by the last get request. The context also contains the vector clock. When processing the put request, a coordinator generates a vector clock for a new key-value pair and broadcasts the new key-value pair and the vector clock to the N servers. For a get operation, the coordinator compares the vector clocks of key-value pairs in R replies returned from the N servers, and if a causal relationship exists between key-value pairs of the R versions, the latest key-value pair and the corresponding vector clock are returned to the client. If no causal relationship exists between the key-value pairs of some versions, all such key-value pairs and corresponding vector clocks are returned to the client. The client resolves the conflict through its own logic, determines a correct version, and writes this version back to the Dynamo system. To reduce the storage space consumed by vector clocks, Dynamo has made some optimizations: Values that are 0 are not stored in vector clocks, that is, using a compact list instead of an array of server lengths. In addition, Dynamo limits the maximum number of elements in each vector clock and removes elements that are older than the predefined threshold.

8.3.1.2 RAMCloud

RAMCloud is a distributed key-value store system proposed by researchers at Stanford University. It is designed to achieve low latency. Therefore, RAMCloud maintains all data in dynamic random-access memory (DRAM) and relies on high-speed NICs and user-mode network stacks for inter-server communication.

Figure 8.3 shows the RAMCloud system architecture. RAMCloud runs in a single data center and consists of multiple storage servers, one coordination server, and

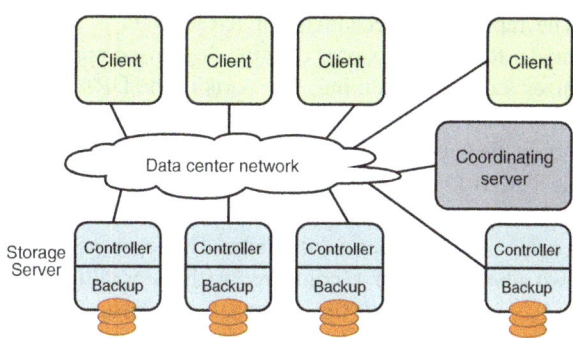

Fig. 8.3 The RAMCloud cluster architecture [3]

multiple client servers. Each storage server consists of a master component and a backup component. The master stores key-value pairs in DRAM and serves client requests. The backup stores key-value pairs owned by other masters into disks or flash memory for redundancy. The coordination server stores the cluster configuration information, along with the location mapping from data partitions to storage servers, in a highly reliable manner. Application servers cache the location mapping and can directly send a storage request to storage servers.

System Interfaces and Semantics

A RAMCloud cluster stores multiple tables with unique IDs and names, and each table contains multiple data partitions. Key-value pairs in a data partition are stored in one master and multiple backups as a whole. RAMCloud has the following basic access interfaces:

createTable(name): creates a table named **name** and returns the table ID.
getTableId(name): returns the ID of the table named **name**.
dropTable(name): deletes the corresponding table.
read(tableId, key): queries the key-value pair whose primary key is **key** in the table corresponding to **tableId** and returns the version of the key-value pair.
write(tableId, key, value): inserts or updates <key, value> to the table corresponding to **tableId** and returns the version of the key-value pair.
delete(tableId, key): deletes the key-value pair whose primary key is **key** from the table specified by **tableId**.

RAMCloud also provides other interfaces to cope with different application scenarios. For example, the batch processing interface supports simultaneous reading and writing of multiple key-value pairs, the atomic operation interface supports atomic conditional updates and auto-increment operations, and the management interface of data partitions supports the splitting of a data partition into multiple partitions and the migration of a data partition to another master.

Replication Mechanism and Read/Write Process

The replication mechanism of RAMCloud is closely related to its local storage mode. RAMCloud stores data in a log-based manner. Specifically, the master organizes a data partition into two parts in the DRAM: a log area for appending and a hash index. The log area stores the key-value pairs, and the hash index is used to query the key-value pairs in the log area. The log area consists of fixed-length (such as 8 MB) log segments and is the basic unit of the copy mechanism. Given that DRAM is expensive and data will be lost after a power failure, RAMCloud stores copies of log segments into disks. The log-based storage mode in RAMCloud has two advantages. First, compared with the traditional DRAM allocator, RAMCloud reduces the DRAM fragments of the master and improves the DRAM usage.

8.3 Distributed Key-Value Store Systems

Second, it makes the data format of DRAM the same as that of disks, greatly simplifying system storage management. However, the log-based storage manner involves garbage collection overheads. If a large quantity of invalid key-value pairs exist in a log segment (denoted as A), RAMCloud uses a clearing thread to move valid key-value pairs in A to a new log segment and then releases the space previously occupied by A.

RAMCloud uses primary-backup replication to manage data copies. Figure 8.4 shows how RAMCloud processes write operations. The client sends the key-value write request to the corresponding master. The master first appends the new key value to the corresponding log segment, then updates the hash table, and broadcasts the key-value pair to the corresponding backups. When receiving the key-value pair, the backups write it to the local nonvolatile buffer (such as persistent memory) and reply to the master. When the master has received all replies, data can be reliably stored in all copies. In this case, the master can reply to the client, indicating that the write operation is successfully completed. When the local nonvolatile buffer of a backup is full, it writes the log segments accumulated in the buffer to the disks. RAMCloud introduces nonvolatile buffers to simultaneously achieve low latency and persistence of replication operations. For a read operation, the client sends the request to the corresponding master. The master then queries the hash table of the corresponding data partition, locates the key-value pair in the log area, and returns the key-value pair to the client.

When a storage server breaks down, DRAM data in its master will be lost. In this case, RAMCloud must recover the log segment corresponding to the crashed master from the backup, identify the latest key-value data, and reconstruct the hash table in the DRAM. To maintain high availability, recovery needs to be completed quickly. Therefore, RAMCloud designs a mechanism for fast parallel recovery. Specifically, the master of a data partition tries to distribute different log segments across a large number of backups. When the data partition is recovered, different log segments can be read from different backups. This allows the aggregate bandwidth of numerous disks to be fully utilized. Furthermore, RAMCloud divides the data stored on a crashed master into multiple partitions, and each partition is recovered and

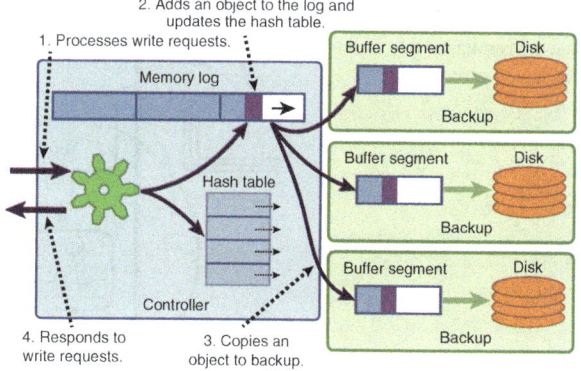

Fig. 8.4 RAMCloud write operation processing flowchart [3]

reconstructed by a normal storage server. This allows the network and CPU resources of different storage servers to be fully utilized.

8.3.2 Key Technologies of Distributed Key-Value Store Systems

This section describes the key technologies of distributed key-value store systems, including the distributed replication technology and distributed secondary index technology.

8.3.2.1 Distributed Replication Technology

Distributed key-value store systems use distributed replication technology to provide high availability. In addition to the quorum-based replication mechanism used by Dynamo and the primary-backup replication mechanism used by RAMCloud, the replication mechanism based on consensus protocols is also commonly used.

As shown in Fig. 8.5, in the replication mechanism based on consensus protocols, each copy has command logs and state machines. Command logs record the requests sent by users, such as key-value pair updates. State machines then execute the requests in the command logs in sequence. As long as the command logs of each copy are the same, the state machines of each copy are also the same. Consensus protocols ensure that command logs of different copies are the same. Common consensus protocols include Paxos [4] and Raft [5]. The following information briefly describes the execution process of the Raft protocol.

In the Raft protocol, each server plays one of the following roles: leader, follower, or candidate. The leader initiates the command log replication process, with the client sending a request to the leader. The leader then appends the request to the local command log and broadcasts it to the followers. The followers write the received request to the corresponding location in the command log. When the leader receives responses from most of the followers, it confirms that the request has been

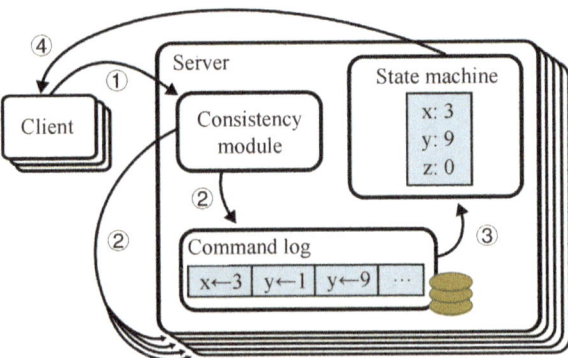

Fig. 8.5 Replication mechanism based on consensus protocols [5]

8.3 Distributed Key-Value Store Systems

successfully submitted, and the leader can execute the request to the state machine and reply to the client. If a follower detects that the leader has failed through the timeout mechanism, it becomes a candidate and attempts to elect itself as the new leader. The Raft protocol divides time into ascending numbers (denoted as Term) and uses a specific election mechanism to ensure that each Term has only one leader. When multiple leaders exist in a cluster due to network partitions (the leaders' Terms are different), the Raft protocol ensures that a log item is not be submitted with different content across copies.

The primary-backup replication mechanism is easier to implement than the quorum-based replication mechanism and the replication mechanism based on consensus protocols. However, in the event of a crashed server, primary-backup replication requires a special recovery protocol, resulting in lower availability. Finally, the primary-backup replication has higher latency because it must wait for replies from all secondary copies, while the other two replication technologies only need replies from some copies.

8.3.2.2 Distributed Secondary Index Technology

Distributed key-value store systems can use secondary indexes to access key-value data with secondary keys. Secondary indexes maintain the mapping from secondary keys to primary keys. Similar to primary indexes, secondary indexes can be distributed across multiple storage servers to improve scalability. Secondary indexes have two distribution modes, each with distinct secondary index query performance and primary index modification performance.

In the first mode, secondary keys and their corresponding primary keys are stored on the same server, as shown in Fig. 8.6a. This mode is used by distributed key-value store systems such as Cassandra [6]. Here, one secondary key corresponds to multiple primary keys (the primary keys corresponding to the secondary key "1997" exist on two servers). Therefore, for a secondary index query operation, the client must send requests to all servers in parallel, wasting a large number of network resources. However, in terms of a primary key modification operation, this mode

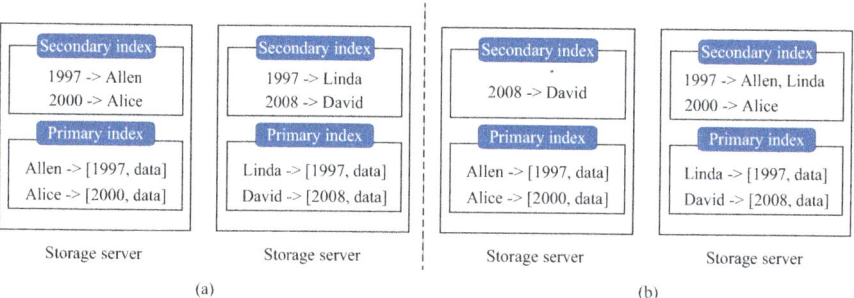

Fig. 8.6 Two distribution modes of secondary indexes. (**a**) Mode 1. (**b**) Mode 2

involves only one server, resulting in low latency and requiring no distributed coordination.

In the second mode, used by RAMCloud, secondary indexes are distributed to multiple machines according to secondary keys, as shown in Fig. 8.6b. In this mode, a secondary index query operation requires two network round trips: First, the server where the corresponding secondary index is located is queried to obtain the corresponding primary key list. Then, the server in which the corresponding primary index is located is queried based on these primary keys to obtain the target key-value pair data. The modification of a primary index requires multiple network interactions: Initially, the modification request is first sent to the server (denoted as A) corresponding to the primary index to query the old secondary key. Then, A sends data to the secondary index server where the new secondary key is located, requesting the creation of a new secondary index entry. Afterward, A writes the new key-value pair into the primary index. Lastly, A sends data to the secondary index server where the old secondary key is located, requesting the deletion of the old secondary index entry.

8.4 Distributed Object Storage Systems

Distributed object storage systems store objects across numerous storage servers and provide object access interfaces. Distributed object storage systems allow users to create buckets for storing multiple objects. Each object can vary in size up to several terabytes and is identified by a globally unique ID. A globally unique ID can be used to simulate the path of a file system (such as directory 1/directory 2/file name). Unlike a file system, object storage does not provide object renaming services. Distributed object storage systems offer read/write operations on objects as well as access to a certain range of objects. To reduce metadata maintenance, distributed object storage systems make some sacrifices in terms of interfaces. To update an object, the entire object needs to be rewritten at once, and atomic updates of multiple objects are not supported. In addition, although distributed object storage systems provide interfaces for metadata operations, such as the LIST operation (when a start ID is given, other objects in a bucket are listed in the dictionary order of IDs), the performance of such interfaces is poor.

In addition to the traditional advantages of cloud storage (such as pay-as-you-go and economies of scale), distributed object storage systems provide users with compute and storage resources that can be independently expanded. Currently, these systems are commonly used to store unstructured data, which is data without a predefined data model, such as office documents, pictures, audios, and videos. Such systems are commonly applied to scenarios like static website hosting, data backup and archiving, Video On Demand (VOD), big data analytics, and enterprise cloud disks. In recent years, some cloud-native data lake systems (such as Delta Lake) have started using object storage as a base and have implemented a set of high-performance table storage that supports transaction semantics based on object storage.

8.4.1 Typical Distributed Object Storage Systems

This section describes two typical distributed object storage systems: Swift and Ceph RADOS systems.

8.4.1.1 Swift

The Swift distributed object storage system is one of the projects of OpenStack, an open-source project dedicated to developing and maintaining cloud computing services. Swift leverages consistent hashing technology [7] to ensure the scalability and load balancing of distributed systems and uses the replication protocol to ensure high system availability. Swift provides users with read and write operations on containers and objects, and the consistency level is eventual consistency. In addition, Swift allows data to be shared among multiple tenants.

Cluster Architecture

Swift consists of proxy servers, object servers, container servers, account servers, and auditors. To ensure that distributed object storage systems remain consistent even when a temporary fault occurs, Swift replicates data to multiple servers. Swift also uses the audit service to check the consistency status of the system.

The proxy server functions as the entry for other Swift components. For each request, it queries the specific location of the account, container, or object and routes the request accordingly. The return value of the request is also returned to the client through the proxy server.

The object server is used to store objects in Swift. It can store, retrieve, and delete objects stored on local devices. Objects are stored in the file system as binary files, and object metadata is stored in the extended attributes (xattrs) of the files. Therefore, Swift requires extended attributes of files to be supported by the underlying file system of the object server.

The container server is used to store object lists of Swift. It does not store the location of objects in a container, only the object lists in a specific container. The lists are stored as an SQLite database file. Similar to object data, there are multiple copies in the cluster. The container server collects container information, including the total number of objects in the container and the total storage usage of the container.

The account server is used to store container lists of Swift. The function and organization of a container list are similar to those of an object list in the container server, but the content of the container list changes from object information to container information.

Auditors are used to crawl server content to check the integrity of objects, containers, and accounts. If a fault is detected, the incorrect object is isolated, and the object is copied from another copy to replace the incorrect data. If other errors are detected, the audit service records them in logs.

Data Organization

Swift uses a three-level (account/container/object) flat organization mode that facilitates storage cluster expansion. Each storage service user can have multiple accounts. Each account can include multiple containers, and each container can contain multiple objects. To support scalability and load balancing, Swift must distribute objects across different storage nodes. Swift uses a consistent hashing algorithm to distribute object data.

The consistent hashing algorithm is an improvement on common hashing algorithms. Common hashing algorithms hash an object ID and perform a modulo operation with the hash results and the number of storage nodes. The modulo operation result is then used as the storage node ID of the object. This involves huge overhead for the system whenever nodes are added or deleted. Specifically, when a node is added or deleted, the original mapping relationship between an object and the storage node becomes invalid. Therefore, systems need to migrate a large amount of data, blocking the storage node from serving the object request of the user.

The consistent hashing algorithm, first proposed by Karger of the Massachusetts Institute of Technology in 1997, can solve this problem. It is implemented through the data structure of a hash ring, with each hash value corresponding to a location on the hash ring. Storage nodes are evenly distributed on the hash ring. The hash value of an object is calculated using hash functions and incrementally searched on the hash ring for the storage node closest to the hash value of an object, and this node is used as the storage node of the object. When a storage node is added to the hash ring, only some objects in the storage node that is adjacent to the new storage node in an anticlockwise direction need to be moved to the new node, and the mapping of other objects remains unchanged. As shown in Fig. 8.7, there are three storage nodes (1, 2, and 3) and four objects (A, B, C, and D) in an existing hash ring, where A belongs to storage node 2, B belongs to storage node 3, and C and D belong to storage node 1. When node 4 is added as a new node, it is calculated to be located between storage nodes 3 and 1. Therefore, the object (C) located between nodes 3 and 4 will be migrated to the new storage node. Other objects (A, B, and D) do not need to be migrated. When a storage node is deleted, only object data in the deleted node needs to be moved to a storage node that is adjacent to the deleted node in an anticlockwise direction in the hash ring, and data on other nodes does not need to be migrated. As shown in Fig. 8.8, assume that storage node 3 needs to be deleted, then object (B) originally managed by storage node 3 is moved to storage node 4, and other objects (A, C, and D) do not need to be migrated.

The consistent hashing algorithm uses virtual nodes to maintain load balancing after nodes are added or deleted. Specifically, each storage node is split into multiple virtual nodes, and all the virtual nodes are then evenly distributed on the hash ring. The hash range of an object actually managed by each storage node consists of

8.4 Distributed Object Storage Systems

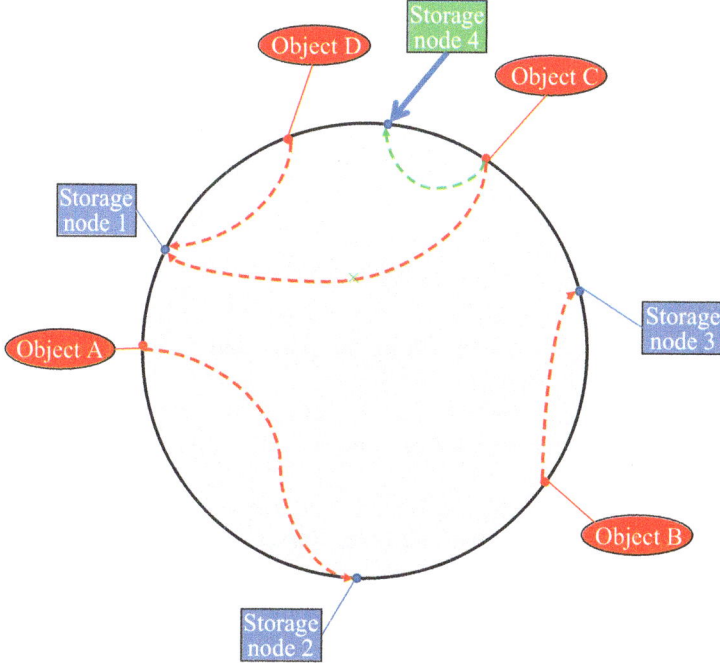

Fig. 8.7 Consistent hashing algorithm: adding a node

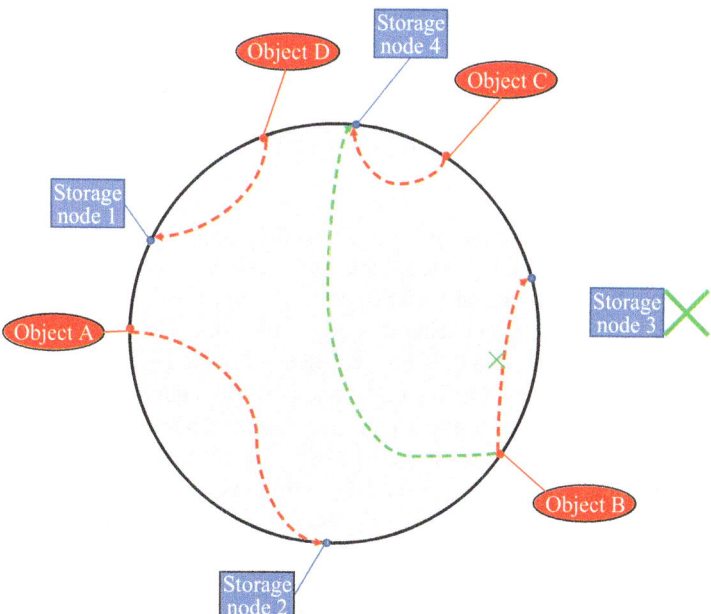

Fig. 8.8 Consistent hashing algorithm: deleting a node

several discontinuous intervals on the hash ring. Through the use of virtual nodes, objects stored on a storage node when that node is deleted can be distributed to other remaining storage nodes to the greatest extent. When a storage node is added, the new storage node stores objects from other existing storage nodes.

Data Read/Write Process

The following example describes the Swift data read/write process using the writing of an object:

1. The proxy server provides an API for the client, and the client initiates a write request to the proxy server.
2. Before accessing the Swift service, the client must obtain the access token through the authentication service and add the authentication token to the message.
3. The proxy server sends the write request to the object server.
4. If the write operation succeeds, the object server sends a write success message to the proxy server.
5. The proxy server forwards the write operation result to the client.

8.4.1.2 Ceph RADOS

Reliable Autonomic Distributed Object Store (RADOS) is the object storage system of Ceph. It provides reliable object storage interfaces and supports dynamic and heterogeneous storage nodes.

Cluster Architecture

Ceph RADOS consists of monitoring nodes and Object Storage Daemon (OSD) storage nodes. The interfaces provided by Ceph RADOS are compatible with Swift object storage APIs and Amazon S3 APIs.

Monitoring nodes maintain cluster mappings, including storage node status and mappings between objects and data nodes. Clients of storage clusters need to read the cluster mapping from monitoring nodes and perform object addressing based on the cluster mapping. The monitoring nodes are deployed across multiple servers to ensure high system availability in cases when a single monitoring node breaks down. Multiple monitoring nodes form a monitoring node cluster, and data in each monitoring node is synchronized through a distributed consistency protocol.

Storage nodes are used to store objects in RADOS. In addition, OSD, the daemon process, is deployed on each storage node and is responsible for checking the status of the storage node itself and other storage nodes and reporting the status to monitoring nodes. OSD actively sends the storage node status to monitoring nodes.

Data Organization

Storage nodes use a local file system to store objects. Each object corresponds to a file in the file system. In contrast to Swift, which uses extended attributes of files to store object metadata, Ceph RADOS can store object metadata as key-value pairs in a local key-value store engine (such as LevelDB) to break the limit of file extended attributes on the amount of object metadata.

Ceph RADOS uses two components to distribute object data: storage pool and placement group (PG). A storage pool contains multiple PGs, and objects in each storage pool use the same data redundancy policy (replication policy or EC policy). A PG contains multiple objects. Objects in a PG can be distributed across multiple storage nodes, and each storage node can store objects from different PGs.

To support scalability and load balancing, Ceph RADOS uses the Controlled Replication Under Scalable Hashing (CRUSH) algorithm [8] to manage the mapping between objects and storage nodes.

The CRUSH algorithm manages the mapping between objects and storage nodes based on PGs. The input of the algorithm is the PG_ID, the hierarchical cluster map, and placement rules. The output of the algorithm is a list of storage nodes that can be selected.

The hierarchical cluster map describes the topology structure of a storage cluster. A leaf node in the topology corresponds to a storage device (OSD), and a non-leaf node in the topology corresponds to a container (called a bucket) of a device. Common bucket types include data centers, rooms, racks, and hosts. A data center contains multiple rooms, each room contains multiple racks, each rack contains multiple hosts, and each host can contain multiple OSDs. A user can specify a weight for an OSD to control the amount of data to be stored. Therefore, the load of an OSD is directly proportionate to its weight. The weight of a bucket is the sum of the weights of its sub-buckets. The placement rules provide an interface for a user to freely define a replication placement rule for an object in the PG.

The CRUSH algorithm consists of two steps. In the first step, the CRUSH algorithm maps each object to a PG. When the number of PGs is a power of 2, the CRUSH algorithm obtains the PG ID (PG_ID) by performing a modulo operation with the object hash value and the number of PGs (PG_NUM). When the number of PGs is not a power of 2, the CRUSH algorithm performs a modulo operation with the object hash value and two powers of 2 close to the number of PGs, that is, [power 1 less than the number of PGs (PG_NUM1) and power 2 greater than the number of PGs (PG_NUM2)], and obtains two hash values (hash1 and hash2). If hash2 is less than PG_NUM, the CRUSH algorithm uses hash2 as the PG_ID of the object. If hash2 is greater than PG_NUM, there is no corresponding PG, and the CRUSH algorithm uses hash1 as the PG_ID of the object. This method aims to minimize the migration of objects when the number of PGs changes. The following example better explains this method. Assume that there are two objects, A and B. Hash (A) = 25 and Hash (B) = 29. Also, suppose that there are eight PGs (PG_NUM = 8) at the beginning. In this case, it is calculated, through the CRUSH algorithm, that PG_IDA = 1 and PG_IDB = 5. When the number of PGs increases from

8 to 12 (PG_NUM = 12), PG_NUM1 is 8 and PG_NUM2 is 16. The CRUSH algorithm first calculates hash2A and hash2B for A and B, and the values are 9 and 13, respectively. Because hash2A is smaller than PG_NUM, hash2A is directly used as the PG_ID of object A. Since hash2B is greater than PG_NUM, the CRUSH algorithm calculates hash1B of object B, and the result (hash1B = 5) is used as the PG_ID of object B. Comparing the mapping before and after the PG number increase shows that the mapping of object B remains unchanged.

In the second step, the CRUSH algorithm selects a corresponding OSD list for each PG. The CRUSH algorithm allocates the hierarchical cluster map layer by layer from the root node. The allocation policy is determined by the user-defined placement rules. Specifically, the placement rule specifies the quantity of buckets selected at each layer and the selected random algorithm. The CRUSH algorithm uses a bucket as the input, selects a specified quantity of sub-buckets from the available buckets through the random algorithm, and continues the execution with the selected sub-buckets as the input until a corresponding OSD is selected. The random algorithm selects sub-buckets based on the weight of each sub-bucket. Currently, the CRUSH algorithm has five random algorithms (Uniform, List, Tree, Straw, and Straw2). Different random algorithms use different formulas to select sub-buckets. The Straw2 random algorithm is the most popular because it has the minimum migration overhead when the cluster mapping or the number of PGs is changed.

Data Read/Write Process

The following example describes the write process of Ceph RADOS, as shown in Fig. 8.9. The read process is basically the same:

1. The client requests cluster mapping from the monitoring node.
2. The monitoring node returns the cluster mapping to the client, which will be used as the input for the CRUSH algorithm.
3. The client calculates the mapping between the object and the OSD.
4. The client sends a write request to the active OSD node.

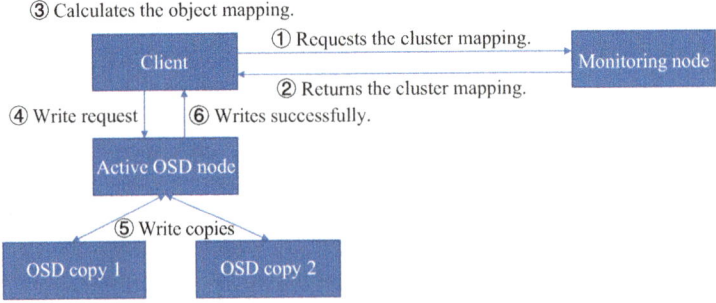

Fig. 8.9 Ceph RADOS write process

5. The OSD on the active OSD node is responsible for synchronizing object data in the copies.
6. If the write operation succeeds, the active OSD node sends a write success message to the client.

8.4.2 Key Technologies of Distributed Object Storage Systems

This section describes the key technologies involved in distributed object storage systems, including data distribution and load balancing.

8.4.2.1 Data Distribution

Generally, data distribution technologies must achieve the following three outcomes:

1. Ensure good load balancing. The data stored on each server and the requests received by each server should not vary greatly.
2. Minimize the amount of data to be migrated when servers are added or deleted.
3. If copies are considered, ensure the distribution of copies provides high availability.

Common data distribution technologies used in distributed object storage systems include the consistent hashing algorithm used by Swift and the CRUSH algorithm used by Ceph RADOS. There are also some extensions to these algorithms, such as MapX [9].

MapX addresses the issue of uncontrollable data migration during cluster capacity expansion in the traditional CRUSH algorithm. The main method employed by MapX is to add the time dimension mapping mechanism to the original CRUSH algorithm. Specifically, MapX records the creation time of objects and the time of each cluster capacity expansion. During each capacity expansion, MapX creates a new virtual layer in the hierarchical cluster map and allocates new PGs that contain timestamp information for the capacity expansion. The newly created objects are then mapped to the new virtual layer, while the distribution locations of the original objects remain unchanged, avoiding data migration. Furthermore, the original PGs can be easily re-mapped to the new virtual layer by adjusting the timestamps of the original objects, thereby solving the potential load imbalance problem.

8.4.2.2 Load Balancing

Different objects have different access frequencies and can become hotspots, which change over time. Therefore, distributed object storage systems must use load balancing technology to prevent some servers from becoming overloaded and affecting overall system performance. Common load balancing technologies are as follows:

Data migration: When some servers are overloaded, this technology migrates some objects from the overloaded servers to other idle servers. During migration, access to the migrated objects may be blocked. In addition, the storage bandwidth and network bandwidth used for migration can affect normal tasks.

Data caching: This technology uses a distributed memory system to cache hotspot objects, reducing the access traffic to the distributed object storage systems and achieving load balancing. The main issue with this technology is that it cannot handle write operations for hotspot objects. Since distributed memory systems cannot perform data persistence, back-end distributed object storage systems are still required to process write operations for hotspot objects.

Data replication: This technology replicates objects to multiple servers based on their hotspot degree. In this way, different servers can serve requests for hotspot objects, preventing some servers from becoming overloaded. The main disadvantage is the need to handle consistency between copies.

Data splitting: This technology splits an object into multiple fixed-length data blocks and stores them on different servers. The read/write overhead for hotspot objects is evenly distributed across multiple servers. However, this technology has two disadvantages. First, it is only effective for large objects; for small objects, much network throughput is wasted. Second, it increases object read/write tail latency because it needs to wait for responses from all servers.

8.5 Distributed Block Storage Systems

Distributed block storage systems use block interfaces to expose the storage resources of server clusters to external applications. Common protocols used in these systems include Fibre Channel over Ethernet (FCoE), Fibre Channel (FC), and Internet Small Computer System Interface (iSCSI). Distributed block storage systems divide data into blocks and store each block of data in a different location. When clients request data, the systems reassemble the blocks and return the data to the clients.

Distributed block storage systems ensure low latency, high availability, simplified management, and high scalability. These systems are typically used to provide virtual disks for virtual machines (VMs), with existing single-node applications running on the virtual disks. For example, Linux file systems can be created on the virtual disks, and LevelDB can run over the file systems to provide key-value store services. In addition, distributed block storage systems are widely used in database and mail system applications.

8.5.1 Typical Distributed Block Storage Systems

This section describes two major distributed block storage systems: Amazon Elastic Block Store (EBS) and Blizzard [10].

8.5 Distributed Block Storage Systems

8.5.1.1 Amazon EBS

Amazon EBS is a cloud distributed block storage system designed for use with Elastic Compute Cloud (EC2) instances. Amazon EBS works as system disks or data storage disks for EC2 instances and can be accessed as easily as using host disks. Compared with conventional cloud storage, Amazon EBS emphasizes storage availability and consistency. Availability refers to quick response to client requests, and consistency refers to compliance of stored data with the execution semantics of users.

There are two types of EBS block storage services: network-attached storage (NAS) and instance storage. Amazon EBS involves the following basic operations: creating an EBS volume, attaching an EBS volume to an EC2 instance, and taking a snapshot for an EBS volume.

When Amazon EBS is used as NAS, you need to format EBS volumes and initialize the corresponding file systems to use the volumes like traditional disks. However, you cannot attach EBS volumes to servers. You can take snapshots for fast backup. The snapshots are incremental, and the new snapshots save only the data that has changed since the last snapshots.

When Amazon EBS is used as instance storage, it is regarded as a block device and does not require formatting. However, the instance storage must be attached to instances for use and cannot be used independently.

Amazon EBS has the following features:

High scalability: Amazon EBS uses snapshots for flexible scaling of diverse workloads. Snapshots can back up existing EBS volumes and create new EBS volumes.
Data backup: Incremental snapshots back up Amazon EBS data to the Amazon S3 platform, and existing snapshots can create new EBS volumes.
Flexible deployment: EBS volumes can be increased or decreased in size on demand.

8.5.1.2 Blizzard

Blizzard is a distributed block storage system optimized for POSIX and Win32 applications. It provides virtual disk abstractions for upper-layer applications. Traditional applications (POSIX and Win32 applications) and cloud-native applications have significantly different I/O patterns and consistency requirements, creating many challenges in Blizzard design. For example, cloud-native applications (such as MapReduce) typically issue large, sequential I/Os, whereas traditional applications mainly issue small, random I/Os in large quantities with poor locality. Similarly, in terms of consistency, cloud-native applications can run with relatively low levels of consistency and can tolerate some data loss, whereas traditional applications that experience even a small amount of data loss can cause metadata inconsistencies or other catastrophic consequences, such as inability to index data. Moreover, traditional applications use fsync() system calls to control the ordering and durability of writes. This prevents subsequent writes from completing before all previous writes are durable, limiting disk parallelism. Therefore, Blizzard needs to reduce the adverse impact of fsync() on distributed block storage performance.

To achieve this, Blizzard builds itself on Flat Datacenter Storage (FDS) [11], a datacenter-class blob storage system. FDS breaks blobs, which are unstructured objects, into 8 MB data fragments called tracts. Specifically, FDS splits a blob into multiple tracts and scatters the tracts among different disks. Blizzard virtual disks are backed by FDS blobs. POSIX application I/Os are generally smaller than tracts, which are 8 MB in size. To enable POSIX applications to exploit the parallelism of multiple disks, Blizzard uses a new striping scheme called nested striping. Blizzard divides the address space of virtual disks into multiple segments, with each segment containing multiple tracts. Data in a segment is striped in blocks and stored in these tracts. Figure 8.10 shows an example of Blizzard's nested striping, specifically, the mapping of a segment with two tracts. Access to contiguous data blocks in the segment is converted into access to two tracts, or to two disks in particular. Nested striping enables Blizzard to leverage the aggregate bandwidth of multiple disks.

Blizzard provides three types of consistency semantics: write-through, flush epoch, and out-of-order (listed in descending order of durability and ascending order of performance).

Write-through consistency ensures the durability of associated data on FDS disks when a virtual disk write is acknowledged to the relevant client. This ensures minimal data loss in the event of a system crash but provides the lowest performance.

Flush epoch consistency divides writes into multiple periods (called epochs) based on flush operations initiated by users. In this scheme, when writes from an epoch become durable on disks, writes from all previous epochs are also durable. This consistency scheme allows users to issue requests faster because Blizzard immediately acknowledges a write or flush request upon receiving the request without waiting for the durability process to complete. To maintain the order of writes from different epochs, Blizzard uses the cache and write queues to temporarily store writes and issues the writes to disks for durability in order of epochs. When a read arrives, Blizzard reads data from the cache first. If no data is hit, Blizzard then reads data from disks.

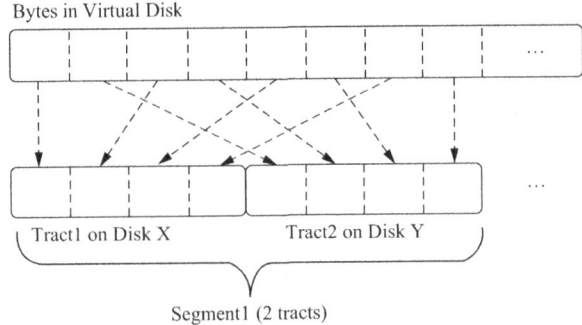

Fig. 8.10 Nested striping of Blizzard [10]

8.5 Distributed Block Storage Systems

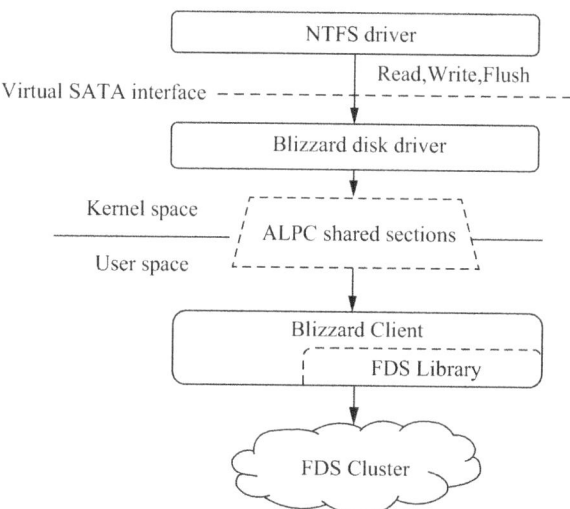

Fig. 8.11 Architecture of Blizzard on Windows [10]

Out-of-order consistency allows writes to be acknowledged immediately and issued for durability in any order, maximizing performance. For consistency semantics purposes, Blizzard uses a log structure instead of nested striping to avoid updating blocks in place. This ensures that a consistent version of the relevant virtual disk can be recovered in the event of a system fault.

Figure 8.11 shows the Blizzard architecture on Windows. A virtual disk mainly consists of two parts: a kernel-mode Blizzard disk driver and a user-mode Blizzard client that comprises the FDS library. File systems or applications send I/O requests to the SATA driver, which forwards the requests to the Blizzard client. The client then converts them into FDS requests and sends them to the FDS cluster. To reduce the overhead of data exchange across the user-kernel boundary, Blizzard uses Windows Advanced Local Procedure Calls (ALPC) to provide zero-copy communications with shared memory.

8.5.2 Key Technologies of Distributed Block Storage Systems

This section describes caching optimization and parallelism technologies used by distributed block storage systems.

8.5.2.1 Caching Optimization

In cloud environments, VMs typically use distributed block storage systems to provide virtual disks. When a large number of VMs simultaneously run I/O-intensive workloads, the I/O performance of the distributed block storage systems is usually limited by the network. A common approach for minimizing network traffic is to use local disk devices as the block-level cache for remote disks. However, this approach has drawbacks: First, blocks adjacent on remote disks may not be adjacent anymore in the local cache, which slows access performance. Second, cache metadata management results in additional overheads. Third, this approach cannot ensure data consistency for virtual disks in the event of a VM breakdown.

To address these issues, vStore [12] manages a block-level cache at the hypervisor layer. The hypervisor intercepts upper-layer I/O requests, a process that is transparent to VMs. The vStore block-level cache is a set-associative cache that is divided into multiple rows based on the number of remote virtual disks. This prevents the caches of remote virtual disks from affecting each other. When evicting cache blocks, vStore considers the impact of the eviction on cache sequentiality and prefers to evict nonsequential or modified cache blocks. vStore records the metadata and hash value of each cache block on adjacent disk space. If a VM breaks down and restarts, vStore compares cache blocks with hash values. If there is a mismatch between a cache block and its hash value, the cache block is incomplete and is discarded.

8.5.2.2 Parallelism

Parallelism is the process of allocating random small I/Os to different physical disks to exploit the aggregate bandwidth of multiple storage devices. Parallelism, such as Blizzard's nested striping, is used in most distributed block storage systems. URSA [13] further defines three levels of parallelism: on-disk parallelism, inter-disk parallelism, and network parallelism.

On-disk parallelism: Primary replicas are stored on solid-state drives (SSDs) in URSA. To fully explore SSD parallelism, URSA runs multiple block management processes on an SSD, and each process can issue parallel I/O requests via an asynchronous I/O library.
Inter-disk parallelism: It is a combination of multiple kinds of parallelism, including striping, out-of-order execution, and out-of-order completion.
Network parallelism: URSA processes requests in a pipeline for each network connection. Network delay is masked by data processing, further minimizing request delay.

8.6 Distributed File Systems

As a scale-out extension of single-node file systems, distributed file systems manage storage resources from multiple servers and provide users with file system access interfaces and unified file system namespaces. Besides file data, distributed file systems also store directory metadata and file metadata. Directory metadata contains directory names, IDs, permissions, timestamps, and entry lists (i.e., names of files and subdirectories under each directory), and file metadata contains file names, IDs, and permissions.

8.6 Distributed File Systems

Distributed file systems typically use client/server architecture. The client provides standard file system access interfaces for applications, and the server stores all data and metadata in file systems. Generally, distributed file systems use the design architecture of separating data from metadata. File data is stored on multiple data servers in the form of data blocks, and file metadata and directory metadata are stored on one or more metadata servers. Distributed file systems are widely used because of advantages in management, scalability, reliability, and availability. For example, big data applications store datasets and intermediate results in distributed file systems to fully exploit the high aggregate bandwidth and mass storage capacity of the systems.

8.6.1 Typical Distributed File Systems

This section describes two major distributed file systems: Google File System (GFS) [14] and InfiniFS [15].

8.6.1.1 GFS

To address rapidly increasing data processing requirements, Google designed GFS, a distributed file system, to provide better back-end storage support for the large-scale parallel data computing model MapReduce [16]. GFS forms a high-performance data storage cluster built from a large number of inexpensive servers. The design of GFS is optimized based on the data storage and access characteristics of the Google search engine. In the Google search engine, files are relatively large. New data is written in append mode, and file overwriting and random writing are rare. GFS thus needs to support high-speed sequential reads and writes on large files. In the Google MapReduce computing framework, one computational process is scheduled to multiple servers in a cluster for parallel data reads, writes, and computing. Therefore, GFS needs to support parallel access from a large number of clients. Because of the use of large-scale inexpensive servers, component failures are the norm. At any time when the distributed file system is running, any server may fail or even break down due to human errors, software faults, disk errors, memory errors, or network errors. Therefore, GFS needs to exploit mechanisms such as cluster monitoring, error detection, automatic recovery, and fault tolerance to ensure high system availability.

Cluster Architecture

GFS has two types of servers: metadata server (called master) and data server (called chunkserver), as shown in Fig. 8.12. A single GFS instance consists of a single master and multiple chunkservers. Multiple applications can access the same GFS instance in parallel and share the same file system namespace. GFS service programs run in user mode.

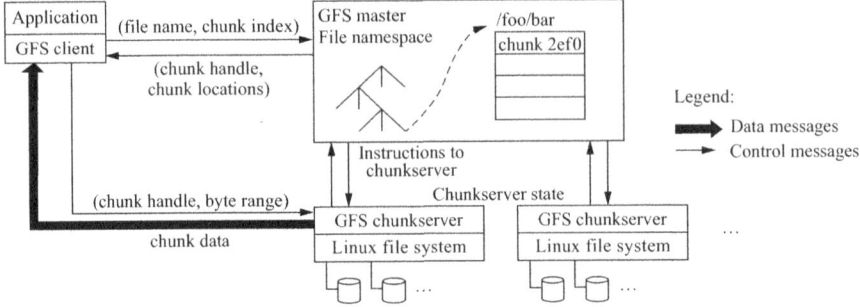

Fig. 8.12 GFS architecture [14]

GFS uses a single master to manage the metadata of all files in a cluster. This metadata includes file system directory trees, access control information, mapping from files to chunks, and mapping from chunks to chunkservers that store the chunks. The master translates a read/write operation from an application on a file path name into a read/write operation performed on the relevant chunk. The master periodically sends heartbeat messages to all chunkservers to monitor the state of each chunkserver in the cluster.

GFS divides file data into chunks and distributes the chunks to chunkservers. The default chunk size in GFS is 64 MB. When a chunk is created, the master assigns a 64-bit chunk handle to identify the chunk. In a storage cluster, each chunk handle is globally unique. Chunkservers use local file systems to store and manage chunks. After obtaining chunk handles and chunkserver addresses, clients can bypass the master and interact directly with chunkservers for data reads and writes. GFS leverages a multi-replica scheme to ensure chunk reliability and exploits a primary-secondary replica mechanism to maintain consistency across replicas. By default, each chunk has three replicas stored on different chunkservers.

Clients provide access interfaces of the GFS distributed file system for applications and interact with the master and chunkservers for file reads and writes. Clients interact with the master to process file metadata requests from applications. For file data read and write requests from applications, clients first obtain relevant chunk handles and chunkserver addresses from the master. Then, the clients interact directly with the relevant chunkservers for subsequent reads and writes. GFS clients do not provide file system interfaces with complete POSIX semantics, and applications need to use specified file operation primitives to access the GFS distributed file system. GFS clients do not cache file data because GFS is designed for large-file storage and sequential reads/writes. Client caching has little benefit for these workloads.

Directory Tree Structure

In GFS, all file metadata is stored in the memory of the master. GFS uses logs to maintain persistent file system directory tree and file-to-chunk mapping information in local disks and uses a multi-replica mechanism to back up the logs in the cluster

8.6 Distributed File Systems

to provide highly reliable file system metadata services. The master does not store the mapping from chunk handles to server addresses. Instead, the master simply asks each chunkserver about its chunk information during cluster startup. By doing so, the master reconstructs a complete mapping from chunk handles to server addresses.

The objective of GFS is to optimize storage when the number of files is moderate but individual files are large. Therefore, GFS uses a single master as the metadata server for centralized metadata management and loads all metadata information to the memory of the server, remarkably simplifying the design of file metadata servers. However, the use of a single master limits the scalability of GFS. As the number of files increases, the pressure on the master in GFS rises significantly. Due to limited memory, the master may be unable to store all the file metadata of the file system in memory. The performance and running time of the entire cluster are limited by the performance and running time of the master. If the master breaks down, the entire storage cluster becomes inaccessible. During startup, the master needs to query all chunkservers to obtain the mapping from chunks to servers. As the number of files increases, the startup time of the master becomes longer and longer.

File Creation Process

In GFS, an application creates a file as follows:

1. The application calls the file creation request function provided by the client to specify the path and permissions of the file to be created.
2. The client sends the file creation request to the master.
3. The master creates the file name in the directory tree, initializes file metadata according to permissions, and logs the creation operation in local disks.

File Read/Write Process

In GFS, an application reads file data as follows:

1. The application calls the read request function provided by the client to specify the path name, offset, and read length of the file to be read.
2. Based on the chunk size, the client converts the offset and length information into a chunk index, intra-chunk offset, and length within the file. The client then sends the file read request to the master.
3. The master queries the directory tree to obtain the chunk handle and the addresses of chunkservers that store the primary and secondary replicas according to the file path name and chunk index. Then, the master returns the obtained information to the client.
4. The client selects the nearest chunkserver from all the chunkservers that store the primary and secondary replicas to establish a connection. Within a certain period of time, the client will interact directly with the chunkserver for any subsequent reads/writes on the relevant chunk without the involvement of the master.

8.6.1.2 InfiniFS

With the rapid growth of data, modern large-scale data centers often manage tens or even hundreds of billions of files, generating massive amounts of file metadata that far exceeds the capacity of metadata servers in current distributed file systems (such as GFS). Currently, data centers are usually divided into smaller clusters, with each cluster running a distributed file system instance separately. However, managing the entire data center using a single distributed file system instance is much more conducive to achieving global data sharing, high resource utilization, and low maintenance costs.

Scalable and efficient distributed metadata services are critical to distributed file systems. Modern data centers usually have tens or even hundreds of billions of files. Using an extremely large-scale file system to manage all the files creates serious challenges in metadata services. First, directory trees of distributed file systems expand continuously, and application workloads are diverse. It is difficult to partition the directory trees to multiple metadata servers while ensuring high metadata access locality and sound load balancing. Second, files are deep in extremely large-scale file systems. As a result, the latency of path resolution can be high. Third, because an extremely large-scale file system usually serves a huge number of clients concurrently, the overhead of coherence maintenance for client caches can be significant.

InfiniFS, an efficient metadata service for extremely large-scale distributed file systems, addresses these challenges. First, InfiniFS decouples directory metadata structures, groups associated directory metadata and file metadata, and evenly distributes the metadata across metadata servers for both metadata locality and load balancing. Second, InfiniFS generates IDs for directories using a cryptographic hash and conducts path resolution in parallel with speculation, reducing metadata access latency. Third, to alleviate near-root read hotspots caused by path resolution, InfiniFS designs an optimistic access metadata cache. By checking the correctness of access operations on metadata servers, InfiniFS reduces the overhead of coherence maintenance for the cache.

Cluster Architecture

The metadata architecture of InfiniFS consists of three parts: clients, metadata servers, and a coordinator, as shown in Fig. 8.13.

- **Clients**: InfiniFS provides a global file system directory tree shared by multiple clients. InfiniFS clients offer two access modes: user-space library and FUSE user-level file system. Clients minimize latency through parallel path resolution and mitigate read loads on near-root directories through optimistic metadata caching.
- **Metadata servers**: The file system directory tree is decoupled and distributed across metadata servers for high metadata locality and good load balancing.

8.6 Distributed File Systems

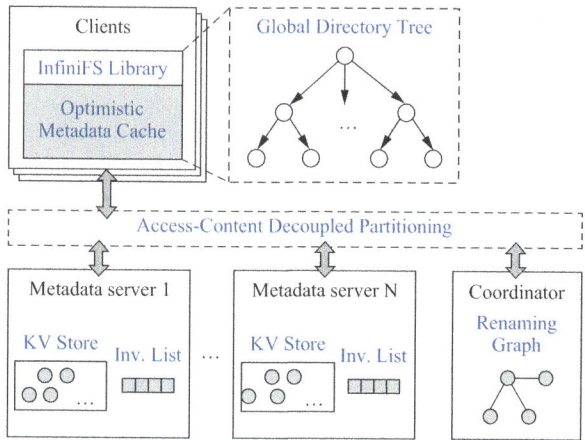

Fig. 8.13 Metadata architecture of InfiniFS [15]

Each server stores metadata in the local key-value store (KV Store) system. Servers typically cache metadata in memory and log updates to disks for high performance. Metadata servers use an invalidation list (Inv. List) to record directory rename operations and validate client metadata requests.

Coordinator: InfiniFS uses a central coordinator to handle directory rename operations. The coordinator first checks concurrent directory rename operations to prevent orphaned loops and then broadcasts modification information to invalidation lists of metadata servers.

Directory Tree Structure

In distributed file systems, directories are organized into a single-root inverted tree structure, as shown in Fig. 8.14a. Files consist of metadata and data. Metadata of the file system directory tree includes directory metadata and file metadata. A directory within a directory is called a subdirectory.

As shown in Fig. 8.14b, InfiniFS decouples directory metadata in the directory tree into two parts: directory access metadata and directory content metadata. Directory access metadata contains the directory metadata relevant to directory access, such as path resolution and permission checking. During path resolution, directory access metadata is used to index files and check permissions on directories and files. Directory access metadata includes directory IDs, names, and permissions (perm), which encompass user IDs, group IDs, and access permissions, such as read-only, read and write, and execute. Directory content metadata contains the directory metadata related to directory children, which are the files and subdirectories within a directory. The directory content metadata is updated when directory contents are listed, or files and subdirectories within directories are created or deleted. Directory content metadata includes directory timestamps (times), such as atime, mtime, and ctime, and the entry list (dirent), which is the name list of files and subdirectories.

As shown in Fig. 8.14c, InfiniFS groups the decoupled metadata and partitions the groups to different metadata servers (MSn). For each directory in the file system directory tree, the access metadata of the directory is grouped with the content metadata of its parent directory, and the content metadata of the directory is grouped with the access metadata of its subdirectories and the metadata of its files. This approach forms per-directory fine-grained metadata groups. These fine-grained metadata groups are then partitioned to different metadata servers with consistent hashing. Each metadata group has only one parent directory. The metadata groups are partitioned to metadata servers by hashing the IDs of parent directories. All metadata objects in metadata groups are then stored on the relevant metadata servers.

As shown in Table 8.1, InfiniFS structures metadata into key-value pairs for division and storage. Key-value pairs are classified into three types: directory access metadata, directory content metadata, and file metadata. The directory access metadata contains directory IDs and permissions. Character strings consisting of parent directory IDs and directory names are used as the keys. The directory content metadata contains directory timestamps and the entry list. Directory IDs are used as the keys. The file metadata uses character strings consisting of parent directory IDs and file names as the keys. Key-value pairs of directory access metadata are partitioned to target metadata servers by consistent hashing on parent directory IDs. Key-value

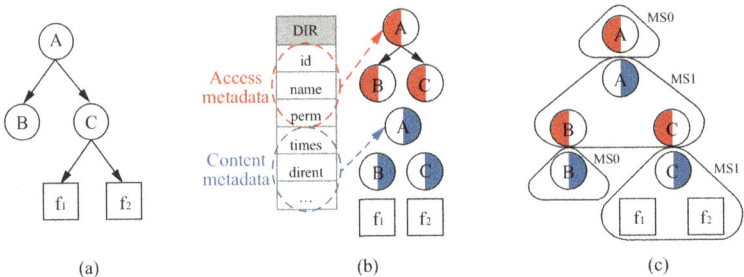

Fig. 8.14 Directory tree decoupling and partitioning in InfiniFS [15]. (**a**) Single-root inverted tree structure. (**b**) Metadata decoupling. (**c**) Metadata grouping and partitioning

Table 8.1 Key-value structure of metadata

Metadata type	Metadata value	Metadata key	Key-value pairs partitioned to servers by
Directory access metadata	Directory ID and permission	ID of the parent directory and name of the directory	Consistent hash function (ID of the parent directory)
Directory content metadata	Timestamp and entry list	ID of the directory	Consistent hash function (ID of the directory)
File metadata	All file metadata	ID of the parent directory and name of the file	Consistent hash function (ID of the parent directory)

8.6 Distributed File Systems

pairs of directory content metadata are partitioned to target metadata servers by consistent hashing on directory IDs. Key-value pairs of file metadata are partitioned to target metadata servers by consistent hashing on parent directory IDs.

InfiniFS introduces the optimistic metadata cache on the client side. Clients cache only the directory access metadata, including directory names, IDs, and permissions. Based on the file system hierarchy, clients organize cache entries in a tree structure and link the leaf entries as a least-recently-used (LRU) list. When cache replacement occurs, the least recently used leaf entry according to the LRU list is evicted. This ensures that access metadata of near-root directories remains cached on clients. Directory rename and permission modification operations can cause a large number of cache entries to become stale. InfiniFS exploits a lazy cache invalidation mechanism to reduce the overhead of coherence maintenance for the cache.

As shown in Fig. 8.15a, InfiniFS first sends directory rename operations (op) to the coordinator to detect whether the operations would result in orphaned loops, which lead to metadata loss. The coordinator then assigns an incremental version number (ver) to each directory rename operation. Second, InfiniFS broadcasts the version numbers and operation information of the directory rename operations to invalidation lists (Inv. List) of all metadata servers (MSn). The operation information includes the operation type, source path (src), and destination path (des). Third, InfiniFS processes the target metadata. That is, it deletes the access metadata key-value pairs of the directories on the metadata servers of the source paths and creates the access metadata key-value pairs of the directories on the metadata servers of the destination paths.

As shown in Fig. 8.15b, metadata servers detect whether client metadata requests involve any stale cache entries when processing the requests. Specifically, client metadata requests contain path names and cache version numbers. The metadata

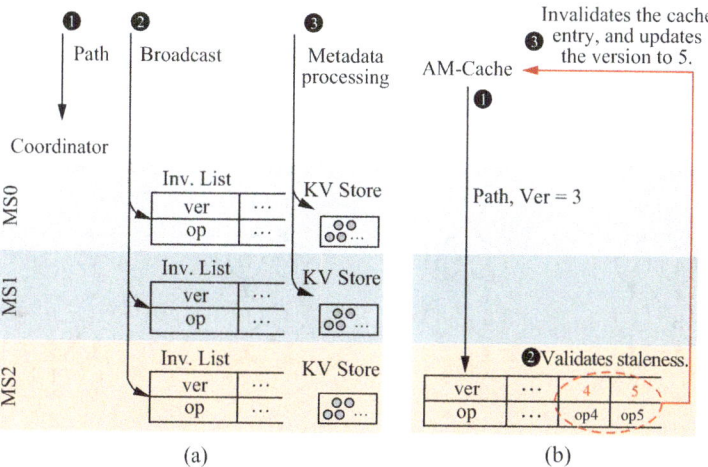

Fig. 8.15 Lazy cache invalidation mechanism. (**a**) rename(src,des). (**b**) stat(path)

servers validate staleness by comparing the path names in the requests against rename operations in the invalidation lists. The metadata servers only need to compare the operation information with version numbers greater than the cache version numbers. If the metadata servers find that the requested path names are invalid, the metadata requests involve stale cache entries. In this case, the servers abort the requests and return the new operation information in the invalidation lists to the clients. If the metadata servers find that the requested path names are valid, the metadata requests do not involve stale cache entries. In this case, the servers process the requests and return the processing results and the new operation information in the invalidation lists to the clients. With the operation information, the clients then invalidate local stale cache entries and update the cache version numbers to the latest numbers.

InfiniFS leverages a globally unique coordinator to prevent orphaned loops caused by concurrent directory rename operations. Figure 8.16 shows an orphaned loop caused by two directory rename operations. As shown in Fig. 8.16a, directories in the file system directory tree should be connected and not form a loop. In Fig. 8.16b, c, client 1 attempts to rename directory E to be a child of directory C, and client 2 attempts to rename directory B to be a child of directory F. The metadata (in the blue boxes) required for the two rename operations does not overlap. Therefore, the two operations can be performed concurrently. But the two operations result in an orphaned loop, as shown in Fig. 8.16d. Files and directories in the orphaned loop cannot be accessed. To address this problem, the InfiniFS coordinator locally maintains a renaming graph to record the source and destination paths of current in-flight directory rename operations. Before executing a new directory rename operation, a client needs to contact the coordinator to check whether the new operation will produce an orphaned loop with the in-flight operations. If an orphaned loop will be produced, the directory rename operation is terminated. If no loop will be produced, the directory rename operation is allowed to proceed. The source and destination paths of directory rename operations are kept in the renaming graph throughout the rename procedure and deleted after the directory rename operations are complete.

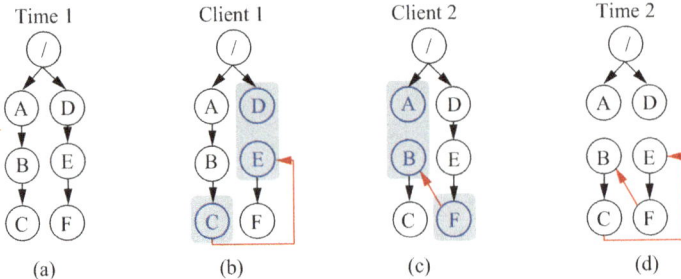

Fig. 8.16 Orphaned loop caused by concurrent directory rename operations. (**a**) No orphaned loop. (**b**) Rename E to be a child of C. (**c**) Rename B to be a child of F. (**d**) Form an orphaned loop

8.6 Distributed File Systems

File Creation Process

Figure 8.17 shows the process for an application to create a directory in InfiniFS. The application calls the directory creation operation to specify the path name for the new directory. An ID is required for the directory during creation. InfiniFS generates a candidate ID for the new directory by hashing the parent directory ID, the directory name, and the version number (0 by default). InfiniFS then checks whether the candidate ID collides with any existing directory. If there are no collisions, the candidate ID becomes the ID of the new directory. Otherwise, InfiniFS increases the version number by one, generates a new candidate ID, and checks for collisions again until the ID of the new directory is determined. After that, InfiniFS initializes key-value pairs for directory access metadata and directory content metadata. Then it partitions the key-value pairs of directory access metadata to the relevant metadata server using the parent directory ID and the directory name, and partitions the key-value pairs of directory content metadata to the relevant metadata server using the directory ID. The key-value pairs are stored in key-value stores of the servers.

Data Read and Write Parsing

In InfiniFS, an application reads or writes a file as follows: The application calls the file read/write operation to specify the path name, offset, and read/write length of the file. InfiniFS first conducts path resolution in parallel with speculation. Starting from the root directory (the root directory ID is known) with 0 (default value) as the version number, InfiniFS predicts the IDs of all intermediate directories in the path name based on the cryptographic hash. Then, InfiniFS reconstructs the keys to the metadata of all intermediate directories and the metadata of the target file using the speculated directory IDs and directory names. After that, InfiniFS sends network requests in parallel to access the metadata corresponding to these keys. If the

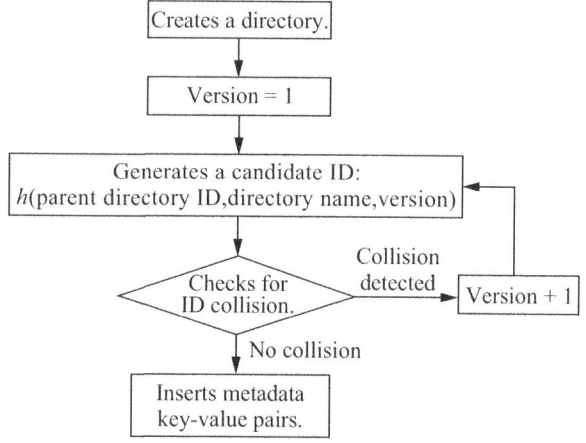

Fig. 8.17 Process of creating a directory

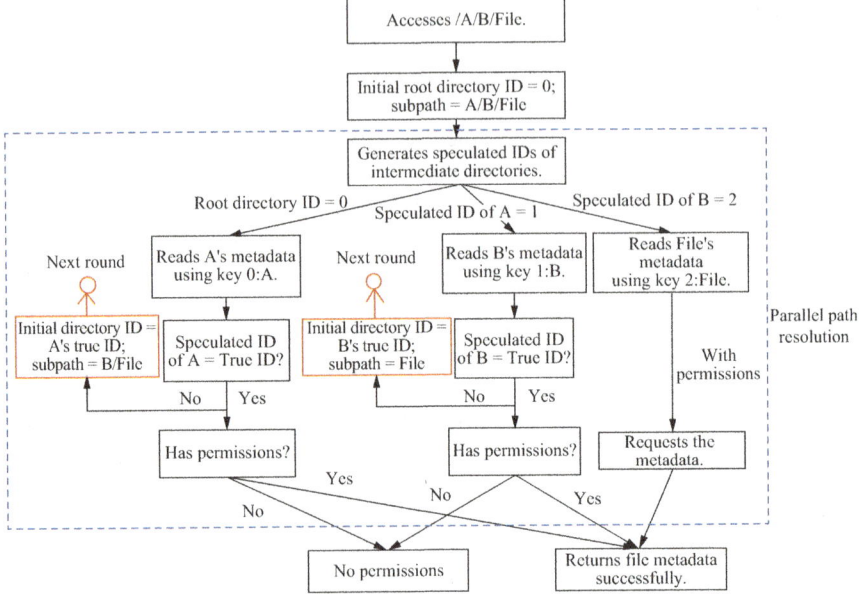

Fig. 8.18 Parallel path resolution

speculated ID of an intermediate directory is incorrect, the true ID of the directory is returned from the key-value store based on the ID of the parent directory and the name of the directory. With the true ID, InfiniFS continues to resolve the subpath in parallel under the directory until the permissions of all intermediate directories are checked and the metadata of the target file is returned. Figure 8.18 shows one round of parallel path resolution. Clients obtain identifiers of data blocks and locations of data servers from the file metadata and interact directly with the data servers for reads and writes.

8.6.2 Key Technologies of Distributed File Systems

This section describes directory tree management and fast path resolution technologies used in distributed file systems.

8.6.2.1 Directory Tree Management

As the number of files increases rapidly, distributed management of file system directory trees is critical. Currently, major directory tree management technologies used in distributed file systems include the following:

Static partitioning: Used by HDFS Federation, Lustre DNE, and PVFS, static partitioning is the most widely used directory tree partitioning approach. Static partitioning typically requires an administrator to assign a subtree of the directory tree to a specified metadata server for management. File and directory rename operations across different partitions are not allowed. Static partitioning can achieve high metadata access locality. For applications with high access locality, experienced administrators can exploit static partitioning to effectively balance loads between servers. However, as the scale of data and the use of cloud computing continue to expand, one platform may now run thousands of different applications simultaneously. The static partitioning approach obviously cannot properly handle these complex workloads.

Dynamic subtree partitioning: Dynamic subtree partitioning is another widely used approach. One of the major systems that use this approach is CephFS. CephFS migrates hotspot directory subtrees to non-busy nodes dynamically based on the system access load, thereby achieving load balancing. CephFS allows an administrator to dynamically add and remove metadata servers based on the access load. Each metadata server can dynamically take over metadata access under a subtree, providing high scalability. However, if access hotspots jitter back and forth, the system has to frequently and repeatedly migrate subtrees. This eventually results in performance jitters. Moreover, during migration, subtrees are frozen and their file data cannot be accessed.

Hash-based partitioning: Hash-based partitioning is also widely used in file systems, such as LocoFS [17, 18], DeltaFS, and IndexFS. These systems are mainly dedicated to processing access requests to a single directory containing a huge number of files in supercomputing scenarios. As a single directory in a supercomputing program may have millions of files, using index-based metadata management creates huge query overheads. To address the issue of large-scale file access requests to a single directory, the hash-based partitioning approach partitions the namespace to multiple metadata servers by hashing.

8.6.2.2 Fast Path Resolution

Distributed file systems support the following three path resolution approaches:

Centralized path resolution: Some distributed file systems, such as LocoFS, GFS, and HDFS, store all directory metadata on a central metadata server to reduce network access due to path resolution, thereby minimizing latency. However, this approach suffers from single-node bottlenecks.

Serial path resolution: Some distributed file systems, such as IndexFS and Tectonic, partition the file system directory tree to multiple metadata servers to prevent performance bottlenecks caused by a single node. However, this approach requires access to multiple metadata servers in serial order to obtain the locations of next-level directories and to check corresponding permissions during path resolution for metadata requests, resulting in high data access latency.

Parallel path resolution: Some distributed file systems, such as BetrFS, Giraffa, and CalvinFS, leverage full path names or hashing on full path names for file indexing. BetrFS uses full path names to index file metadata in the local file system. Giraffa uses full path names as primary keys to store file metadata in databases. CalvinFS locates file metadata by hashing full path names. Therefore, the systems can send network requests in parallel during path resolution to check the permissions of each intermediate directory. However, this approach makes it difficult to implement hierarchy semantics of the directory tree. For example, the overhead of directory rename operations is huge because the operations change the full path names of all descendants, resulting in the need to migrate the metadata of all descendants to new locations.

8.7 Summary

Distributed storage systems scatter data across a large number of servers and aggregate the servers' compute, storage, and network resources to provide large-capacity high-performance storage services. Compared with local storage systems, distributed storage systems have stricter requirements on scalability, consistency, and availability. A wide variety of distributed storage systems are available, such as distributed key-value store systems, distributed object storage systems, distributed block storage systems, and distributed file systems. They expose different access interfaces to upper-layer applications. The semantics of these interfaces greatly impact the design of the relevant distributed storage systems. For example, distributed file systems need to consider how to partition metadata to different servers and provide a globally unified directory tree abstraction. As applications develop, new distributed storage systems will emerge, such as distributed vector storage systems for artificial intelligence (AI) workloads.

8.8 Practice Questions

1. **Scalability is critical to distributed storage systems. What are the common methods for improving the scalability of distributed storage systems?**

    Answer: (1) Avoid introducing centralized components, such as coordination servers, to data access paths. (2) Avoid coordination between multiple servers, for example, executing distributed transactions.

2. **Dynamo, a distributed key-value store system, provides semantic assurance for eventual consistency. Compared with a system with strong consistency, Dynamo may cause exceptions. Please provide an example of such possible exceptions.**

    Answer: In a social media scenario, user A in one country publishes a tweet and then calls user B in another country located on a different continent to check the tweet. However, user B does not see the tweet after logging in to the website.

3. **For a distributed key-value store system, does the secondary index function affect the performance of common write operations? Explain your answer.**

Answer: The performance of common write operations will deteriorate. This is because the mapping from secondary keys to primary keys needs to be added to the secondary indexes when common write operations are performed.

4. **Compared with consistent hashing, what are the advantages of Ceph's CRUSH algorithm?**

    Answer: The CRUSH algorithm constructs a hierarchical cluster map, can detect the physical topology and performance characteristics of storage clusters, and allows users to customize allocation policies. Therefore, compared with consistent hashing, the CRUSH algorithm has higher fault tolerance and can adapt to the heterogeneity of storage clusters.

5. **How do distributed block storage systems improve the aggregate bandwidth?**

    Answer: (1) Parallel technology is used to distribute data to multiple storage devices, maximizing the aggregate bandwidth of storage and network hardware. (2) The local cache technology is also used to reduce network access.

6. **GFS, a distributed file system, has only one centralized metadata server (called master). Which methods are used to prevent the server from becoming the system performance bottleneck?**

    Answer: (1) After obtaining chunk handles and chunkserver addresses, clients can bypass the master and interact directly with chunkservers for data reads and writes. (2) The master does not persistently store the mapping from chunk handles to server addresses. Instead, the master reconstructs the mapping by simply asking each chunkserver during cluster startup.

7. **GFS organizes file data in the unit of chunks. How does the selection of data chunk sizes affect system performance?**

    Answer: When the chunk size is larger, the number of interactions between the client and the master decreases, and the total amount of metadata stored on the master declines. However, the system load is more likely to be unbalanced due to the existence of some hotspot files.

8. **InfiniFS decouples directory metadata into two parts: directory access metadata and directory content metadata. What is the reason for this design?**

    Answer: For both access locality and load balancing. On one hand, for each directory in the file system directory tree, the access metadata of the directory is grouped with the content metadata of its parent directory, and the content metadata of the directory is grouped with the access metadata of its subdirectories and the metadata of its files. In this way, most metadata operations involve only one metadata server, which achieves sound access locality. On the other hand, InfiniFS can use hashing to distribute directory metadata to different metadata servers and thus achieve excellent load balancing.

9. **Can the dynamic subtree partitioning and hash-based partitioning approaches be used together to improve the metadata processing efficiency of distributed file systems?**

    Answer: Yes. The dynamic subtree partitioning approach is used for the entire file system directory tree to achieve better access locality. In addition, for a directory that contains massive files, the hash-based partitioning approach can be used separately for that directory, so that metadata access to the directory does not become a system hotspot bottleneck.

References

1. Herlihy MP, Wing JM. Linearizability: a correctness condition for concurrent objects. ACM Trans Program Lang Syst. 1990;12(3):463–92.
2. Decandia G, Hastorun D, Jampani M, et al. Dynamo: Amazon's highly available key-value store. ACM SIGOPS Oper Syst Rev. 2007;41(6):205–20.
3. Ousterhout J, Gopalan A, Gupta A, et al. The RAMCloud storage system. ACM Trans Comput Syst. 2015;33(3):1–55.
4. Lamport L. Paxos made simple. In: ACM SIGACT news (distributed computing column) 32, 2001, vol 4(121). p. 51–8.
5. Ongaro D, Ousterhout J. In search of an understandable consensus algorithm. In: USENIX. 2014 USENIX annual technical conference (Usenix ATC 14). Berkeley: USENIX Association; 2014. p. 305–19.
6. Lakshman A, Malik P. Cassandra: a decentralized structured storage system. ACM SIGOPS Oper Syst Rev. 2010;44(2):35–40.
7. Karger D, Lehman E, Leighton T, et al. Consistent hashing and random trees: distributed caching protocols for relieving hot spots on the world wide web. In: ACM. Proceedings of the twenty-ninth annual ACM symposium on theory of computing. New York: Association for Computing Machinery; 1997. p. 654–63.
8. Weil SA, Brandt SA, Miller EL, et al. CRUSH: controlled, scalable, decentralized placement of replicated data. In: IEEE. SC'06: proceedings of the 2006 ACM/IEEE conference on supercomputing. New York: Association for Computing Machinery; 2006. p. 31.
9. Wang L, Zhang Y, Xu J, et al. {MAPX}: controlled data migration in the expansion of decentralized {object-based} storage systems. In: USENIX. 18th USENIX conference on file and storage technologies (FAST 20). Berkeley: USENIX Association; 2020. p. 1–11.
10. Mickens J, Nightingale EB, Elson J, et al. Blizzard: fast, cloud-scale block storage for cloud-oblivious applications. In: USENIX. 11th USENIX symposium on networked systems design and implementation (NSDI 14). Berkeley: USENIX Association; 2014. p. 257–73.
11. Nightingale EB, Elson J, Fan J, et al. Flat datacenter storage. In: USENIX. 10th USENIX symposium on operating systems design and implementation (OSDI 12). Berkeley: USENIX Association; 2012. p. 1–15.
12. Tak B, Tang C, Chang RN, et al. Block-level storage caching for hypervisor-based cloud nodes. IEEE Access. 2021;9:88724–36.
13. Li H, Zhang Y, Li D, et al. Ursa: hybrid block storage for cloud-scale virtual disks. In: ACM proceedings of the fourteenth EuroSys conference 2019; 2019. p. 1–17.
14. Ghemawat S, Gobioff H, Leung ST. The Google file system. In: Proceedings of the nineteenth ACM symposium on operating systems principles. New York: Association for Computing Machinery; 2003. p. 29–43.
15. Lv W, Lu Y, Zhang Y, et al. {InfiniFS}: an efficient metadata service for {large-scale} distributed filesystems. In: USENIX. 20th USENIX conference on file and storage technologies (FAST 22). Berkeley: USENIX Association; 2022. p. 313–28.
16. Dean J, Ghemawat S. MapReduce: simplified data processing on large clusters. Commun ACM. 2008;51(1):107–13.
17. Li S, Liu F, Shu J, et al. A flattened metadata service for distributed file systems. IEEE Trans Parallel Distrib Syst. 2018;29(12):2641–57.
18. Li S, Lu Y, Shu J, et al. LocoFS: a loosely-coupled metadata service for distributed file system. In: ACM. The international conference for high performance computing, networking, storage and analysis (SC). Denver: Association for Computing Machinery; 2017. p. 1–12.

Chapter 9
Storage Reliability

The rapid development of information technology has put data at the heart of modern production, and data is now experiencing a period of tremendous growth. In 2019, the International Data Corporation (IDC) predicted that the Global Datasphere (a measure of all new data that is captured, created, and replicated in any given year across the globe) will grow from 32 ZB (zettabytes) recorded in 2018 to 175 ZB by 2025 [1]. Traditional data centers match the storage demands stemming from all this data creation either by beefing up installed devices or setting up extra devices. But a storage system with more devices succumbs more easily to failures and disruptions. Recently, over half of data centers have reported service or access disruptions, each incurring million-dollar losses or more. In light of these risks, a data storage system must provide high reliability through robust fault tolerance policies and mechanisms that are designed based on the hardware and software failure modes.

Businesses and individual users might want to know exactly how their data storage solutions can live up to reliability expectations. This chapter delves into this question in detail, from general knowledge about storage reliability to specific aspects like the reliability of disks and flash memory media, erasure coding technology, and distributed storage reliability.

9.1 Overview of Storage Reliability

9.1.1 Reliability Metrics and Calculation

To clearly understand storage reliability, one concept is key—the quantification of a system's storage reliability against another's. The industry has defined a set of metrics to describe the reliability of a storage component.

The mean time to failure (MTTF) measures the average amount of time a storage component operates before it fails.

The mean time to repair (MTTR) measures the average time required to repair a failed component. For a storage device, the MTTR is relevant to device capacity. The larger the device capacity, the longer the MTTR.

The mean time between failures (MTBF) measures the average time between two failures of a storage component, represented in hours or years. The longer the MTBF, the fewer failures that a storage system experiences within an observation window.

The MTBF of a storage component can be calculated (1) by accumulating the MTTF and MTTR values or (2) by dividing the length of the observation window by the number of failures that occur within the window. If the MTBF is 1, a storage component will fail once a year on average.

The mean time to data loss (MTTDL) is usually calculated through a Markov chain. The MTTDL of a storage system can be determined by describing the system's error states and the transition probability across each state. For example, a RAID 5 storage system that tolerates the failure of a single storage device has three error states: 100% health of storage devices, failure of a single storage device, and data loss. The MTTDL of the RAID 5 storage system can be obtained by constructing a chain that possesses the Markov property.

The annual failure rate (AFR) measures the average number of times a storage device fails in a year. The AFR is the reciprocal of the MTBF and can therefore be calculated as follows:

$$\text{AFR} = \frac{1}{\text{MTBF}(\text{in years})} = \frac{8760}{\text{MTBF}(\text{in hours})} \quad (9.1)$$

The availability of a component measures the ratio of system uptime to the interval between two consecutive chunks of uptime.

$$\text{Availability} = \frac{\text{MTTF}}{\text{MTTF}+\text{MTTR}} = \frac{\text{MTTF}}{\text{MTBF}} \quad (9.2)$$

In general, if the value of availability approximates more to 1, the ratio of the uptime of a component is higher, and higher component availability corresponds to shorter downtime. For example, 99.9999% (6 nines) availability means that component downtime is just 31.5 s/year. The reliability and availability of data are often mistakenly used interchangeably, and their differences will be covered in Sect. 9.1.3.

9.1.2 Layered Reliability Design

A storage system is designed in (1) a hardware layer that concerns storage devices (such as disks, memories, and flash solid-state disks) and (2) a software layer that involves a storage management system. Accordingly, system reliability is defined at

9.1 Overview of Storage Reliability

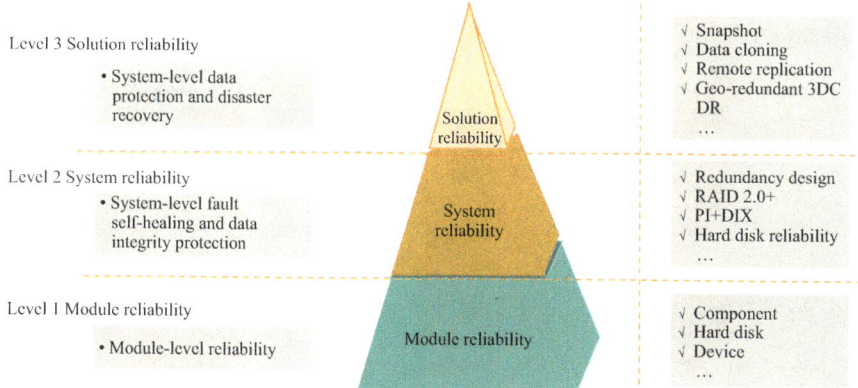

Fig. 9.1 Logical map of three-level reliability

three levels from the bottom up: module reliability, system reliability, and solution reliability, as illustrated in Fig. 9.1.

At the module level, storage hardware is backed by engineering and process technologies to make media less prone to failures and more able to tolerate faults. At the system level, certain redundant data is saved, and the corresponding input/output (I/O) processing logic is modified. This allows invalid data to be restored through a pre-defined operation when a storage device fails. At the solution level, data is protected system-wide through data protection and disaster recovery (DR) technologies like snapshot, data cloning, remote replication, and geo-redundant 3DC DR. As solution-level reliability is covered in Chap. 11, this chapter will focus on module- and system-level reliability.

9.1.3 Reliability vs. Availability

Availability and reliability are two different notions in regard to a data storage system but are often used incorrectly. As explained in Fig. 9.2, availability is measured by the ratio of time a system stays operational under normal circumstances to the total elapsed time and represents system uptime during a given time window; meanwhile, reliability is the probability that a system will run without experiencing failures and describes the system failure frequency. For example, if a storage system is down for 30 ms/h, it is deemed extremely unreliable, albeit with high availability of up to 99.999% (5 nines) (see Table 9.1), because the maximum uptime is only 1 h.

Fig. 9.2 Availability vs. reliability [2]

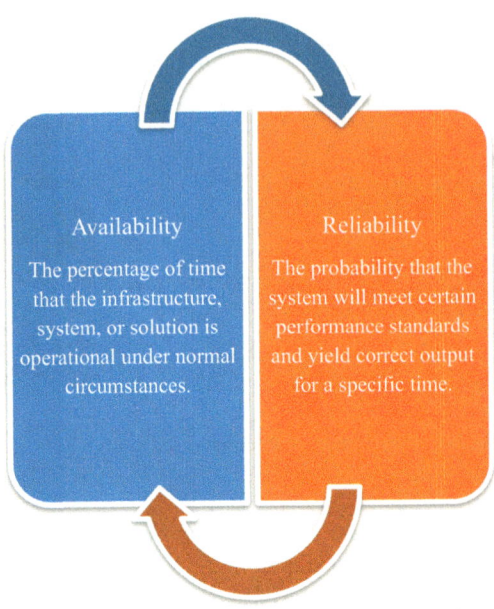

Table 9.1 Availability vs. service downtime

Availability (%)	Downtime per year	Downtime per month	Downtime per week
90	36.5 days	72 h	16.8 h
99	3.65 days	7.2 h	1.68 h
99.9	8.76 h	43.8 min	10.1 min
99.99	52.56 min	4.32 min	1.01 min
99.999	5.26 min	25.9 s	6.05 s
99.9999	31.5 s	2.59 s	0.605 s
99.99999	3.15 s	0.250 s	0.0605 s

9.2 Disk Reliability

Storage system components are subject to durability and reliability tests before delivery. Once in service, components are likely to err at a probability that changes with their service time.

9.2.1 Analysis of Disk Error Characteristics

This section discusses the error characteristics of two typical storage devices—hard disk drive (HDD) and solid-state drive (SSD)—and names a few module-level reliability technologies.

9.2.1.1 HDD Error Characteristics

The AFR of HDDs takes on a classic bathtub curve as their service time increases. Figure 9.3 illustrates the AFR curve of HDDs at three phases of the HDD lifespan [3].

Infancy: The AFR is typically high on new HDDs over the first 3–5 months after they first enter use in real-life production environments.
Useful life: After infancy, the AFR of HDDs drops and remains relatively stable for 3–4 years.
Wearout: This is the last phase of the HDD lifespan where the AFR starts to climb.

9.2.1.2 SSD Error Characteristics

An SSD works by storing data on flash memory cells, which are made of floating-gate metal oxide semiconductors (FGMOSs). Electric charge is applied to the cells to generate different threshold voltages, which help represent data. Figure 9.4 demonstrates four possible states of a multi-level cell (MLC) and their corresponding threshold voltages.

Because of the update-out-place write mechanism, SSDs frequently perform garbage collection to free up space. During garbage collection, the flash controller performs the erase operation, which involves expelling the electric charge stored in cells so that the cells are usable again. When the erase operation is executed repeatedly, the cells' ability to store electric charge is irreversibly compromised. As a result, electric charge easily escapes from the cells, leading to represented charge

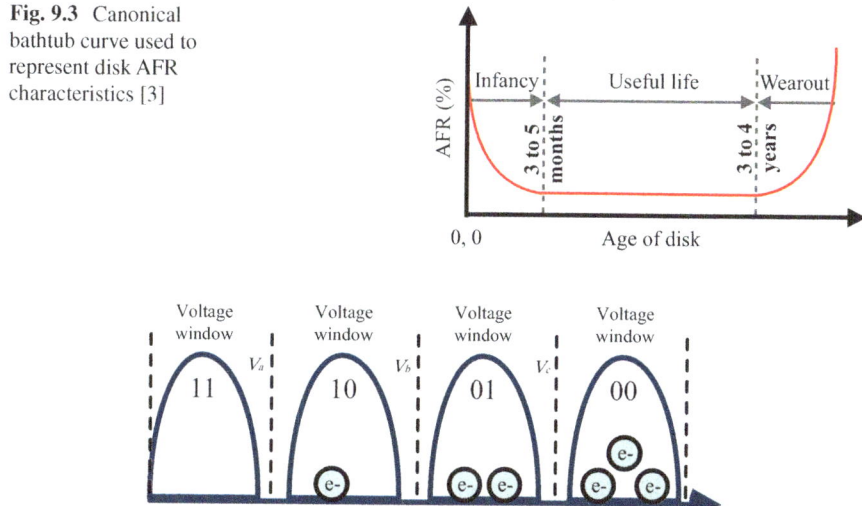

Fig. 9.3 Canonical bathtub curve used to represent disk AFR characteristics [3]

Fig. 9.4 The four states in an MLC flash cell [4]

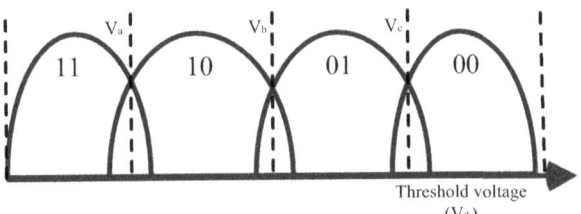

Fig. 9.5 Threshold voltage distribution for cell-stored data showing shifts after multiple erase operations [4]

shift. As illustrated in Fig. 9.5, after multiple erase operations, the threshold voltage distribution corresponding to the data stored in a cell shifts, and even the threshold voltage windows overlap, increasing the probability of data identification errors.

To ensure the reliability of storage devices, error correction code (ECC) is configured within firmware by default in order to detect and correct errors in stored data. ECC implements this capability of error detection and correction by adding redundancy bits (or parity bits). The most commonly used ECCs are the Bose-Chaudhuri-Hocquenghem (BCH) code, the Hamming code, and the low-density parity-check (LDPC) code. In terms of HDDs, user data and the corresponding ECC redundancy information are usually stored in the data area of an HDD sector. Regarding SSDs, user data is generally stored in a flash memory page, whereas corresponding parity information is stored in an out-of-band area of the page.

9.2.2 Disk Failure Warning and Monitoring

Misoperation or occurrences in the external environment may cause disks to fail, causing disk read/write to slow down, damaging file data, or, worse still, leaving the entire storage system unable to start normally. To ensure data security and promptly rectify disk failures, the system must monitor and warn off such failures.

9.2.2.1 S.M.A.R.T. Attribute Values

The advent of the big data era is causing data security to emerge as a major concern. Self-Monitoring, Analysis, and Reporting Technology (S.M.A.R.T.), first proposed by Compaq together with IBM and Fujitsu, requires that a complete set of reliability metrics be defined at the same time products are being made so that the system can predict possible disk failures by monitoring related metrics and comparing them against the preset safe levels.

The technology provides some S.M.A.R.T. attributes for determining the disk reliability state and sets a raw value for each attribute. When the disk state changes, the value of the corresponding attribute also changes (increases or decreases). Each drive manufacturer defines its own set of S.M.A.R.T. attributes and chooses some of

9.2 Disk Reliability

Table 9.2 Commonly used S.M.A.R.T. attributes

ID	Attribute	Description
0x01 (001)	Read Error Rate	Rate of disk data read errors
0x04 (004)	Start/Stop Count	Count of disk start/stop cycles
0x05 (005)	Reallocated Sectors Count	Count of reallocated sectors on a disk
0x09 (009)	Power-On Hours	Length of time that electrical power is applied to a disk
0x0C (012)	Power Cycle Count	Count of disk power on/off cycles
0xC2 (194)	Temperature	Disk temperature
0xF1 (241)	Total LBAs Written	Total size of all logical block address (LBA) write count
0xF2 (242)	Total LBAs Read	Total size of all LBA read count

them as its own industry standards. This inevitably leads to disparities in attribute settings with other manufacturers, notwithstanding some key attributes that are universally adopted by many manufacturers. Table 9.2 lists the more commonly used S.M.A.R.T. attributes, where the ID item identifies the specific attributes.

S.M.A.R.T. technology sets a raw value, a current value, a worst value, and a threshold for each attribute.

The raw value is individually defined by the manufacturer and is usually the real test value observed when a disk is running. For example, the disk temperature is represented in degrees Celsius, and its raw value changes when the temperature changes. The raw value can also be calculated. For example, the read error rate of a disk can be obtained based on the total amount of data read and the total amount of data read with errors within a given period of time.

The current value is the raw value normalized to a range of 1–253 based on a formula. This is used to measure the disk health status. The higher the value, the healthier the disk, and vice versa. The greater the disk service time, the lower the disk reliability, and naturally, the lower the current value. However, this value cannot measure the health of different types of disks, as one manufacturer may use a different formula from other manufactures and a manufacturer may even use different formulas for different types of disks.

The worst value records the worst attribute value recorded by a disk since delivery. Considering disk wearout, the current value is equal to the worst value, with exceptions such as the temperature attribute.

The threshold is individually defined by each manufacturer and measures whether a device is safe. The threshold maps to the current value. If the current value falls below the threshold, a disk may have failed, and data may be lost.

S.M.A.R.T. provides basic information, namely three states, so that users can analyze disk health: Good, Warning, and Bad. (These names may differ by software.) The S.M.A.R.T. states can be measured by comparing the current or worst value against the threshold. If the current or worst value is lower than the threshold, the observed disk is unreliable, and its S.M.A.R.T. state is Bad. If the current or worst value is close to the threshold, the disk is unreliable, and its S.M.A.R.T. state is Warning. If the current or worst value is far greater than the threshold, the disk is reliable, and its S.M.A.R.T. state is Good, as demonstrated in Fig. 9.6.

9.2.2.2 ML-Based Disk Fault Prediction Algorithm

While most disks nowadays are equipped with S.M.A.R.T., the technology is increasingly unable to satisfy users' demands for reliable data storage by relying on its predictive capability alone. Luckily, disk fault prediction algorithms based on machine learning (ML) have appeared right on time. Not only are the algorithms able to learn the many causes of disk faults, but they can also analyze disk status and promptly send results to the disk monitoring hardware or software for repair or tuning.

Interestingly, most ML-based disk fault prediction algorithms still rely on S.M.A.R.T. attributes one way or another. That is, a fault prediction model is created to analyze the correlation between attributes and disk faults and the weight of each attribute on disk faults so that disk health can be analyzed based on the S.M.A.R.T. attribute values.

9.2.2.3 Bad Sector Detection and Repair

Bad sectors are a common disk fault and fall into logical and physical bad sectors. A logical bad sector is essentially a software-repairable logic error, that is, a mismatch between the sector content and the error correction code. Physical bad sectors

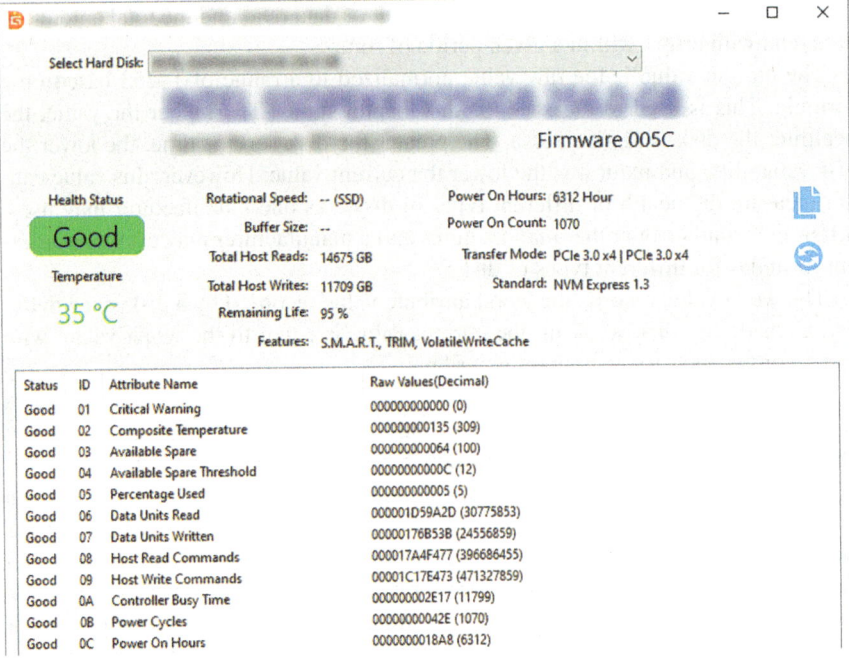

Fig. 9.6 S.M.A.R.T. state evaluation on the DiskGenius

9.2 Disk Reliability

arise from physical damages due to the external environment (such as overheating and physical impacts). Some physical bad sectors can also be repaired by software (for example, by isolating the damaged area). There are often telltale signs when a disk has bad sectors, which include the following:

1. Data read or write is slow and sometimes accompanied by errors or strange noises.
2. Running programs load slowly and often stop responding.
3. The system starts up or runs slowly, a "blue screen of death" often occurs, and sometimes, the system may even fail to start up altogether.
4. Error messages prompting that the disk is inaccessible are displayed.

To ensure data security, bad sectors need to be detected and repaired in a timely manner. Some common causes of bad sectors include the following:

Physical wear: All disks have finite lifespans and experience physical wear-and-tear and aging of components, which increase the more they are used.

Virus or malware infection: Viruses may falsely mark good sectors as bad.

Disk overheating: When a disk runs for a long time without stopping, its temperature may exceed a safe level, which will damage components and may result in physical bad sectors.

Dust particles: During operation, a disk rotates at a very high speed. If any dust ingresses into the disk housing and comes into contact with the disk, the disk can easily be damaged.

External impact: When a disk is subject to impact from strong external forces, its components may become deformed, which may result in physical bad sectors.

File system damage: If the file system is damaged, the system cannot correctly identify the boundaries of logical partitions, and the partition table may also be damaged. As a result, the disk cannot be accessed normally.

There are some simple but effective methods for detecting and repairing bad sectors. These include the following:

Data backup: Data backup is the easiest way to deal with bad sectors. Multiple copies of data are stored so that, in the event of data loss or damage due to bad sectors, an undamaged copy of the data can be used instead. A major downside of this method is that it creates a storage overhead that is several times the size of the original data.

Error detection: All files stored on a disk are scanned and then compared against the error correction code or the MD5 value calculated based on the file data to locate any files with data errors. If the detection process fails due to errors, it indicates that the disk may have bad sectors.

Surface scan: Disks are scanned for read errors to identify bad sectors. An error means that a sector is unreliable, which can then be marked so that the file system skips this sector when processing read/write operations.

S.M.A.R.T. test: Disk health is evaluated based on its S.M.A.R.T. state.

Software repair: Bad sectors are detected and repaired using software such as DiskGenius and CHKDSK in Windows 10.

Despite the existence of all the aforementioned methods for detecting and repairing bad sectors, disks should always be used in a rational and healthy fashion to safeguard storage reliability. The following are some methods for preventing bad sectors from occurring on disks:

1. Keep the computer safe from static electricity. Static electricity may build up in computers, leading to electrostatic breakdown that damages device components.
2. Do not turn off a disk while it is running. Suddenly turning off a running disk causes it to abruptly lose power, which may cause structural damage to the disk.
3. Regularly clean up the disk and back up important data. Though some disk faults are predictable, data loss may occur due to a myriad of unpredictable reasons. Backing up important data can help minimize losses due to sudden faults.
4. Do not continue to use disks that have overheated. When a disk is under heavy load for an extended period of time, the temperature of the disk will remain high for a long period of time, which may result in structural damage to the disk.

9.2.3 Environment-Aware Disk Reliability Design

Because storage systems consist of both storage media and controllers, the overall reliability of their hardware is of the utmost importance. Storage systems are deployed in various locations (with wide ranges in temperature and environmental conditions), which have different impacts and risks to hardware health. For this reason, storage systems need to be designed with enhanced overall reliability in order to withstand the most unfavorable environments they may be deployed in. Some common solutions to improving overall reliability are described below.

9.2.3.1 Anti-corrosion

A storage system may be used in corrosive environments (for example, by the seaside or near an industrial facility). To ensure normal functioning in such environments, core hardware (such as disks and controllers) must be designed with corrosion-resistance in mind. For example, an anti-corrosion coating may be applied to the disk drive. Corrosion protection may also be designed in a way that takes into account temperature and voltage distribution to improve the overall durability and reliability of the storage system.

9.2.3.2 Cooling

Storage devices generate heat while they are operating due to reasons such as friction caused by the rapidly spinning disk. This causes the device temperature to rise, which increases the device failure rate. For SSDs, overly high temperatures can

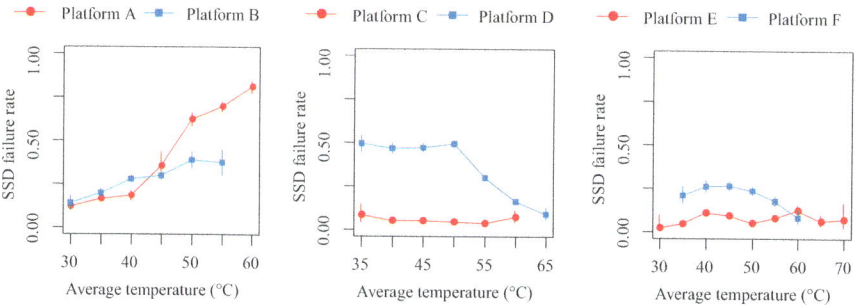

Fig. 9.7 SSD failure rate vs. temperature [5]

easily result in the Arrhenius effect, which leads to faster wear of flash memory cells. Figure 9.7 shows the SSD failure rates at different temperatures. Generally, the system monitors component temperatures in real-time (for example, by using temperature sensors built into flash cards). If the temperature rises sharply, fans are activated to cool the system and maintain the temperature at a safe level.

If the temperature is unusual or rises to a level whereby the storage system can no longer function properly or data reliability is jeopardized, users should stop the currently running service or power off the system altogether.

9.2.3.3 Vibration-Resistant Design

Strong vibrations may loosen the connectors on storage devices, causing data access errors and even failures. To protect the storage system from damage caused by vibrations, its design needs to feature greater rigidity, resonance resistance, and more securely fastened components. Rubber dampers and springs can also be used to isolate vibrations and protect components from damage caused by strong impacts.

9.3 Flash Media Reliability

In order to increase the capacity of SSDs, manufacturers use high-density flash media as basic flash memory cells whose data representation is based on the threshold voltages represented by the amount of electric charge stored in the cells. The number of bits that can be stored in a high-density cell, such as a triple-level cell (TLC) or quad-level cell (QLC), increases with the cell density. Each TLC and QLC can store 3 and 4 bits of data, respectively. To represent all possibilities of multiple bits of data, the TLC (or QLC) divides the threshold voltage into 8 (or 16) states. As the storage density increases, the voltage fault-tolerant space between different voltage states decreases. For this reason, when a cell experiences varying degrees of interference from error sources, it is more likely to show a threshold voltage distribution shift. This causes an incorrect read of stored data and affects storage reliability.

SSDs usually measure flash media reliability through the raw bit error rate (RBER), which is the ratio of error bits to the total bits after page data reads. The four main error sources affecting flash media reliability are program/erase (P/E) cycle, retention time, program interference, and read disturb [6]. Increasing P/E cycles cause flash memory cells to experience irreversible wear, and consequently, the charge is trapped in the cells and fails to be detrapped promptly. As the retention time increases, the charge trapped in the cells leaks. Through program interference and read disturb, electric charge is mistakenly applied to immediately adjacent cells of the target cells subject to data write and read operations, respectively. These error sources will cause a shift in the threshold voltage distribution, leading to incorrect data read.

9.3.1 Flash Media Error Sources

9.3.1.1 P/E Cycle

As the P/E cycle count increases, electric charge is trapped and builds up in the tunnel oxide layer of flash memory cells. As a result, the erase operation cannot reset the threshold voltage to the initial state, or the program operation cannot set the threshold voltage to the preset state, leading to incomplete erasing or program errors. Without considering other error sources, growing P/E cycles cause each state's threshold voltage distribution to systematically (1) shift to the right and (2) widen. The threshold voltage distribution shift occurs because the charge trapped in the tunnel oxide layer causes the charge of subsequent program operations to tunnel through the oxide more easily. This leads to higher threshold voltages than the preset values, i.e., a shift of the distribution to the right [7]. The threshold voltage distribution of each state widens as a result of process variation between cells. As more P/E cycles occur, process variation increases; that is, threshold voltage values presented by different cells differ more notably, i.e., the distribution widens.

Data errors caused by P/E cycles become more serious as the density of flash memory cells increases, because the fault-tolerant space between cell states becomes smaller. Therefore, the maximum count of tolerable P/E cycles of a cell decreases as the cell density increases. Table 9.3 lists the maximum tolerable P/E cycles for

Table 9.3 Maximum tolerable P/E cycles of flash memory cells with different storage densities

Flash memory cell	Number of stored bits	Maximum P/E cycles
SLC	1 bit	~100,000
MLC	2 bits	~10,000
TLC	3 bits	~1000
QLC	4 bits	~150

four major cells with different storage densities. The single-level cell (SLC) and QLC have the largest and smallest maximum tolerable P/E cycles, respectively. When the P/E cycles experienced by a cell exceed the given range, the storage device becomes unreliable, meaning data storage reliability becomes an issue.

9.3.1.2 Retention Time

As the retention time of data stored in flash memory cells elapses, the electric charge stored in cells leaks, causing the cell voltage to drop. This results in data errors. Charge trapped in the tunnel oxide layer due to P/E operations damages the oxide insulation. Errors caused by long retention times are positively correlated with P/E cycles. Given the same retention time, the larger the P/E cycle count, the more serious the charge leakage. Without considering other error sources, as the retention time elapses, each state's threshold voltage distribution systematically (1) shifts to the left, (2) widens, and (3) shifts more sharply at high-voltage states. Threshold voltage distribution shifts occur when the electric charge stored in cells leaks from the tunnel oxide layer, causing the cell voltage to drop, i.e., a shift of the distribution to the left. The threshold voltage distribution of each state widens as a result of process variation between cells. The process variation causes varying levels of charge leakage, which leads to varying levels of impact of retention time on cells, i.e., widening of the distribution. The high-voltage distributions shift more sharply because the higher strength of the electric field formed inside high-voltage cells causes the charge to leak more easily.

9.3.1.3 Program Interference

Program interference stems from a parasitic capacitance coupling effect between adjacent flash memory cells; that is, the voltage rise of a cell causes the voltage rise of immediately adjacent cells. Figure 9.8 shows the eight immediately adjacent cells (or neighbors) of a victim cell, including two wordline neighbors, two bitline neighbors, and four diagonal neighbors. The amount of interference that program

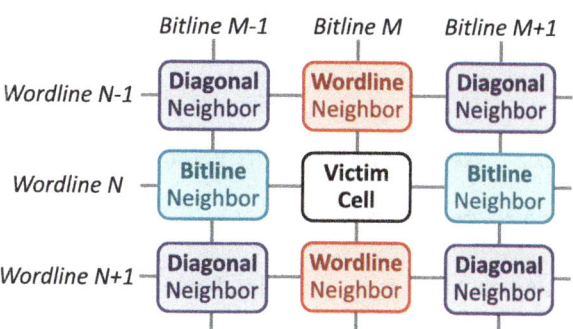

Fig. 9.8 Program interference range [6]

operations to the immediately adjacent cells can induce on the victim cell is expressed as $\Delta V_{\text{victim}} = \sum_X K_X \Delta V_X$, where ΔV_{victim} is the change in voltage of the victim cell due to cell-to-cell program interference, K_X is the coupling coefficient between neighbor X and the victim cell (the wordline neighbor has the highest coupling coefficient and accordingly the biggest program interference), and ΔV_X is the threshold voltage change of neighbor X during programming (the higher the neighbor's program state, the greater the impact on the victim cell). Program interference is the main cause of rightward shifts of threshold voltage distributions of cells; that is, the rising of the threshold voltage as more charge is applied.

9.3.1.4 Read Disturb

Read operations of a flash memory cell apply a high voltage to the immediately adjacent wordlines to mask the read behaviors of these wordline immediately adjacent cells. This causes extra charge to be applied to these cells and accordingly affects their threshold voltage states; that is, each state's threshold voltage distribution systematically (1) shifts to the right and (2) shifts more sharply at low-voltage states. The threshold voltage distribution shifts to the right because the threshold voltage of immediately adjacent cells rises due to the extra charge applied to them. The low-voltage distributions shift more sharply because the greater difference between the high voltage applied by read operations and the low voltage of cells strengthens the tunneling effect, making the cells more susceptible to additional charge in the low-voltage state.

Studies have been conducted on these error sources and the reliability issues they cause. Four optimization solutions are shadow program sequence [8], refresh [9, 10], read-retry [11, 12], and voltage optimization [13, 14]. Through shadow program sequence, the program operations of different bits of different wordline flash memory cells are mixed to minimize program interference. Through refresh, data is read into the flash controller, which corrects the bits with errors and rewrites them to a new cell, thereby solving bit errors caused by long retention times. Through read-retry, different read reference voltages are used for data read operations so that RBER is controlled within the ECC range. Through voltage optimization, the read reference voltage is tuned based on data error characteristics to lower the error rate during ECC operations and improve data read reliability. The following section provides detailed explanations of these four techniques.

9.3.2 Key Flash Reliability Technologies

9.3.2.1 Shadow Program Sequence

Using an MLC as an example, the conventional program sequence is shown in Fig. 9.9a. Data is programmed strictly according to the wordline sequence. First, the least significant bit (LSB) and most significant bit (MSB) of wordline 0 are

9.3 Flash Media Reliability

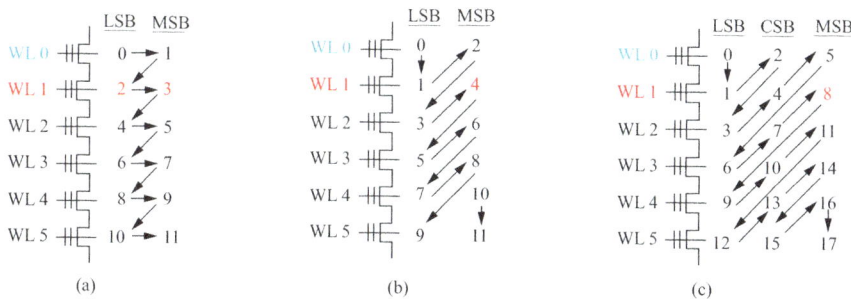

Fig. 9.9 Shadow program sequence [8]. (**a**) Conventional MLC program sequence. (**b**) MLC shadow program sequence. (**c**) TLC shadow program sequence

programmed in sequence, followed by those of wordline 1, and so on. This causes the flash memory cells of wordline 0 to be interfered with twice by data programming of wordline 1, i.e., LSB and MSB programming of wordline 1, as shown in numbers 2 and 3 in Fig. 9.9a. In order to minimize this program interference, the shadow program sequence allows the programs to be mixed, as shown in Fig. 9.9b. Sequential programming is first done for the LSBs of wordline 0 and wordline 1 and then for the MSBs of wordline 0 and wordline 1. The LSB and MSB programming of wordline 0 is still affected by the programming of wordline 1, but the MSB programming of wordline 0 can correct interference with its LSB programming, which means less interference with wordline 0. With the shadow program sequence, only the MSB programming is affected by the programming of the next wordline, as shown in number 4 in Fig. 9.9b. This means improved reliability in the data programming process and reduced probability of data errors caused by program interference. The shadow program sequence can also work with TLCs, as shown in Fig. 9.9c.

9.3.2.2 Refresh

Flash media data errors occur when the quantity of electrons stored in flash memory cells exceeds or falls below the fixed threshold voltage beyond the ECC correction capability. With the refresh mechanism, erroneous data or predicted erroneous data is read into the flash controller, which corrects data errors within the ECC correction capability and writes the data back to the flash media. For SSDs, two refresh mechanisms are available: out-of-place refresh [9, 10, 15] and in-place refresh [16]. With out-of-place refresh, data is first read into the flash controller, which corrects data errors and then rewrites the data to a new flash page. With in-place refresh, after data errors have been corrected, the data is refreshed and rewritten in the place where it was originally stored based on the error bit information. Out-of-place refresh can be actively or periodically triggered or triggered at an idle time. In active mode, flash read latency is observed to determine whether data stored in a flash page has a high error rate tendency. If it does, refresh is triggered to correct the data and rewrite it to a new flash page. In periodic mode, a trigger threshold is set so that refresh is performed in a fixed time period. In idle mode, out-of-place refresh is performed whenever an SSD is idle. Compared with periodic and active refresh, idle mode has a smaller impact on host data read and write due to extra write operations caused by out-of-place refresh. It therefore reduces performance interference. Unlike out-of-place refresh, in-place refresh does not entail extra data writes, which significantly reduces write amplification. However, in-place refresh can only handle data reliability issues due to electron leakage, which is detected by whether a flash memory cell's threshold voltage shifts to the left. If electron leakage is detected, a verifying voltage is applied to finely rewrite data to the affected cell to restore it to the preset threshold voltage stage. Because of this limitation, in-place refresh cannot be used to handle program interference and read disturb. This is where hybrid refresh [9, 16] is useful; that is, in-place refresh or out-of-place refresh is performed based on the threshold voltage shift of flash memory cells.

9.3.2.3 Read-Retry

In flash reads, reference voltages are applied to flash memory cells to determine their threshold voltage states. However, interference from different error sources will cause the threshold voltage distributions to shift to the left or right. As a result, the threshold voltage states cannot be correctly determined based on the preset reference voltages. With read-retry, different reference voltages are used for data reads to assist ECC correction. Specifically, the data content of flash memory cells is read by using a preset reference voltage and then fed back to the flash controller for ECC correction. If the error is corrected, the read operation ends. If the error persists, a different reference voltage is applied to trigger another data read and ECC correction. This process continues until the data error is corrected or the preset upper limit of error correction times is reached. Multiple read-retries increase the latency of single-page reads, which hinders system responsiveness. Research suggests that the rangeability of reference voltages should be dynamically tuned based on the error state of the threshold voltage of flash memory cells. For example, for data with a high error rate, the rangeability of reference voltages can be increased to achieve fast exertion of the ECC correction capability, thereby reducing read-retries [12].

9.3.2.4 Voltage Optimization

Read-retry gets the optimal read reference voltage after applying the reference voltage multiple times, but the resulting large read overhead affects the device's overall performance. Voltage optimization, on the other hand, records the optimal reference voltage to reduce the number of reads in the read-retry process. Since the threshold voltage state of flash memory cells changes over time and depending on the memory access behavior, obtaining and recording the optimal reference voltage are also processes that constantly change. The two main methods of voltage optimization are sampling [17, 18] and modeling [19]. In the sampling method, each flash memory block is periodically sampled to obtain the optimal reference voltage for the flash memory cells within that block. This method works because the error rate of flash media is stable for a short period of time (for example, 1 day), and cells within the same block have similar error rates. This process involves frequent data sampling reads, which increases data storage overheads. The alternative is the modeling method, which predicts the threshold voltage distributions by recording the memory access and reliability characteristics of flash memory devices and then applies the optimal reference voltage.

There are other technologies that can also optimize the reliability of flash media, such as hybrid ECC [20, 21] and hybrid flash media [22, 23]. Hybrid ECC identifies the read/write characteristics of different data and then uses an ECC with a high error correction capability for data that has a high error rate and an ECC with fast decoding for data that has a low error rate. This improves data read/write performance while ensuring data reliability. Hybrid flash media technology stores data that requires high reliability in SLCs or other flash media with a high fault tolerance and data with lower reliability requirements in TLCs or QLCs to minimize the extra overheads associated with ensuring reliability as the data error rate increases.

9.4 Erasure Coding

We can use multi-copy or erasure coding technologies to store data in a way that ensures system-level reliability. These technologies will have different overheads arising from storage, computing, and restoration tasks.

9.4.1 Multi-copy

Copy is the simplest redundancy technology that can be used to ensure reliable data storage. In principle, the system makes multiple copies of every piece of data and stores these copies in different places (such as in storage devices, nodes, and even cabinets) so that the system can retrieve another copy from another place if the original data is lost. With r-copy technology ($r > 1$), the storage system can tolerate the failure of any $r - 1$ copies of the data, but it results in storage overheads of r times the stored data. Three-copy technology is a popular choice within the industry. For example, Google File System (GFS) [24] uses the three-copy solution to handle storage reliability and access load balancing.

9.4.2 Erasure Coding

While multi-copy can deliver a required level of fault tolerance, the storage overhead is prohibitive. Erasure coding [25–28], on the other hand, offers high fault tolerance with low storage overheads. The technology has two simple parameters:

k, which represents the storage efficiency, and m, which represents the fault tolerance. It inter-converts raw data and redundant data through encoding (i.e., calculating redundant data) and decoding (i.e., repairing raw data). During encoding, k fixed-size data blocks are used as inputs so that m redundant blocks (i.e., parity blocks) of the same size can be obtained, thereby obtaining a stripe made of $k + m$ blocks. In the case of a data failure, any k blocks from the same stripe can be used for decoding to repair the original k blocks. This means that erasure coding can tolerate the failure of any m blocks in any stripe. Let us look at the most representative Reed-Solomon (RS) code [25]. RS code can flexibly set the k and m parameters based on the system's storage capacity and fault tolerance requirements. Because of this advantage, it is widely used in open-source storage systems (e.g., Hadoop 3.0 [29] and Ceph [30]) and commercial storage systems (e.g., Facebook f4 [31] and Windows Azure Storage [32]). Figure 9.10 shows how RS code is deployed in a data center where $k = 2$ and $m = 2$. Each stripe has two data blocks ($k = 2$) and two parity blocks ($m = 2$), meaning that the storage system can tolerate the failure of any two blocks. By storing each stripe in $k + m$ nodes, the storage system can tolerate the failure of any m nodes. This means, however, that in order for the data center to be able to tolerate the failure of a single cluster, no more than m blocks from each stripe can be stored in the same cluster.

The encoding operation used in erasure coding can be mathematically expressed as the matrix multiplication operation of the finite field. Assuming that there are k data blocks, erasure coding constructs a data vector $\boldsymbol{D}$ of length k, where $\boldsymbol{D} = (D_0, D_1, ..., D_{k-1})$. This data vector is multiplied by a code matrix $\boldsymbol{G}_{(k+m) \times k}$ to obtain $k + m$ coded chunks, which can be written as $\boldsymbol{C} = (C_0, C_1, ..., C_{k+m-1})$. The existing erasure codes fall into one of the following two categories based on their encoding effect:

Systematic erasure code: A systematic erasure code is one in which the first k chunks of the $k + m$ coded chunks remain consistent with the original data chunks even

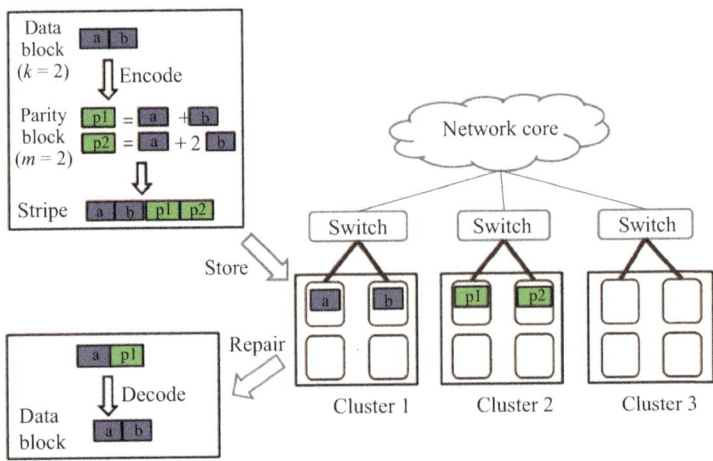

Fig. 9.10 Erasure coding being deployed in a data center

9.4 Erasure Coding

after encoding. The original data chunks can be accessed directly after the encoding operation without having to be decoded. In this sense, in systematic erasure codes, the first $k \times k$ submatrix of the code matrix $G_{(k+m) \times k}$ is an identity matrix. Figure 9.11 illustrates the encoding operation for a systematic erasure code.

Non-systematic erasure code: The $k + m$ coded chunks generated by a non-systematic erasure code are completely different from the original data chunks. For this reason, the data chunks need to be decoded before they can be accessed. This consumes certain storage I/O and network I/O resources.

Although erasure coding introduces only lightweight computing overheads, the repair operation easily consumes a lot of network bandwidth and storage I/O resources. For example, to repair a block, the erasure code needs to retrieve k available blocks, which means that both the network bandwidth overhead and the storage I/O overhead increase by a factor of k. According to a study on one of Facebook's warehouse clusters, the median number of block failures per day in the cluster was 95,500. This meant that a median of 180 TB of data was being transferred across racks per day [33]. To reduce such network and storage I/O overheads, several new erasure code technologies have been proposed since 2010. Two of them, locally repairable code (LRC) and regenerating code, have become mainstream approaches. The next section will describe their respective construction and repair strategies.

Unlike RS code, LRC [32] has three parameters: k, l, and m. In principle, k data blocks are first encoded to generate m global parity blocks and then further divided into l groups so that each group has k/l data blocks. A local parity block is maintained for each group. If one data block fails, LRC can retrieve the remaining k/l blocks from the group corresponding to the failed data block and perform the decoding operation to recover the invalid data. While RS code needs to retrieve k data blocks for repair, LRC can reduce network traffic to $1/l$ of the original repair traffic. LRC is used in Microsoft Azure storage systems. Figure 9.12 illustrates how a (6, 2, 2) LRC divides six data blocks into two groups (each with three data blocks) and maintains a parity block for each group (e.g., by generating a local parity block p_x

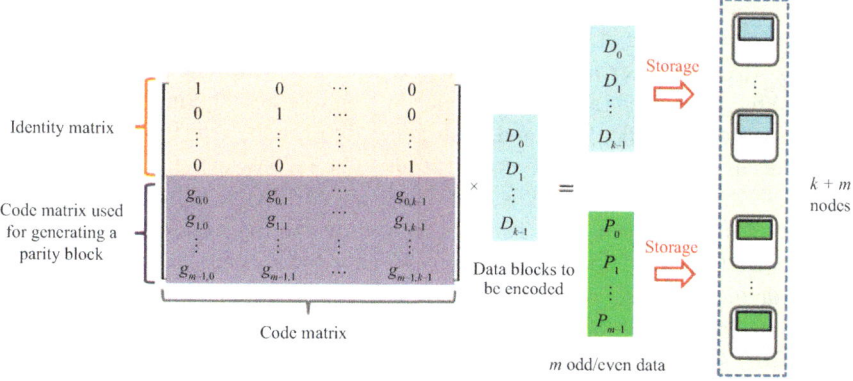

Fig. 9.11 Systematic erasure code—encoding operation

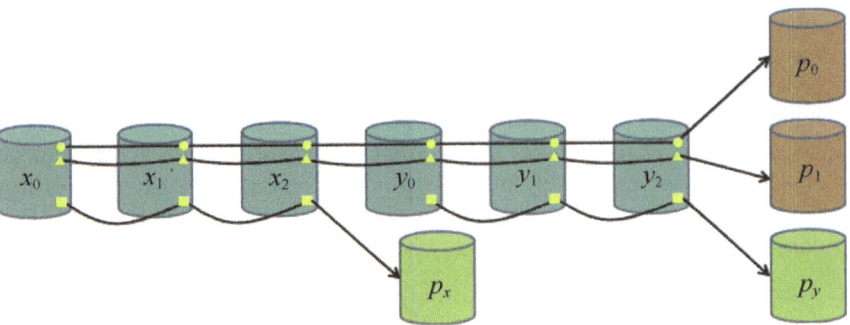

Fig. 9.12 (6, 2, 2) LRC [32]

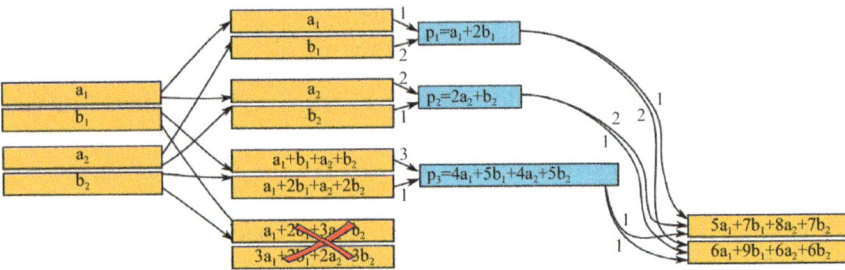

Fig. 9.13 MSR code [34]

for group $\{x_0, x_1, x_2\}$). When one data block (e.g., x_0) fails, data can be restored by retrieving the three blocks (e.g., $\{x_1, x_2, p_x\}$) in the corresponding group.

Regenerating code [34] reduces the network traffic used for the repair process by having data computed via the storage node before it is transmitted and by retrieving data from more nodes (than k nodes with RS code). Assume that α data symbols are stored in each storage node and one storage node fails. During the repair process, regenerating code can contact any d available storage nodes and download β data symbols from each contacted storage node for data reconstruction. To reconstruct each data symbol, data of size $d \cdot \beta$ needs to be downloaded. Regenerating code can be categorized as either minimum storage regeneration (MSR) code or minimum bandwidth regeneration (MBR) code depending on whether it optimizes the storage capacity or repair traffic. MSR and MBR codes differ in the selection of $\{\alpha, \beta,$ and $d\}$. Let us assume that a file of size M needs to be reconstructed. MSR code will choose $\alpha = M/k$ and $\beta = M/(k \cdot (d - k + 1))$, whereas MBR code will choose $\beta = 2M/(k \cdot (2d - k + 1))$ and $\alpha = d \cdot \beta$. This is illustrated in Fig. 9.13: Assuming that $k = 2$ and $m = 2$, the size of each data block is $M/2$ for MSR code. Each data block is further divided into two data fragments, a_1 and b_1. When a parity block fails, the MSR code can contact the remaining three available storage nodes for repair (i.e., $d = 3$). Each storage node transmits a data fragment of size $M/4$ to aid the repair (i.e., $\beta = M/4$). The network traffic required for repairing a block is $3M/4$.

9.4 Erasure Coding

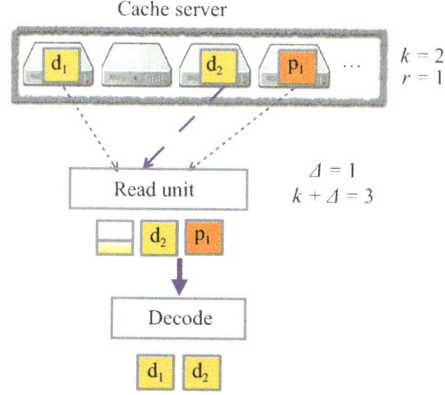

Fig. 9.14 EC-Cache for fast object access using erasure coding [35]. r represents the number of parity blocks

Compared with multi-copy, erasure coding consumes much more network traffic and is therefore generally used to back up cold data for long-term storage. Windows Azure Storage and Facebook f4 are the main systems for this. They usually use blocks sized between 64 MB and 1 GB. Recently, erasure coding has also been used in memory object storage systems to reduce the tail latency of system access. For example, EC-Cache [35] divides an object into k data blocks and creates m additional parity blocks. When an object is accessed, EC-Cache retrieves $k + \Delta$ ($\Delta \geq 1$) of the available blocks from the same stripe. After receiving the first k blocks, EC-Cache immediately discards the next Δ blocks that do not arrive and performs a decoding operation to obtain the object that is to be accessed. This is how it reduces the tail latency. Figure 9.14 shows an example of EC-Cache where $k = 2$ and $\Delta = 1$. When an object is accessed, EC-Cache reads three blocks of the stripe in which the object is located and uses the first two blocks to arrive for data reconstruction. This process enables it to promptly obtain the object that is to be accessed and respond to the request.

9.4.3 Typical Erasure Codes

Aside from RS code, RAID 6 is another erasure coding technology that can tolerate the failure of any two disks [27]. With RAID 6, two independent parity information algorithms generate parity data which is stored on two different parity disks or on all member disks. If two disks fail at the same time, RAID 6 resolves binary equations to reconstruct the data on the two failed disks. To improve encoding and decoding performance, RAID 6 mainly uses exclusive-OR (XOR) and cyclic shift (CS) operations. RAID 6 has the following characteristics:

- RAID 6 is based entirely on the XOR operation, which is easy to realize and offers fast operation speed.
- RAID 6 does not need the parameter distribution matrix to aid the operation.

- RAID 6 is not as general-purpose as RS code, because it is specifically developed to meet the fault tolerance requirements.
- RAID 6 can achieve optimal or near-optimal update efficiency.
- With RAID 6, horizontal code can be used for any number of storage devices (e.g., by adding virtual devices), while vertical code features comparatively poorer scalability because the stripe size is limited.

9.4.3.1 EVENODD Code

EVENODD code is the earliest array code technology used in RAID 6. Proposed by Blaum et al. in 1995, EVENODD code is a horizontal code that can tolerate the failure of up to two strips and is also a standard maximum distance separable (MDS) code [36]. The size of an EVENODD stripe is a two-dimensional array of $(p - 1) \times (p + 2)$, where p is a prime number.

EVENODD code stores data information in $(p - 1) \times p$ arrays and parity information in the last two columns. The parity information in two columns is obtained by XORing the data information in the same row and the information on the diagonal of a given slope. The formulas for generating the parity information are as follows:

$$a_{i,p} = \bigoplus_{p-1}^{t=0} a_{i,t} \qquad (9.3)$$

$$S = \bigoplus_{p-1}^{t=1} a_{p-1-t,t} \qquad (9.4)$$

$$a_{i,p+1} = S \oplus \left(\bigoplus_{p-1}^{t=0} a_{(i-t)\%p,t} \right) \qquad (9.5)$$

where $a_{i,j}$ represents data information in row i column j, i is ranged within $0 \leq i \leq p - 2$, $a \% b$ represents $a \bmod b$, and S is a parity factor.

EVENODD coding process: Based on Fig. 9.15, assuming $p = 5$, formula (9.3) shows that the parity information for column 6 is the sum of the XOR value of each row of data information, formula (9.4) shows how to generate a parity factor, and formula (9.5) shows how to generate the parity information for column 7. Geometrically, S is obtained by XORing the data information with a slope of 1 starting from column $p - 1$. The parity information for column 5 is the sum of the XOR value of p pieces of data information in the row, while the parity information for column 6 is the sum of the XOR value of the data information with a slope of 1 or is generated by XORing S.

If data becomes invalid in a column, EVENODD code repairs the invalid data by (1) directly coding if it is a parity column or (2) horizontally XORing based on the horizontal parity columns and surviving data columns if it is a data column. When data becomes invalid in two strips, data becomes invalid in (1) all parity columns, (2) one data column and one parity column, or (3) all data columns. The following

9.4 Erasure Coding

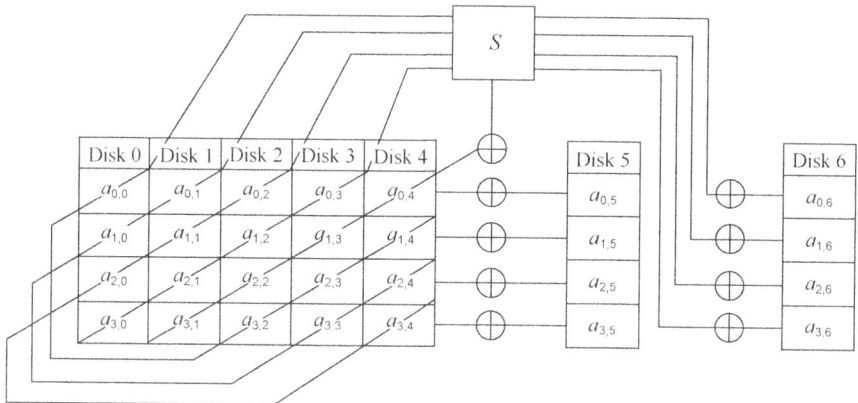

Fig. 9.15 EVENODD coding process

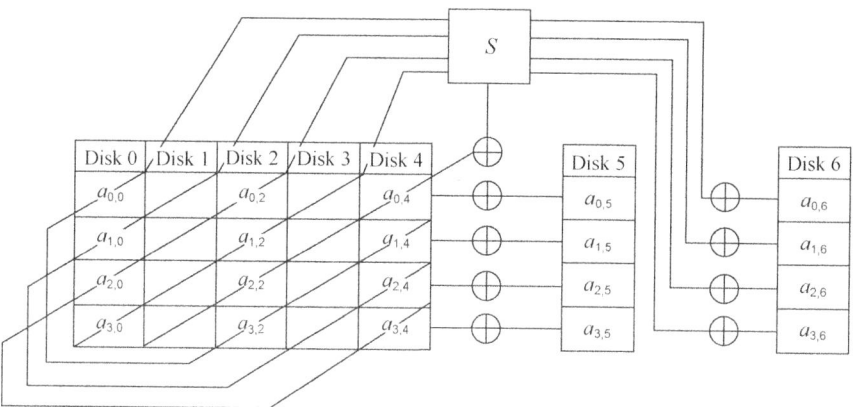

Fig. 9.16 EVENODD decoding process

information will describe the decoding process in situations 1 and 2, for they are comparatively simpler.

EVENODD decoding process: Based on Fig. 9.16, assuming that data becomes invalid in columns 1 and 3, the common factor is calculated first:

$$S = \bigoplus_{p-2}^{i=0} \left(a_{i,5} + a_{i,6} \right) \quad (9.6)$$

Then, S is XORed with $a_{2,6}$ and $a_{0,6}$ in the diagonal parity columns respectively to obtain

$$a_{1,1} = a_{2,6} \oplus S \oplus a_{0,2} \oplus a_{2,0} \oplus a_{3,4} \quad (9.7)$$

$$a_{2,3} = a_{0,6} \oplus S \oplus a_{0,0} \oplus a_{3,2} \oplus a_{1,4} \quad (9.8)$$

At this point, each piece of parity information in rows 0 and 3 of the horizontal parity columns contains two pieces of unknown information, and each piece of parity information in rows 1 and 2 contains one piece of unknown information. That said, $a_{1,3}$ and $a_{2,1}$ can be calculated by using $a_{1,5}$ and $a_{2,5}$ in the horizontal parity columns. The calculation formulas are as follows:

$$a_{1,3} = a_{1,5} \oplus a_{1,0} \oplus a_{1,1} \oplus a_{1,2} \oplus a_{1,4} \tag{9.9}$$

$$a_{2,1} = a_{2,5} \oplus a_{2,0} \oplus a_{2,2} \oplus a_{2,3} \oplus a_{2,4} \tag{9.10}$$

The diagonal information $a_{3,1}$ is calculated by using the common factor:

$$a_{3,1} = S \oplus a_{0,4} \oplus a_{1,3} \oplus a_{2,2} \tag{9.11}$$

Then, $a_{0,3} = a_{3,6} \oplus S \oplus a_{3,0} \oplus a_{2,1} \oplus a_{1,2}$ is obtained. $a_{0,1}$ can be solved by $a_{0,5}$, $a_{0,0}$, $a_{0,2}$, $a_{0,3}$, and $a_{0,4}$. $a_{3,3}$ can be solved by using $a_{3,5}$, $a_{3,0}$, $a_{3,1}$, $a_{3,2}$, and $a_{3,4}$. By analogy, all lost data information in columns 1 and 3 can be solved.

9.4.3.2 RDP Code

Proposed by NetApp, row-diagonal parity (RDP) code is widely used in real storage systems [37]. It is also a RAID 6 code that can tolerate the failure of two disks. Assuming p is a prime number greater than 2, RDP code can be represented by a two-dimensional array of $(p - 1) \times (p + 1)$, with the first $(p - 1)$ columns being data columns and the other two columns being parity columns. Given the same p value, the RDP code has one fewer data column than the EVENODD code.

RDP coding process: Use $a_{i,j}$ to represent data information j in row i, where $i \in [0, p - 2]$ and $j \in [0, p]$. For horizontal parity columns, RDP code uses the same coding method as EVENODD code. In calculating the diagonal parity columns, the common factor does not need to be calculated, but horizontal parity columns participate in the calculation as data columns. The coding formulas are as follows:

$$a_{i,p-1} = \bigoplus_{p-2}^{j=0} a_{i,j} \tag{9.12}$$

$$a_{i,p} = \bigoplus_{p-1}^{j=0} a_{(i-j)\%p,j} \tag{9.13}$$

Formula (9.12) can calculate horizontal parity columns, and formula (9.13) can calculate diagonal parity columns. Figure 9.17 shows the coding process on the assumption of $p = 5$.

The generation of parity information using RDP code is interrelated, meaning the update efficiency is not optimal. The diagonal data information can be updated only after the horizontal parity information is updated.

9.4 Erasure Coding

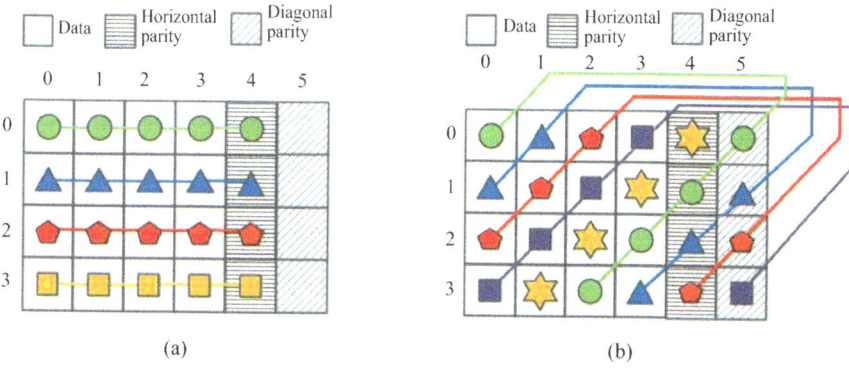

Fig. 9.17 RDP coding process [37]. (**a**) Horizontal parity information obtained by performing an XOR operation on the data information of each row. (**b**) Diagnol parity information obtained by performing an operation on the diagnol data information

RDP decoding process: When data becomes invalid in a column, RDP code repairs the invalid data by (1) recoding if it is a parity column or (2) horizontally XORing based on the horizontal parity columns and surviving data columns if it is a data column. When data becomes invalid in two strips, data becomes invalid in (1) all parity strips, (2) one data strip and one parity strip, or (3) all data strips. In situation 1, the RDP code repairs the invalid data by coding. In situations 2 and 3, the RDP code repairs the invalid data through the interleaved participation of two types of parity information. As shown in Fig. 9.18a, assume that data is invalid in columns 0 and 1. As shown in Fig. 9.18b, the first step is to restore information in row 0 column 0 based on the diagonal parity block.

$$a_{0,0} = a_{0,5} \oplus a_{1,4} \oplus a_{2,3} \oplus a_{3,2} \qquad (9.14)$$

Then, as shown in Fig. 9.18c, restore information in row 0 column 1 based on the horizontal parity block.

$$a_{0,1} = a_{0,0} \oplus a_{0,2} \oplus a_{0,3} \oplus a_{0,4} \qquad (9.15)$$

Then, as shown in Fig. 9.18d, restore information in row 1 column 0 based on the diagonal parity block.

$$a_{1,0} = a_{1,5} \oplus a_{2,4} \oplus a_{3,3} \oplus a_{0,1} \qquad (9.16)$$

Then, as shown in Fig. 9.18e, restore information in row 1 column 1 based on the horizontal parity block.

$$a_{1,1} = a_{1,0} \oplus a_{1,2} \oplus a_{1,3} \oplus a_{1,4} \qquad (9.17)$$

As shown in Fig. 9.18f, continue the restoration process this way till the data on the two failed disks is restored.

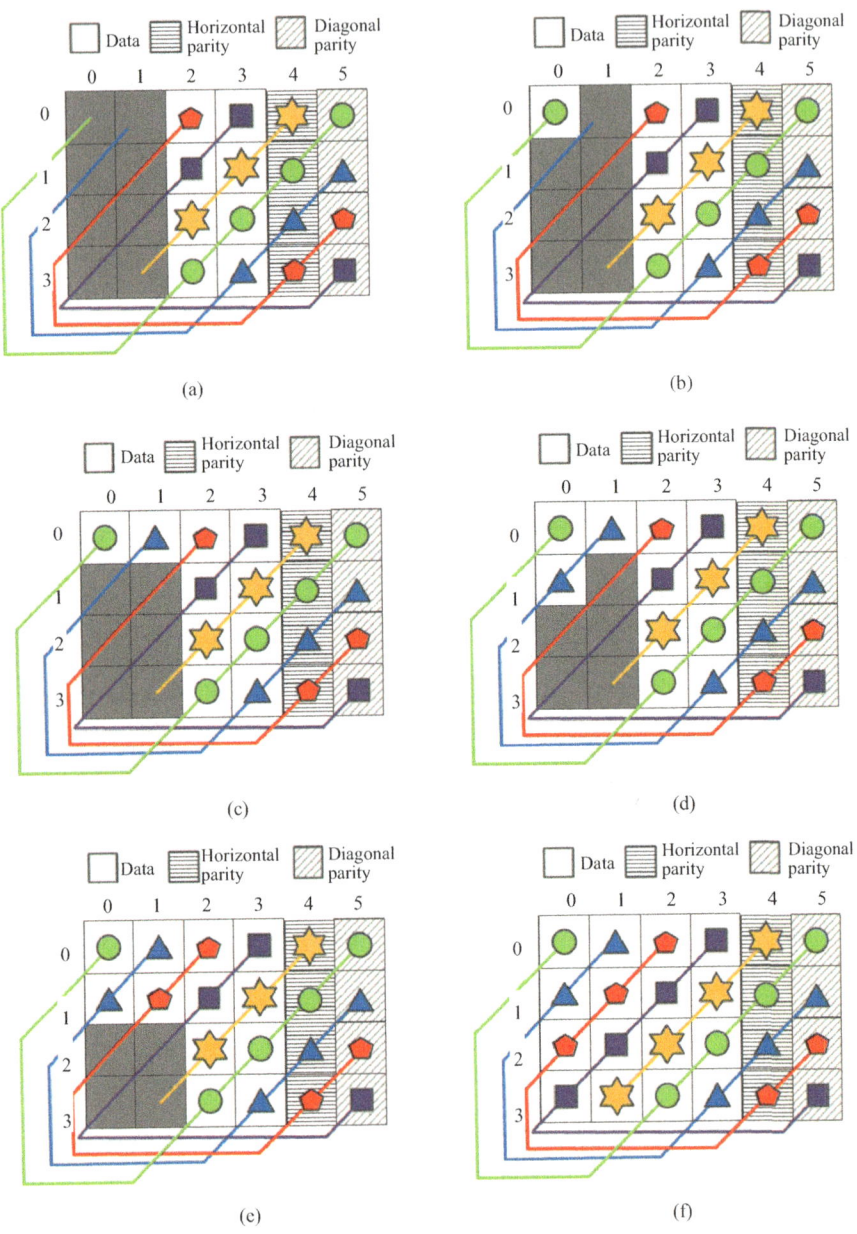

Fig. 9.18 RDP code repair process (columns 0 and 1 failed, $p = 5$). (**a**) Columns 0 and 1 failed. (**b**) Repairing $a_{0,0}$. (**c**) Repairing $a_{0,1}$. (**d**) Repairing $a_{1,0}$. (**e**) Repairing $a_{1,1}$. (**f**) Repairing residual data

9.4.3.3 X-Code

Proposed by Xu and Bruck in 1999, X-code is a typical RAID 6 vertical code [38] that tolerates the failure of two disks. Similar to EVENODD and RDP codes, X-code is constructed in a two-dimensional array of $p \times p$ (where p is a prime number greater than 2). Different from RDP code, X-code stores parity information in rows, with the first $(p-2)$ rows being data rows and the other two rows being parity rows.

X-code coding process: Different from horizontal codes, X-code uses two types of parity rows: diagonal parity row and back-diagonal parity row. Use $a_{i,j}$ to represent data block j in row i, with $i \in [0, p-1]$ and $j \in [0, p-1]$. The formulas for generating the parity row information are as follows:

$$a_{p-2,i} = \bigoplus_{p-3}^{k=0} a_{k,(i-k-2)\%p} \qquad (9.18)$$

$$a_{p-1,i} = \bigoplus_{p-3}^{k=0} a_{k,(i+k+2)\%p} \qquad (9.19)$$

Formulas (9.18) and (9.19) show how to generate information for the first and second parity rows, respectively. Geometrically, the two parity rows are the sum of the XOR values of the data blocks along the diagonals with a slope of 1 and -1, respectively. Figure 9.19 shows the X-code coding process on the assumption of $p = 5$.

X-code decoding process: If data becomes invalid in two columns, X-code iteratively repairs the invalid data by using diagonal and back-diagonal parity. Figure 9.20 explains the data repair process assuming that data in columns 0 and 1 becomes invalid. First, as shown in Fig. 9.20b, $a_{0,1}$ and $a_{4,0}$ are repaired based on the back-diagonal parity block.

$$a_{0,1} = a_{4,4} \oplus a_{2,3} \oplus a_{1,2} \qquad (9.20)$$

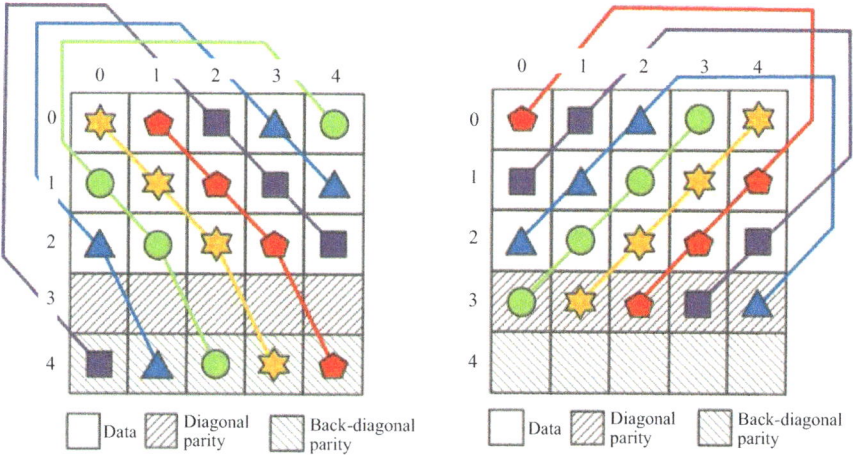

Fig. 9.19 X-code coding process [38] (**a**) Diagonal parity row. (**b**) Back-diagonal parity row

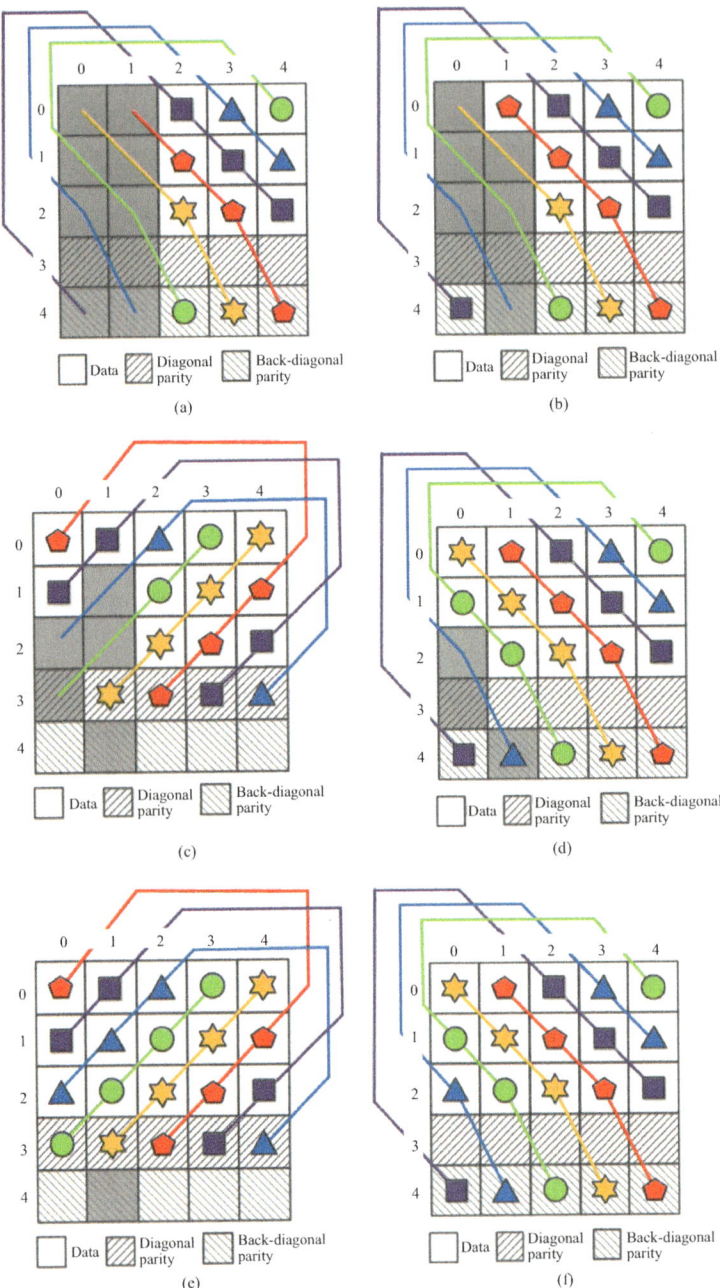

Fig. 9.20 X-code repair process (columns 0 and 1 failed, $p=5$). (**a**) Columns 0 and 1 failed. (**b**) Repairing $a_{0,1}$ and $a_{4,0}$. (**c**) Repairing $a_{0,0}$, $a_{1,0}$, and $a_{3,1}$. (**d**) Repairing $a_{1,1}$ and $a_{2,1}$. (**e**) Repairing $a_{2,0}$ and $a_{3,0}$. (**f**) Repairing $a_{4,1}$

9.4 Erasure Coding

$$a_{4,0} = a_{2,4} \oplus a_{1,3} \oplus a_{0,2} \tag{9.21}$$

Then, as shown in Fig. 9.20c, $a_{0,0}$, $a_{1,0}$, and $a_{3,1}$ are repaired based on the diagonal parity block.

$$a_{0,0} = a_{3,2} \oplus a_{2,3} \oplus a_{1,4} \tag{9.22}$$

$$a_{1,0} = a_{3,3} \oplus a_{2,4} \oplus a_{0,1} \tag{9.23}$$

$$a_{3,1} = a_{0,4} \oplus a_{1,3} \oplus a_{2,2} \tag{9.24}$$

Then, as shown in Fig. 9.20d, $a_{1,1}$ and $a_{2,1}$ are repaired based on the back-diagonal parity block.

$$a_{1,1} = a_{0,0} \oplus a_{4,3} \oplus a_{2,2} \tag{9.25}$$

$$a_{2,1} = a_{0,4} \oplus a_{1,0} \oplus a_{4,2} \tag{9.26}$$

Then, as shown in Fig. 9.20e, $a_{2,0}$ and $a_{3,0}$ are repaired based on the diagonal parity block.

$$a_{2,0} = a_{1,1} \oplus a_{0,2} \oplus a_{3,4} \tag{9.27}$$

$$a_{3,0} = a_{0,3} \oplus a_{1,2} \oplus a_{2,1} \tag{9.28}$$

Finally, as shown in Fig. 9.20f, $a_{4,1}$ is repaired based on the back-diagonal parity block.

$$a_{4,1} = a_{0,3} \oplus a_{1,4} \oplus a_{2,0} \tag{9.29}$$

9.4.3.4 HV Code

Proposed by the Tsinghua University team in 2016, horizontal-vertical (HV) code [39] is a RAID 6 code that combines the advantages of horizontal and vertical codes in order to optimize disk writes while retaining the best coding and decoding efficiency. HV code also boasts higher restoration efficiency because its parity chain is shorter than that of other array codes. HV code is constructed in a two-dimensional array of size $(p - 1) \times (p - 1)$ (where p is a prime number greater than 2).

HV coding process: Use $a_{i,j}$ to represent data information j in row i, with $i \in [1, p - 1]$ and $j \in [1, p - 1]$. The formulas for calculating the parity information are as follows:

$$a_{i,2i_p} = \bigoplus_{p-1}^{j=1} a_{i,j} \left(j \neq 2i_p, j \neq 4i_p \right) \tag{9.30}$$

$$a_{i,4i_p} = \bigoplus_{p-1}^{j=1} a_{k,j} \left(j \neq 8i_p, j \neq 4i_p\right) \tag{9.31}$$

where

$$k = \frac{j - 4i}{2}\bigg|_p = \begin{cases} \frac{1}{2}j - 4i_p & \left(j - 4i_p = 2t\right) \\ \frac{1}{2}\left(j - 4i_p + p\right) & \left(j - 4i_p = 2t + 1\right) \end{cases} \tag{9.32}$$

Formulas (9.30) and (9.31) show how to calculate the horizontal and vertical parity information, respectively. Figure 9.21 shows the HV coding process on the assumption of $p = 7$.

HV decoding process: Similar to the data restoration of other array codes, the HV code repair process for a single column of data is simple. The following will describe the repair process of two disks. As shown in Fig. 9.22, assume that columns 1 and 3 need to be repaired. HV code can construct four chains to restore data at the same time. First, four starting points of the restoration operation are determined by using horizontal and vertical parity, namely, data information $a_{2,3}$, $a_{3,3}$, $a_{5,1}$, and $a_{6,1}$. Then, for each piece of the starting data information, simultaneous repair is performed by alternately switching the parity mode. In Fig. 9.22, $\{a_{5,1}, a_{5,3}\}$ and $\{a_{3,3}, a_{3,1}, a_{4,3}, a_{4,1}\}$ are two of the restoration chains.

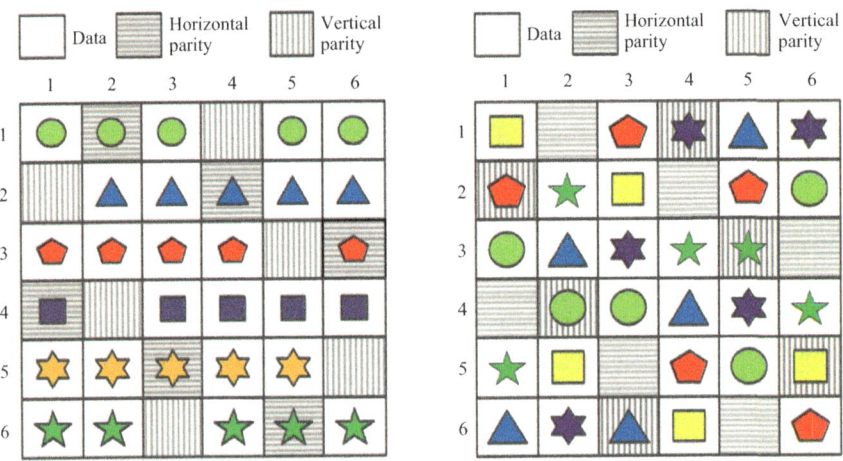

Fig. 9.21 HV coding process ($p = 7$) [39] (**a**) Calculation of horizontal parity information: by XORing data information in the same row, e.g. $a_{1,2} = a_{1,1} \oplus a_{1,3} \oplus a_{1,5} \oplus a_{1,6}$. (**b**) Calculation of vertical parity information: by XORing data information that meets $<2i' + j_p> = j'$. e.g., $a_{1,4} = a_{6,2} \oplus a_{3,3} \oplus a_{4,5} \oplus a_{1,6}$

Fig. 9.22 HV code repair process (columns 1 and 3 failed, $p = 7$) [39]

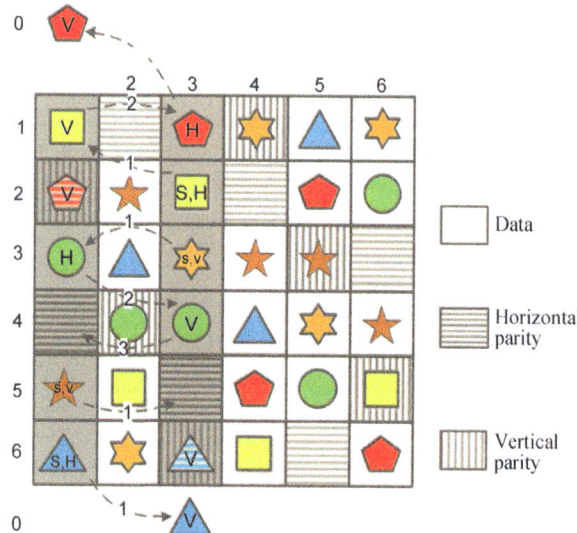

Table 9.4 Parameters comparison of five common array codes

Code	Error tolerance	Update efficiency	Stripe length	Load balancing	Partial write efficiency in a stripe
EVENODD code	2	>3	$p + 2$	Unbalanced	Small number of I/Os
RDP code	2	>3	$p + 1$	Unbalanced	Small number of I/Os
X-code	2	3	p	Balanced	Large number of I/Os
H-code	2	3	$p + 1$	Unbalanced	Small number of I/Os
HV code	2	3	$p - 1$	Balanced	Small number of I/Os

9.4.3.5 Summary and Comparison

As one class of linear erasure code, array code adopts a data layout scheme based on a two-dimensional array. Compared with RS code, array code entails lower computing overhead because it only needs to perform the XOR operation for most processes, and it is easy to implement by software and hardware. Compared with vertical codes, horizontal codes (such as EVENODD and RDP) cannot achieve optimal update efficiency. Horizontal array codes feature better scalability because parity information is stored separately in nodes, in contrast to vertical codes which store parity information evenly among nodes to better balance load.

Still, array code is not without its drawbacks. First, it has poor flexibility, as it cannot be designed for any k and m nodes. The row-column size of the array code usually needs to be a prime, limiting the amount of parity information. Array code is also comparatively poorer in terms of disk fault tolerance.

Table 9.4 compares RAID 6 codes from the perspectives of fault tolerance, update efficiency, stripe length, and load balancing.

9.4.4 Development Trend of Erasure Code Technology

In recent years, as data volumes continue to grow, research into erasure codes for storage systems has expanded from conventional enhanced storage capabilities to other priorities like repair efficiency optimization, repair algorithm construction, active policy selection, redundancy degree conversion, and elastic parameter expansion, leading to many meaningful findings. The following information summarizes current research progress regarding erasure codes.

9.4.4.1 Construction of Erasure Codes with Optimized Repair Efficiency

Conventional Reed-Solomon codes can reduce storage overhead but easily generate a lot of repair traffic. In light of this, LRC [40, 41] and Rotated-RS [42] codes reduce the repair traffic with the storage overhead increased by a small amount. Regenerating codes [34] reduce the repair traffic by making surviving nodes send a linear combination of data or making more surviving nodes participate in the repair (e.g., product-matrix MSR code [43]). Butterfly [44] and Clay [45] codes can directly send locally stored data for repair without data calculation, ensuring consistency between storage I/O and network I/O. Hitchhiker [46] code builds block dependencies across stripes in code construction to reduce repair traffic. Butterfly codes are a typical exact-repair MSR code that can tolerate the failure of two disks or nodes (that is, $n - k = 2$). With the parameter composition of (n, k), Butterfly codes are constructed by a data block matrix D_k which includes $k \times 2^{k-1}$ data blocks and two parity block matrices P_k and Q_k each of which include 2^{k-1} parity blocks. The data block matrix D_k may be represented as a matrix including four elements:

$$D_k = \begin{bmatrix} a_{k-1} & D_{k-1}^1 \\ b_{k-1} & D_{k-1}^2 \end{bmatrix} \quad (9.33)$$

where D_{k-1}^1 and D_{k-1}^2 are two matrices of $(k - 1) \times 2^{k-2}$, elements in the matrices are data blocks, and a_{k-1} and b_{k-1} are two vectors with 2^{k-2} data blocks. Butterfly codes use a recursive design and only require XOR operations for coding/decoding. That is, a (n, k) Butterfly code can be recursively constructed by a $(n - 1, k - 1)$ Butterfly code. Figure 9.23 shows the Butterfly code construction.

9.4.4.2 Erasure Code Repair Algorithm

In addition to building erasure codes with optimized repair efficiency, fast repair algorithms [28, 47–54] are being developed for incumbent erasure codes. Degraded-first scheduling [47] starts a degraded read task with a higher priority so as to leverage the unused network resources for data repair. PUSH [48] transmits requested blocks in a pipelined manner to alleviate network congestion in repair.

9.4 Erasure Coding

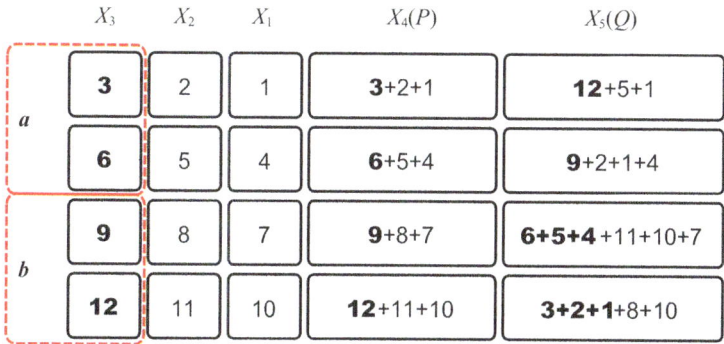

Fig. 9.23 Butterfly code construction $(n, k) = (5, 3)$

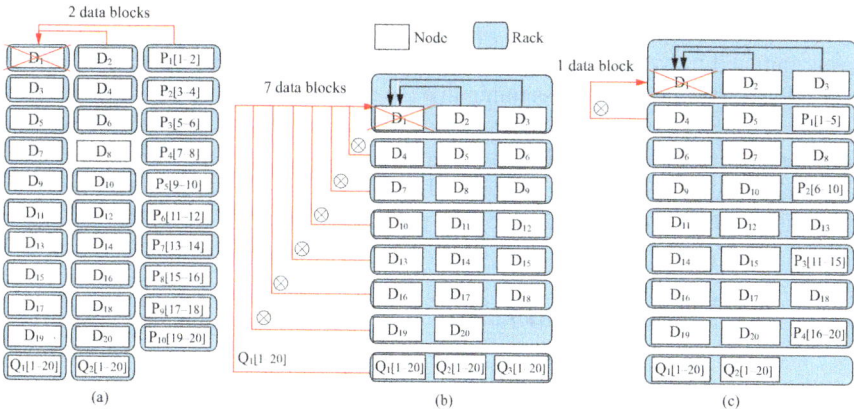

Fig. 9.24 ECWide combined locality data repair [53]. (**a**) Parity locality. (**b**) Topology locality. (**c**) Combined locality

Partial-parallel-repair (PPR) [49] divides the repair scheme into smaller phases to execute in parallel. ECPipe [50] divides a large data block into smaller fragments and pipelines these fragments so that in a homogeneous network environment, the repair time complexity is close to $O(1)$. Considering the bandwidth diversity of tiered data centers, cross-rack-aware recovery (CAR) [51] and ClusterSR [52] reduce and balance repair traffic between clusters. ECWide [53] provides the idea of combined locality, a mechanism that systematically addresses the wide-stripe repair problem through a combination of parity locality and topology locality. RepairBoost [54] is a repair task scheduling framework that accelerates data repair by employing a directed acyclic graph and the topological sorting of repair tasks.

Figure 9.24 visualizes the principle of ECWide. Essentially, LRCs are deployed in tiered data centers to reduce the data repair overhead of wide stripes (those with a large k value). As shown in Fig. 9.24a, with parity locality, although data can be quickly repaired in a rack, an additional local parity block must be stored in each rack. As shown in Fig. 9.24b, with topology locality, although it can reduce redundancy overhead, huge repair traffic is generated because data blocks must be requested from each rack during data block repair. As shown in Fig. 9.24c, ECWide combines parity locality and topology locality, striking a balance between the two locality schemes.

9.4.4.3 Proactive Repair

Incumbent ML-based techniques used for failure prediction typically employ S.M.A.R.T. attributes [55–59] and, given additional system events [60] and performance indicators [61], achieve high prediction accuracy (e.g., at least 95% [56, 57, 59, 60]) and low false positive rates (e.g., only 2.5% [58] and 0.2–0.4% [62]). FastPR [63] combines block migration with erasure code–based repair to accelerate active repair efficiency. Hu et al. [64] proposes that degraded reads can be proactively initiated to bypass hotspots, thereby reducing the tail latency of read operations. As shown in Fig. 9.25, assuming that the popularity of data block a in node S1 is high, degraded reads can be actively initiated when the system receives a request for data block a. In other words, the remaining idle blocks are read from other relatively idle nodes, such as S2, S3, and L1, to repair data block a, so that node S1 becomes less blocked.

9.4.4.4 Redundancy Transitioning

Some researchers are also studying transforming erasure code's parameters to adjust a system's overall storage overheads, without changing its configurations, according to access frequencies or reliability requirements. Hadoop adaptively-coded distributed file system (HACFS) [65] and another similar research [66] propose using fast coding to optimize recovery performance and compact coding to reduce storage overhead and switching between the two codes to dynamically adapt to changing workloads. Popularity-aware redundancy scheme (PaRS) [67] dynamically relocates a block in a redundancy group based on workload popularity to achieve a tradeoff between storage overhead and access parallelism. Heterogeneity-aware

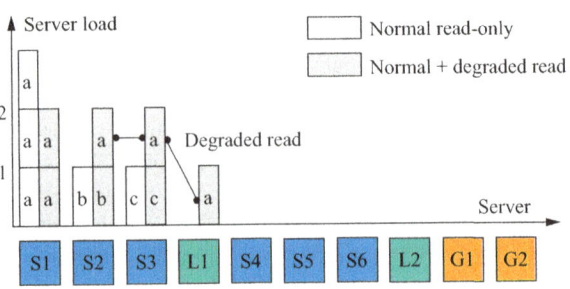

Fig. 9.25 Proactive degraded read [64]

9.4 Erasure Coding

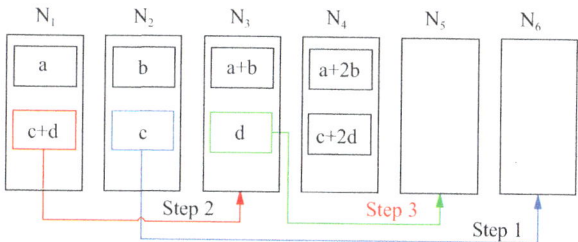

Fig. 9.26 Example of StripeMerge [72]

redundancy tuner (HeART) [68] designs the most space-efficient redundancy scheme to protect data in the face of changing disk failures.

There has been considerable research into reducing the data transfer and parity block update overhead resulting from redundancy transitioning. Stretched Reed-Solomon (SRS) [69] coding extends the RS coding [25] scheme, by applying a larger value of k, to enable a transitioned data block to be directly used for computing a new parity block without any additional data relocation. Elastic Reed-Solomon (ERS) [70] code features a new coordinated design of encoding matrix and data placement to further reduce the data transfer needed for parity block updates while also proposing an optimal theoretical tradeoff between the repair and scaling performance of LRC for tiered storage systems. LRC stripe combination has been further examined as a special redundancy transitioning scheme to form small stripes into a large one [71] for minimal cross-cluster traffic. StripeMerge [72] minimizes the bandwidth overhead for a wide-stripe generation by carefully selecting multiple narrow stripes. Convertible codes [73, 74] have been proposed to achieve the fastest switching of erasure codes with minimal resource cost.

Figure 9.26 illustrates an example of StripeMerge. First, it checks the placement of data blocks and, upon discovering that data blocks b and c are stored in N_2, transfers data block c to N_6 (step 1). Second, to ensure all parity blocks that have the same coding coefficient are in the same node, it transfers data blocks a + b and c + d to the same node and generates a new parity block a + b + c + d (step 2). Third, it transfers data block d from N_3 to an available node N_5 (step 3). So only three transfers of data blocks are performed, meaning a merging cost of just three. This example shows how StripeMerge achieves the highest redundancy transitioning efficiency with the lowest possible overhead.

9.4.4.5 Elastic Parameter Scaling

Scaling storage by adding a new node to a storage cluster increases data blocks in each stripe, which in turn results in extensive data migration. Typical scaling approaches, such as sliding window, lazy updates, and movement scheduling (SLAS) [75]; FastScale [76]; aggregate accesses to data chunks, lazy updates of mapping metadata, and valve-based rate control (ALV) [77]; stripe-based data migration (SDM) [78]; and H-Scale [79], have been proposed to reduce data migration during RAID scaling (data transfers triggered by adding a new disk to a RAID

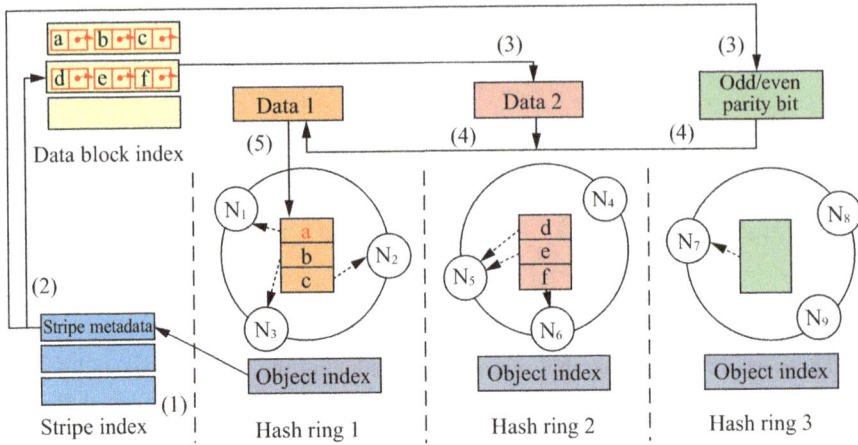

Fig. 9.27 Basic principle of ECHash [82]

volume). Scale-RS [80] is an efficient scaling scheme for Reed-Solomon-coded storage clusters to minimize data movement and guarantee uniform data distribution after cluster scaling. NCScale [81] provides a network coding–based scaling approach to achieving optimal or near-optimal scaling traffic. ECHash [82] is proposed to eliminate parity updates via data grouping and cross coding to markedly reduce scaling traffic. As illustrated in Fig. 9.27, ECHash splits a consistent hash ring into multiple hash rings, with parity blocks forming an extra ring. After a new node is added, ECHash puts it into one of the hash rings, and so only data in the ring is moved. The logical positioning of data blocks means that adding new nodes does not require data to be migrated extensively and parity blocks to be recomputed.

9.5 Distributed Storage System Reliability

Distributed storage uses more complex software and hardware designs than single-node or local storage to scatter data across server nodes. This means that the data reliability between servers is more crucial to them. Normally, the technologies that distributed storage systems use to achieve data reliability include data redundancy, failure recovery, data consistency protocols, and load balancing.

9.5.1 Data Redundancy

In distributed storage systems, multi-copy and erasure coding are typical data redundancy technologies to achieve data reliability. The multi-copy and erasure coding principles in Sect. 9.4 indicate that a multi-copy or erasure coding

9.5 Distributed Storage System Reliability

distributed storage implementation needs to consider the relationship between the number of nodes and the multi-copy or erasure coding configurations in the first place. Specifically, the distributed storage featuring N copies needs N nodes, while the distributed storage based on a $k - m$ erasure coding needs at least $k + m$ nodes. It is important that storage devices in a node are consistent with those in other nodes in terms of capacity provided, because unequal storage capacity between nodes may cause data recovery failures.

The multi-copy and erasure coding schemes differ in usable capacity, reliability, write penalty, and reconstruction performance. Table 9.5 details these differences.

As summarized in the table, multi-copy and erasure coding schemes support the usable capacity of $1/N$ and $k/(k + m)$, respectively. Ideally, a dual-copy scheme has a maximum capacity utilization of 50%, while a 4 + 2 erasure coding scheme can hit the 67% mark. An 8 + 2 erasure coding scheme enables the utilization to reach 80%, showing erasure coding schemes normally support higher capacity utilization than multi-copy ones.

Common multi-copy and erasure coding schemes do not differ much in faulty node tolerance, both allowing just a small number of nodes to fail. For example, three-copy and 4 + 2 erasure coding schemes both allow two nodes to fail. Erasure coding schemes can also further improve data storage reliability by increasing parity blocks, but at the expense of even lower capacity utilization and reconstruction efficiency.

With multi-copy schemes, data block write needs participation from N nodes, causing N write penalties. An erasure coding scheme incurs even more write penalties (at least $2m + 2$), because it requires at least $m + 1$ nodes to complete a data block write, with at least $m + 1$ data and parity block reads and writes involved.

When a tolerable number of nodes are faulty, multi-copy schemes directly read copies for recovery without extra computing overhead to ensure fast reconstruction. On the contrary, erasure coding needs computation based on data and parity blocks for recovery, clearly involving more data reads and writes and extra computing overhead. As a result, its reconstruction performance is weaker than multi-copy implementations.

These differences mean that multi-copy and erasure coding schemes apply in different scenarios. With smaller write penalties and faster reconstruction, multi-copy schemes are best suited to high-performance use cases, such as databases. As it features a higher capacity utilization, there is a preference to deploy erasure coding to implement data backup.

Table 9.5 Differences between multi-copy and erasure coding schemes

Item	Multi-copy (N)	Erasure coding (k, m)
Usable capacity	$1/N$	$k/(k + m)$
Reliability	$N - 1$	m
Write penalty	Relatively low	Relatively high
Reconstruction	Fast	Relatively slow

9.5.2 Failure Recovery

Distributed storage system failures mainly include node faults and network faults, both of which are attributed to hardware or software issues.

Regular detection that quickly identifies and locates faults in distributed storage systems is crucial to high data availability and reliability [83, 84]. Generally, heartbeat mechanisms are used to carry out the detection, which mainly involves a system sending a heartbeat message at a fixed time and receiving feedback from the target host within an expected length of time. If the feedback is received, the target host is active, and the state of the network is good. Otherwise, the target host is defined as faulty. Considering the network may be unstable and the target host may not promptly process the message, the target host is defined as faulty only after heartbeats are continuously lost. A more accurate method uses heartbeat records to determine how faulty the target node or network is. It samples and collects heartbeat feedback records in historical windows of time to calculate the failure probability in the current window based on preset probability thresholds. By providing a probability-based fault determination result for a current window of time, this method reduces the probability of misjudgment.

Failure recovery mechanisms target both node and network faults in distributed storage systems. When a single node fails to respond to data access requests, the primary and secondary strategy applies: one primary node and several secondary nodes are deployed, and when the primary node fails to react to data access requests, a new primary node will be selected from the secondary ones to ensure service continuity. This can boost distributed storage availability without exposing service exceptions to users. Redis clusters are a common example of implementing this primary/secondary strategy. When data fails to reach a destination over a communications network, data replication to synchronize data between primary and secondary nodes is a common recovery mechanism to help ensure data availability and consistency for users both during and after network faults. The details on data replication technology are provided in Chap. 11.

Distributed storage systems have three desirable properties: consistency (C), availability (A), and partition tolerance (P), which is often termed as the CAP theory [85]. CAP basically states that data consistency and availability are contradictory. Data replication technologies differ in their support for fulfilling these CAP properties. This requires data replication technology for failure recovery to be selected based on the specific consistency and availability requirements.

9.5.3 Data Consistency Protocols

Multi-copy and failure recovery schemes ensure that distributed storage systems are fault-tolerant. To maintain data consistency across nodes in multi-copy schemes, consensus protocols are designed.

A multi-copy consistency protocol ensures the command logs and state machines are consistent on different nodes. The command log records the operation commands of a series of nodes, and the state machine executes commands in line with the command log to ensure data consistency. Classic consensus algorithms include Paxos [86] and Raft [87]. Paxos is the first proven consensus algorithm and has been widely practiced in distributed storage systems such as Google's Chubby to maintain data consistency. In recent years, researchers have increasingly turned to Raft due to a number of advantages, including its clearer engineering guidance, simpler sequential log submission, and more efficient election strategy, which enables it to surpass Paxos as the main consensus algorithm in distributed systems [88]. The basic idea behind Raft was described in Sect. 8.3.2.

9.5.4 Load Balancing

Load balancing scatters data across service nodes of distributed storage systems, aiming to prevent failures or crashes that are caused by one or more nodes being overloaded with data access [89, 90].

Load balancing is a twofold approach that covers data placement and data access. Consistent hash algorithms are commonly used to ensure uniform data placement across multiple nodes [91]. They map nodes into a fully connected hash ring, perform hash function calculations on data blocks, and distribute data to nodes based on the results, achieving load balancing. Considering node deletion and addition, these algorithms virtualize nodes to maintain balanced data placement after nodes are deleted or added, as described in Sect. 8.4.1. As a distributed storage system continues to run over time, data access may exhibit evolving characteristics, leading to dynamic workload popularity and changing load distribution between nodes. This shows the necessity of balancing loads during system operation, and the common approaches for this include data migration, caching, replication, and splitting.

9.6 Summary

Reliability is an important metric for estimating the likelihood of a storage system operating without faults. Storage reliability involves the underlying media, storage devices, single-node systems, and distributed systems, which may have different degrees of reliability from application to application. This requires storage reliability to be designed with holistic considerations taken for software and hardware structures and data access requirements, in order to work out user-friendly and hierarchical schemes for storage media, devices, and systems.

In this chapter, we explained the basic concepts of storage reliability metrics and the calculation methods, the core principles of layered reliability design, and the difference between reliability and availability. We provided a detailed analysis of the sources of storage reliability faults at the level of hard disks, flash storage media, and storage systems and described main reliability optimization technologies. We also described the fault warning, fault detection, and design methods targeting different environmental factors of hard disk faults. We studied multi-copy, erasure coding, and other system-level reliability optimization schemes, focusing on the typical technologies and trends of erasure coding. And last but not least, we summarized the major technologies for optimizing the distributed storage system reliability.

Reliability optimization represents a major direction of research in storage systems. New storage devices and architectures present storage reliability research with both challenges and opportunities.

9.7 Practice Questions

1. **What is the key difference between data availability and reliability?**

   Answer: Availability is the ratio of time a system stays operational under normal circumstances to the total elapsed time and indicates system uptime during a given time window. Reliability is the probability that a system will run without failure and indicates the system failure frequency.

2. **Why do high-density flash media, such as TLCs and QLCs, have higher error rates?**

   Answer: As storage density increases, the number of bits that can be stored in a high-density cell, such as a triple-level cell (TLC) or quad-level cell (QLC), increases too, resulting in tighter voltage distributions in the cell. Therefore, when a high-density cell experiences interference from the same error source, the cell is more likely to have a voltage distribution shift, which leads to incorrect reading of stored data and affects storage reliability.

3. **What are the four key technologies for optimizing flash memory reliability and for what error sources are they designed?**

   Answer: The four technologies are shadow program sequence, refresh, read-retry, and voltage optimization. The shadow program sequence is designed to handle program interference. Refresh is designed to handle interference caused by long retention times. Read-retry and voltage optimization are designed to handle interference caused by large P/E cycle counts, long retention times, read disturb, and program interference.

4. **What are the four key technologies for ensuring distributed storage reliability and what are their typical applications?**

   Answer: (1) Data redundancy: Data redundancy is achieved using multi-copy or erasure coding, with multi-copy mainly used for performance-demanding applications such as databases and erasure coding mainly used for data backup, due to its good capacity utilization. (2) Failure recovery: For node failures, the primary and secondary strategy applies. Redis clusters are a common example of implementing this primary/secondary strategy. For network failures, data replication is mainly used for redundancy in data centers and similar applications. (3)

Data consistency protocol: Consensus protocols are designed to maintain data consistency. Classic consensus algorithms include Paxos and Raft. (4) Load balancing: Consistent hash algorithms are used to ensure uniform data placement across multiple nodes. Common approaches for balancing loads upon data access include data migration, caching, replication, and splitting. Load balancing prevents failures and crashes of distributed storage systems.

5. **In what ways are the RDP, EVENODD, X-, and HV codes differently constructed?**

    Answer: Both the RDP and EVENODD codes use horizontal and diagonal parity check. The X-code uses diagonal and back-diagonal parity check. The HV code uses horizontal and vertical parity check.

6. **Why does the LRC code outperform the RS code in reducing repair overhead?**

    Answer: The RS code uses two parameters to set its fault tolerance and storage overhead, and it uses matrix calculation to construct parity information. To repair one block, k blocks need to be requested. However, the LRC code constructs multiple fine-grained groups in a stripe. To repair one block in a group, only the remaining blocks of the group need to be requested.

7. **What are the advantages and disadvantages of multi-copy and erasure coding?**

    Answer: Multi-copy has higher storage overhead but lower repair overhead. Erasure coding has lower storage overhead but higher repair overhead.

8. **Some say that the storage medium is more likely to fail the longer it is used. Is this a true statement?**

    Answer: This statement is incorrect. The failure rate of HDDs generally varies with the bathtub curve, which is high in the infancy and wearout stages but low in the useful life period.

References

1. Coughlin T. 175 Zettabytes by 2025 (2018-11-27) [2023-04-11].
2. Raza M. Reliability vs availability: what's the difference? (2020-05-13) [2023-04-11].
3. Kadekodi S, Rashmi KV, Ganger GR. Cluster storage systems gotta have HeART: improving storage efficiency by exploiting disk-reliability heterogeneity. In: Proceedings of 17th USENIX conference on file and storage technologies (FAST 19). Berkeley: USENIX Association; 2019. p. 345–58.
4. Wu S, Lan S, Zhou J, et al. BitFlip: a bit-flipping scheme for reducing read latency and improving reliability of flash memory. In: Proceedings of international conference on massive storage systems and technology (MSST). Piscataway: IEEE; 2020.
5. Meza J, Wu Q, Kumar S, et al. A large-scale study of flash memory failures in the field. In: ACM SIGMETRICS performance evaluation review, vol. 43(1). New York: ACM; 2015. p. 177–90.
6. Cai Y, Ghose S, Haratsch EF, et al. Error characterization, mitigation, and recovery in flash-memory-based solid-state drives. In: Proceedings of the IEEE, vol. 105(9). Piscataway: IEEE; 2017. p. 1666–704.

7. Mohan V, Siddiqua T, Gurumurthi S, et al. How I learned to stop worrying and love flash endurance. In: Proceedings of the 2nd workshop on hot topics in storage and file systems (HotStorage), vol. 10. Berkeley: USENIX Association; 2010. p. 3.
8. Park J, Jeong J, Lee S, et al. Improving performance and lifetime of NAND storage systems using relaxed program sequence. In: Proceedings of the 53rd annual design automation conference (DAC). Piscataway: IEEE; 2016. p. 1–6.
9. Cai Y, Yalcin G, Mutlu O, et al. Flash correct-and-refresh: retention-aware error management for increased flash memory lifetime. In: Proceedings of the 30th international conference on computer design (ICCD). Piscataway: IEEE; 2012. p. 94–101.
10. Mohan V, Sankar S, Gurumurthi S, et al. reFresh SSDs: enabling high endurance, low cost flash in datacenters. Univeristy of Virginia, Tech. Rep. CS-2012-05, 2012.
11. Cai Y, Haratsch EF, Mutlu O, et al. Threshold voltage distribution in MLC NAND flash memory: characterization, analysis, and modeling. In: Proceedings of the 2013 design, automation & test in Europe conference & exhibition (DATE). Piscataway: IEEE; 2013. p. 1285–90.
12. Li Q, Shi L, Xue CJ, et al. Improving LDPC performance via asymmetric sensing level placement on flash memory. In: Proceedings of the 22nd Asia and South Pacific Design Automation Conference (ASP-DAC). Piscataway: IEEE; 2017. p. 560–5.
13. Cai Y, Luo Y, Haratsch EF, et al. Data retention in MLC NAND flash memory: characterization, optimization, and recovery. In: Proceedings of the 21st international symposium on high performance computer architecture (HPCA). Piscataway: IEEE; 2015. p. 551–63.
14. Jeong J, Hahn SS, Lee S, et al. Lifetime improvement of NAND flash-based storage systems using dynamic program and erase scaling. In: Proceedings of the 12th USENIX conference on file and storage technologies (FAST). Berkeley: USENIX Association; 2014. p. 61–74.
15. Pan Y, Dong G, Wu Q, et al. Quasi-nonvolatile SSD: trading flash memory nonvolatility to improve storage system performance for enterprise applications. In: Proceedings of the 18th international symposium on high-performance comp architecture (HPCA). Piscataway: IEEE; 2012. p. 1–10.
16. Cai Y, Yalcin G, Mutlu O, et al. Error analysis and retention-aware error management for NAND flash memory. Intel Technol J. 2013;17(1)
17. Chen Z, Haratsch EF, Sankaranarayanan S, et al. Estimating read reference voltage based on disparity and derivative metrics. U.S. Patent 9,417,797; 2016-8-16.
18. Wu Y, Cohen ET. Optimization of read thresholds for non-volatile memory. U.S. Patent 9,595,320; 2017-3-14.
19. Luo Y, Ghose S, Cai Y, et al. Enabling accurate and practical online flash channel modeling for modern MLC NAND flash memory. IEEE J Sel Areas Commun. 2016;34(9):2294–311.
20. Huang P, Subedi P, He X, et al. FlexECC: partially relaxing ECC of MLC SSD for better cache performance. In: Proceedings of the 2014 USENIX annual technical conference (ATC). Berkeley: USENIX Association; 2014. p. 489–500.
21. Gao C, Shi L, Wu K, et al. Exploit asymmetric error rates of cell states to improve the performance of flash memory storage systems. In: Proceedings of the 32nd international conference on computer design (ICCD). Piscataway: IEEE; 2014. p. 202–7.
22. Wilson EH, Jung M, Kandemir MT. ZombieNAND: resurrecting dead NAND flash for improved SSD longevity. In: Proceedings of the 22nd international symposium on modelling, analysis & simulation of computer and telecommunication systems (MASCOTS). Piscataway: IEEE; 2014. p. 229–38.
23. Gao C, Ye M, Li Q, et al. Constructing large, durable and fast SSD system via reprogramming 3D TLC flash memory. In: Proceedings of the 52nd annual international symposium on microarchitecture (MICRO). Piscataway: IEEE; 2019. p. 493–505.
24. Ghemawat S, Gobioff H, Leung ST. The Google file system. In: Proceedings of the 19th ACM symposium on operating systems principles (SOSP). New York: ACM; 2003.
25. Reed I, Solomon G. Polynomial codes over certain finite fields. J Soc Ind Appl Math. 1960;8(2):300–4.
26. Li M. Research on erasure coding technology of disk arrays. Beijing: Tsinghua University; 2011.

27. Fu Y. Research on erasure coding mechanisms in storage systems. Beijing: Tsinghua University; 2014.
28. Shen Z. Performance optimization research of erasure code storage systems. Beijing: Tsinghua University; 2015.
29. The Apache Software Foundation. Apache Hadoop 3.0.0 (2017-12-08) [2023-04-11].
30. Ceph. WELCOME TO CEPH; 2016 [2023-04-11].
31. Muralidhar S, Lloyd W, Roy S, et al. f4: Facebook's warm blob storage system. In: Proceedings of the 11th USENIX symposium on operating systems design and implementation (OSDI). Berkeley: USENIX Association; 2014.
32. Huang C, Simitci H, Xu Y, et al. Erasure coding in windows azure storage. In: Proceedings of the USENIX annual technical conference (USENIX ATC). Berkeley: USENIX Association; 2012.
33. Rashmi KV, Nihar BS, Gu D, et al. A solution to the network challenges of data recovery in erasure-coded distributed storage systems: a study on the Facebook warehouse cluster. In: Proceedings of 5th USENIX workshop on hot topics in storage and file systems (hotstorage). Berkeley: USENIX Association; 2013.
34. Dimakis AG, Godfrey PB, Wu Y, et al. Network coding for distributed storage systems. IEEE Trans Inf Theory. 2010;56(9):4539–51.
35. Rashmi KV, Chowdhury M, Kosaian J, et al. EC-Cache: load-balanced, low-latency cluster caching with online erasure coding. In: Proceedings of the 12th USENIX symposium on operating systems design and implementation (OSDI). Berkeley: USENIX Association; 2016. p. 401–17.
36. Blaum M, Brady J, Bruck J, et al. EVENODD: an efficient scheme for tolerating double disk failures in RAID architectures. IEEE Trans Comput. 1995;44(2):192–202.
37. Corbett P, English B, Goel A, et al. Row-diagonal parity for double disk failure correction. In: Proceedings of the 3rd USENIX conference on file and storage technologies. Berkeley: USENIX Association; 2004. p. 1.
38. Xu L, Bruck J. X-code: MDS array codes with optimal encoding. IEEE Trans Inf Theory. 1999;45(1):272–6.
39. Shen Z, Shu J. HV code: an all-around MDS code to improve efficiency and reliability of RAID-6 systems. In: Proceedings of the 44th Annual IEEE/IFIP International Conference on Dependable Systems and Networks (DSN). Piscataway: IEEE; 2014. p. 550–561.
40. Papailiopoulos DS, Dimakis AG. Locally repairable codes. IEEE Trans Inf Theory. 2014;60(10):5843–55.
41. Sathiamoorthy M, Asteris M, Papailiopoulos D, et al. XORing elephants: novel erasure codes for big data. Proc VLDB Endow. 2013;6(5):325–36.
42. Khan O, Burns R, Plank J, et al. Rethinking erasure codes for cloud file systems: minimizing I/O for recovery and degraded reads. In: Proceedings of the 10th USENIX conference on file and storage technologies (FAST). Berkeley: USENIX Association; 2012.
43. Rashmi KV, Shah NB, Kumar PV. Optimal exact-regenerating codes for distributed storage at the MSR and MBR points via a product-matrix construction. IEEE Trans Inf Theory. 2011;57(8):5227–39.
44. Pamies-Juarez L, Blagojevic F, Mateescu R, et al. Opening the chrysalis: on the real repair performance of MSR codes. In: Proceedings of the 14th USENIX conference on file and storage technologies (FAST). Berkeley: USENIX Association; 2016. p. 81–94.
45. Vajha M, Ramkumar V, Puranik B, et al. Clay codes: moulding MDS codes to yield an MSR code. In: Proceedings of the 16th USENIX conference on file and storage technologies (FAST). Berkeley: USENIX Association; 2018. p. 139–54.
46. Rashmi KV, Shah NB, Gu D, et al. A 'Hitchhiker's' guide to fast and efficient data reconstruction in erasure-coded data centers. In: Proceedings of the ACM conference on SIGCOMM (SIGCOMM). New York: ACM; 2014. p. 331–42.
47. Li R, Lee PPC, Hu Y. Degraded-first scheduling for MapReduce in erasure-coded storage clusters. In: Proceedings of the 44th annual IEEE/IFIP international conference on dependable systems and networks (DSN 14). Piscataway: IEEE; 2014. p. 419–30.

48. Huang J, Liang X, Qin X, et al. PUSH: a pipelined reconstruction I/O for erasure-coded storage clusters. IEEE Trans Parallel Distrib Syst. 2015;26(2):516–26.
49. Mitra S, Panta R, Ra M-R, et al. Partial-Parallel-Repair (PPR): a distributed technique for repairing erasure coded storage. In: Proceedings of the 11th European conference on computer systems (EuroSys); 2016. p. 1–16.
50. Li R, Li X, Lee PPC, et al. Repair pipelining for erasure-coded storage. In: Proceedings of the USENIX annual technical conference (USENIX ATC 17). Berkeley: USENIX Association; 2017. p. 567–79.
51. Shen Z, Shu J, Lee PPC. Reconsidering single failure recovery in clustered file systems. In: Proceedings of the 46th annual IEEE/IFIP international conference on dependable systems and networks (DSN 16). Piscataway: IEEE; 2016. p. 323–34.
52. Shen Z, Shu J, Huang Z, et al. ClusterSR: cluster-aware scattered repair in erasure-coded storage. In: Proceedings of IEEE international parallel and distributed processing symposium (IPDPS). Piscataway: IEEE; 2020. p. 42–51.
53. Hu Y, Cheng L, Yao Q, et al. Exploiting combined locality for wide-stripe erasure coding in distributed storage. In: Proceedings of the 19th USENIX conference on file and storage technologies (FAST). Berkeley: USENIX Association; 2021. p. 233–48.
54. Lin S, Gong G, Shen Z, et al. Boosting full-node repair in erasure-coded storage. In: Proceedings of the USENIX annual technical conference (USENIX ATC). Berkeley: USENIX Association; 2021. p. 641–55.
55. Han S, Lee PPC, Shen Z, et al. Toward adaptive disk failure prediction via stream mining. In: Proceedings of the 40th IEEE international conference on distributed computing systems (ICDCS). Piscataway: IEEE; 2020. p. 628–38.
56. Botezatu MM, Giurgiu I, Bogojeska J, et al. Predicting disk replacement towards reliable data centers. In: Proceedings of the 22nd ACM SIGKDD international conference on knowledge discovery and data mining (KDD). New York: ACM; 2016. p. 39–48.
57. Li J, Ji X, Jia Y, et al. Hard drive failure prediction using classification and regression trees. In: Proceedings of the 44th annual IEEE/IFIP international conference on dependable systems and networks (DSN). Piscataway: IEEE; 2014. p. 383–94.
58. Ma A, Douglis F, Lu G, et al. RAIDShield: characterizing, monitoring, and proactively protecting against disk failures. In: Proceedings of the 13th USENIX conference on file and storage technologies (FAST). Berkeley: USENIX Association; 2015. p. 241–56.
59. Zhu B, Wang G, Liu X, et al. Proactive drive failure prediction for large scale storage systems. In: Proceedings of the 29th symposium on mass storage systems and technologies (MSST). Piscataway: IEEE; 2013. p. 1–5.
60. Xu Y, Sui K, Yao R, et al. Improving service availability of cloud systems by predicting disk error. In: Proceedings of the USENIX annual technical conference (USENIX ATC). Berkeley: USENIX Association; 2018. p. 481–94.
61. Lu S, Luo B, Patel T, et al. Making disk failure predictions smarter! In: Proceedings of the 18th USENIX conference on file and storage technologies (FAST). Berkeley: USENIX Association; 2020. p. 151–67.
62. Zhang J, Huang P, Zhou K, et al. HDDse: enabling high-dimensional disk state embedding for generic failure detection system of heterogeneous disks in large data centers. In: Proceedings of the 2020 USENIX annual technical conference (USENIX ATC). Berkeley: USENIX Association; 2020. p. 111–26.
63. Shen Z, Li X, Lee PPC. Fast predictive repair in erasure-coded storage. In: Proceedings of the 49th annual IEEE/IFIP international conference on dependable systems and networks (DSN). Piscataway: IEEE; 2019. p. 556–67.
64. Hu Y, Wang Y, Liu B, et al. Latency reduction and load balancing in coded storage systems. In: Proceedings of symposium on cloud computing (SoCC). Piscataway: IEEE; 2017. p. 365–77.
65. Xia M, Saxena M, Blaum M, et al. A tale of two erasure codes in HDFS. In: Proceedings of the 13th USENIX conference on file and storage technologies (FAST). Berkeley: USENIX Association; 2015. p. 213–26.

References

66. Wang Z, Wang H, Shao A, et al. An adaptive erasure-coded storage scheme with an efficient code-switching algorithm. In: Proceedings of the 49th international conference on parallel processing, (ICPP), New York, NY, USA; 2020. p. 1–11.
67. Zhou P, Huang J, Qin X, et al. PaRS: a popularity-aware redundancy scheme for in-memory stores. IEEE Trans Comput. 2019;68(4):556–69.
68. Kadekodi S, Maturana F, Subramanya SJ, et al. PACEMAKER: avoiding HeART attacks in storage clusters with disk-adaptive redundancy. In: Proceedings of the 14th USENIX symposium on operating systems design and implementation (OSDI). Berkeley: USENIX Association; 2020. p. 369–85.
69. Taranov K, Alonso G, Hoefler T. Fast and strongly-consistent per-item resilience in key-value stores. In: Proceedings of the 13th EuroSys conference (EuroSys); 2018. p. 1–14.
70. Wu S, Shen Z, Lee PPC. Enabling I/O-efficient redundancy transitioning in erasure-coded KV stores via elastic Reed-Solomon codes. In: Proceedings of international symposium on reliable distributed systems (SRDS); 2020. p. 246–55.
71. Wu S, Du Q, Lee PPC, et al. Optimal data placement for stripe merging in locally repairable codes. In: Proceedings of the IEEE infocom 2022—IEEE conference on computer communications (INFOCOM); 2022. p. 1669–78.
72. Yao Q, Hu Y, Cheng L, et al. StripeMerge: efficient wide-stripe generation for large-scale erasure-coded storage. In: Proceedings of the 41st IEEE international conference on distributed computing systems (ICDCS). Piscataway: IEEE; 2021. p. 483–93.
73. Maturana F, Mukka VSC, Rashmi KV. Access-optimal linear MDS convertible codes for all parameters. In: Proceedings of the 2020 IEEE international symposium on information theory (ISIT). Piscataway: IEEE; 2020. p. 577–82.
74. Maturana F, Rashmi KV. Convertible codes: new class of codes for efficient conversion of coded data in distributed storage. In: Proceedings of the 11th innovations in theoretical computer science conference (ITCS); 2020. p. 1–26.
75. Zhang G, Shu J, Xue W, et al. SLAS: an efficient approach to scaling round-robin striped volumes. ACM Trans Storage. 2007;3(1):3-es.
76. Zheng W, Zhang G. FastScale: accelerate RAID scaling by minimizing data migration. In: Proceedings of the 9th USENIX conference on file and storage technologies (FAST). Berkeley: USENIX Association; 2011.
77. Zhang G, Zheng W, Shu J. ALV: a new data redistribution approach to RAID-5 scaling. IEEE Trans Comput. 2010;59(3):345–57.
78. Wu C, He X, Han J, et al. SDM: a stripe-based data migration scheme to improve the scalability of RAID-6. In: Proceedings of the IEEE international conference on cluster computing (CoCC). Piscataway: IEEE; 2012. p. 284–92.
79. Wan J, Xu P, He X, et al. H-Scale: a fast approach to scale disk arrays via hybrid stripe deployment. ACM Trans Storage. 2016;12(3):1–30.
80. Huang J, Liang X, Qin X, et al. Scale-RS: an efficient scaling scheme for RS-coded storage clusters. IEEE Trans Parallel Distrib Syst. 2015;26(6):1704–17.
81. Zhang X, Hu Y, Lee PPC, et al. Toward optimal storage scaling via network coding: from theory to practice. In: Proceedings of the IEEE infocom 2018—IEEE conference on computer communications (INFOCOM); 2018. p. 1808–16.
82. Cheng L, Hu Y, Lee PPC. Coupling decentralized key-value stores with erasure coding. In: Proceedings of the ACM symposium on cloud computing. New York: ACM; 2019. p. 377–89.
83. Yuan D, Luo Y, Zhuang X, et al. Simple testing can prevent most critical failures: an analysis of production failures in distributed data-intensive systems. In: Proceedings of the 11th symposium on operating systems design and implementation (OSDI). Berkeley: USENIX Association; 2014. p. 48–68.
84. Shvachko K, Kuang H, Radia S, et al. The Hadoop distributed file system. In: Proceedings of the 26th symposium on mass storage systems and technologies (MSST). Piscataway: IEEE; 2010. p. 1–10.
85. Abadi D. Consistency tradeoffs in modern distributed database system design: CAP is only part of the story. IEEE Comput. 2012;45(2):37–42.

86. Lamport L. Paxos made simple. ACM SIGACT News. 2001;4(121):51–8.
87. Ongaro D, Ousterhout J. In search of an understandable consensus algorithm. In: Proceedings of the USENIX annual technical conference (ATC). Berkeley: USENIX Association; 2014. p. 305–19.
88. Wang Z, Zhao C, Mu S, et al. On the parallels between paxos and raft, and how to port optimizations. In: Proceedings of the 2019 ACM symposium on principles of distributed computing (PODC). New York: ACM; 2019. p. 445–54.
89. Chou TCK, Abraham JA. Load balancing in distributed systems. IEEE Trans Softw Eng. 1982;4:401–12.
90. Alakeel AM. A guide to dynamic load balancing in distributed computer systems. Int J Comput Sci Inf Secur. 2010;10(6):153–60.
91. Karger D, Lehman E, Leighton T, et al. Consistent hashing and random trees: distributed caching protocols for relieving hot spots on the world wide web. In: Proceedings of the 29th annual symposium on theory of computing (STOC). New York: ACM; 1997. p. 654–63.

Chapter 10
Storage Security

With information technologies permeating more aspects of production and our daily lives, private data, such as that related to locations and consumption records/habits, can be generated anytime and anywhere. Generally, such data is generated by applications and stored on remote servers. This means that the data is beyond the physical control of users, and whether or not this data is being securely stored has become a key concern for users and service vendors. This concern is intensifying with the rapid development of cloud storage, because centralized data storage and management are becoming increasingly popular. In recent years, frequent storage security accidents have resulted in the leakage of massive amounts of private data and, in some extreme cases, even complete loss of service data.

In January 2018, RightScale—a leading cloud computing management company—conducted research into the latest cloud computing trends. The company surveyed 997 technical professionals across a broad cross-section of organizations regarding their adoption of cloud computing and the main challenges posed by cloud computing. As shown in Fig. 10.1, 81% of respondents reported that security was the biggest challenge in cloud computing [1].

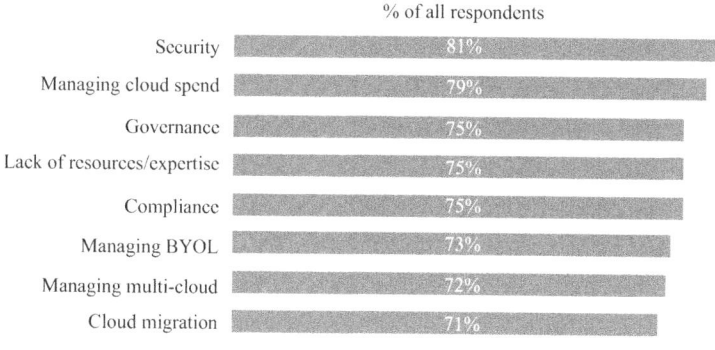

Fig. 10.1 The RightScale report 2018

© Posts and Telecom Press Co., Ltd 2024
J. Shu, *Data Storage Architectures and Technologies*,
https://doi.org/10.1007/978-981-97-3534-1_10

Against the backdrop of the increasing importance of data security, the Data Security Law of the People's Republic of China was officially approved for release at the twenty-ninth meeting of the Standing Committee of the 13th National People's Congress on June 10, 2021. This law aims to regulate data-processing activities and promote data development and use, all while guaranteeing effective data security, thereby protecting the legitimate rights and interests of individuals and organizations and safeguarding national sovereignty, security, and development interests.

10.1 Concept and Security System

Generally, storage systems are built on storage hardware and use system software to efficiently manage data. Users can remotely access data through networks. This means that data storage security is dependent on multiple aspects across hardware, data, and permission management. Here, we will look at the secure storage system Shield, which is designed for the public cloud, as an example (Fig. 10.2) to illustrate design considerations.

The application scenario for the Shield system is that users mount the system locally and then use it to transfer shared files to a storage service provider. During this process, Shield encrypts the shared files, only transmitting and storing ciphertext on the storage service provider's devices. The file owner manages user permissions, granting them read-only or read/write permissions or changing their access permissions (for example, revoking a user's read-only permission so that the user

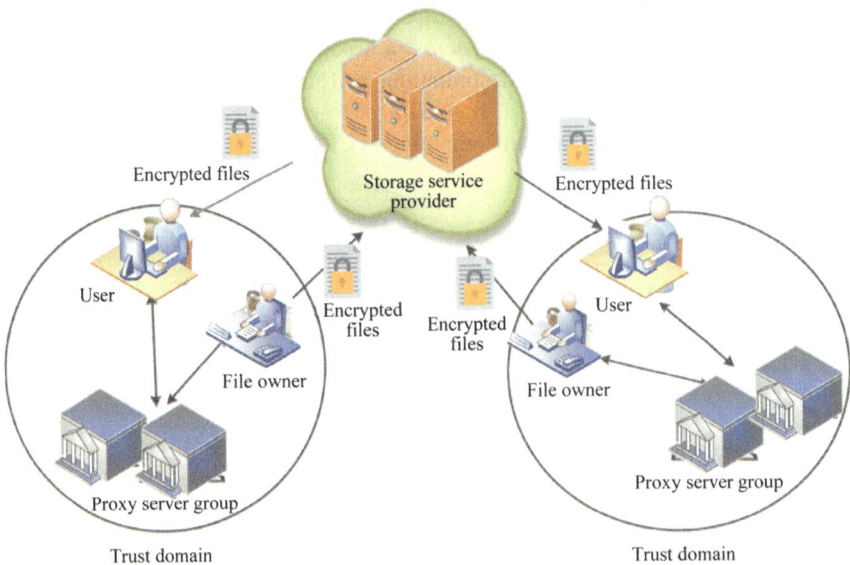

Fig. 10.2 Secure storage system Shield [2] for the public cloud

cannot access certain files). In this storage system, data storage security issues can arise in multiple areas.

System security: The storage system is built upon computer hardware devices and software, meaning that hardware or system attacks can potentially cause a private data breach. For example, a hacker conducting a side-channel attack may access computer information signals (for instance, the electromagnetic radiation of a computer or the sound of running hardware) to attempt to obtain the behavior of users, thus probing user privacy information (such as a password). In addition, certain attackers may attempt to maliciously attack the storage system, causing user data to become inaccessible or even lost.

Data security: Data is transmitted to servers over networks. If attackers listen in on data that is being transmitted over a network or unauthorized users intrude into the systems of storage service providers to spy upon user data, then private data will be at risk of being disclosed. Therefore, unauthorized users must never be able to access data in order to uphold data confidentiality. Additionally, since data is not physically controlled by users and is stored in remote servers, data integrity must be guaranteed to prevent any unauthorized tampering.

Security management: A file owner manages its corresponding files. In actual scenarios, certain users may attempt to upgrade their access permissions, or users whose permissions have been revoked may attempt to access files for which they no longer have access permission. These behaviors threaten data storage security, necessitating the secure and efficient management of user access permissions.

10.2 System Security

This section describes system security, which comprises hardware security, container security, and system resilience.

10.2.1 Hardware Security

Hardware security is the very foundation of system security. Trusted hardware, as shown in Fig. 10.3, is known as a hardware root of trust (RoT). Such hardware lays a computing foundation—built upon cryptography and hardware protection—for trusted environments. Examples of hardware security technologies include keys built in hardware, encryption/decryption algorithms, bottom-layer software built in hardware, and chips that meet the Trusted Platform Module (TPM) standard.

A chain of trust that ensures system security can be built based on the hardware RoT. Piling up underlying trust-foundation layers guarantees effective overall system security, as shown in Fig. 10.4. Chain-of-trust technologies include secure boot, trusted boot, secure storage, secure upgrade, secure transmission, and secure operation.

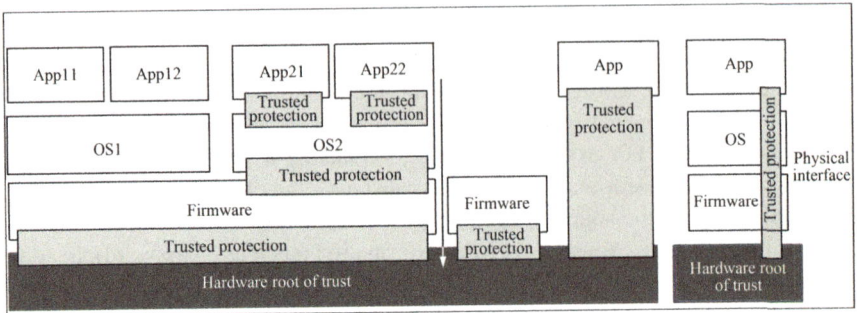

Fig. 10.3 Hardware root of trust

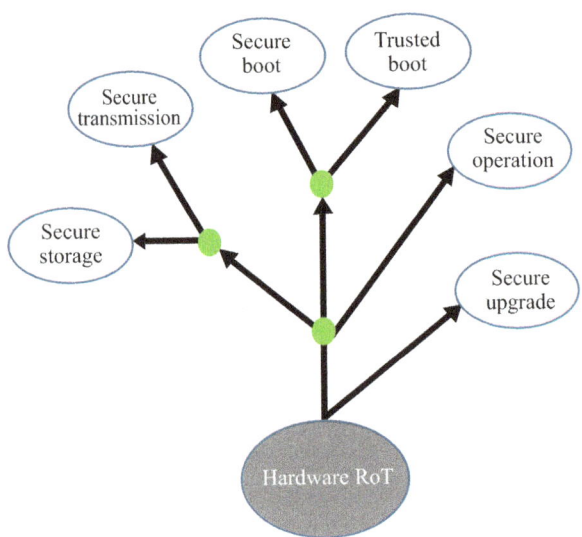

Fig. 10.4 Security technologies built on hardware RoT

Security isolation mechanisms of different levels can be built on trusted hardware through applicable technologies. The most common four are as follows:

Arm TrustZone: An extended technology for hardware security. It divides hardware resources into two execution environments—common and secure environments—to ensure system security. In the common environment, the processor is in a non-secure state, while in the secure environment, the processor is in a secure state.

Intel Software Guard Extensions (SGX): Provides hardware-based security assurance for programs. It stores program code and data in the physical memory space that is protected by an enclave to ensure the secure execution of programs.

AMD Secure Encrypted Virtualization (SEV): Provides security protection for virtual machine (VM) instances. This technology allocates an independent key to each VM instance in order to isolate the guest machine from the hypervisor. The AMD security processor manages the keys. This technology requires the support

of the guest machine and hypervisor. The guest machine specifies the memory pages to be encrypted, while the hypervisor interacts with the AMD security processor to operate the keys of the VM instance.
- Intel Trust Domain Extensions (TDX): Isolates VM instances and hypervisors from other untrusted programs. Intel TDX comprises Secure Arbitration Mode (SEAM), Physical-Address-Metadata Table (PAMT), and Total Memory Encryption-Multi-Key (TME-MK) technologies.

Hardware security is prone to the following three malicious attacks.

- Side-channel attack: This type of attack does not directly target hardware or programs. Instead, such attacks collect and measure physical information (such as power supply, current, power consumption, electromagnetic radiation, and sound/vibration generated during hardware running) that is leaked by encryption software or hardware, so as to analyze such information and launch hardware attacks. Common side-channel attacks include timing attacks, electromagnetic attacks, simple power analysis (SPA), and differential power analysis (DPA).
- Fault injection attack: This type of attack physically causes hardware errors. It bypasses system security protection to launch malicious attacks that change system behaviors, obtain confidential system information, and extract encryption/decryption keys. Common fault injection attacks include voltage glitching, clock glitching, laser injection, and electromagnetic injection.
- Physical attack: Such attacks detect, modify, or damage hardware by physical means. One example is the use of physical tools to detect the bus data and firmware of servers, in order to directly modify or damage key server components.

10.2.2 Container Security

A container is a standard unit of software that packages up code and all its dependencies so that the software runs quickly and reliably from one computing environment to another. Compared with traditional VM technologies that implement virtualization based on hardware resources, containers implement virtualization at the operating system (OS) level. Both virtualization technologies have advantages and disadvantages in terms of performance and security.

VM is the virtualization of physical hardware resources. As shown in Fig. 10.5, a hypervisor or OS runs on the host machine, and guest machine instances run on this set. The guest machines use independent OSs. Examples include Kernel-based Virtual Machine (KVM) and Quick Emulator (QEMU) that run on the host machine OS Linux. In terms of security, VM technology provides strong isolation security for VM instances and facilitates the exclusive use of hardware resources by these VM instances, all owning to the independent running of OSs in the hypervisor and guest machines.

Containers virtualize resources at the OS level. As shown in Fig. 10.6, the resource isolation tool provided by the host machine OS runs to build containers,

Fig. 10.5 Traditional VM technology

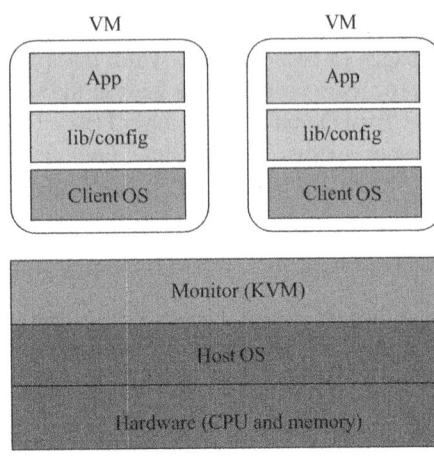

Fig. 10.6 Container technology

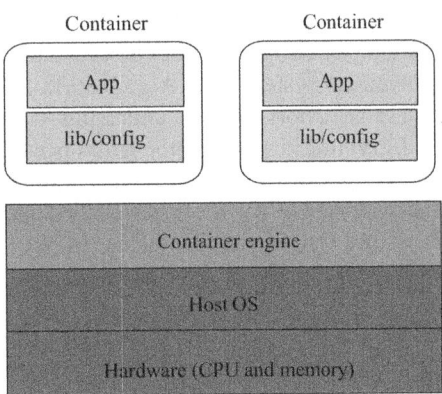

thus enabling software to operate in their own computing environments. We can take Docker container technology as an example. This technology is implemented using the cgroup and namespace tools provided by the Linux kernel. Compared with traditional VM technologies, containers are more efficient and lightweight, and their security is guaranteed by the host machine OS.

Secure containers built on VM technology combine the strong isolation function of VM technology with the efficient and lightweight advantages of container technology. As shown in Fig. 10.7, containers are deployed on VM instances, acting as a layer of virtualization, to implement container isolation. One such example is the Kata Containers.

Figure 10.8 shows a secure container based on hardware security technologies. When a container runs on a host machine, the host machine OS ensures its security. As a result, the security of the container may be affected by the software stack on the host machine. However, hardware-based security technologies, such as Intel SGX and AMD SEV, provide security assurance for containers.

10.2 System Security

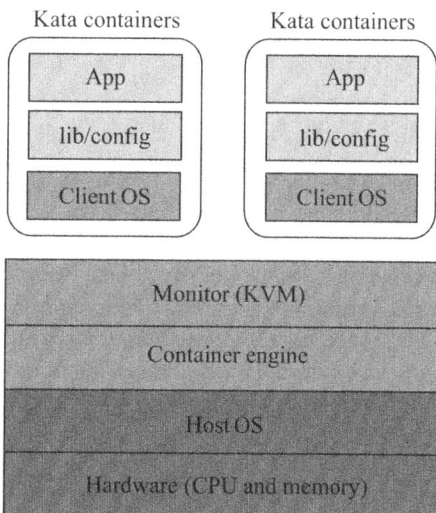

Fig. 10.7 Secure container based on the VM technology

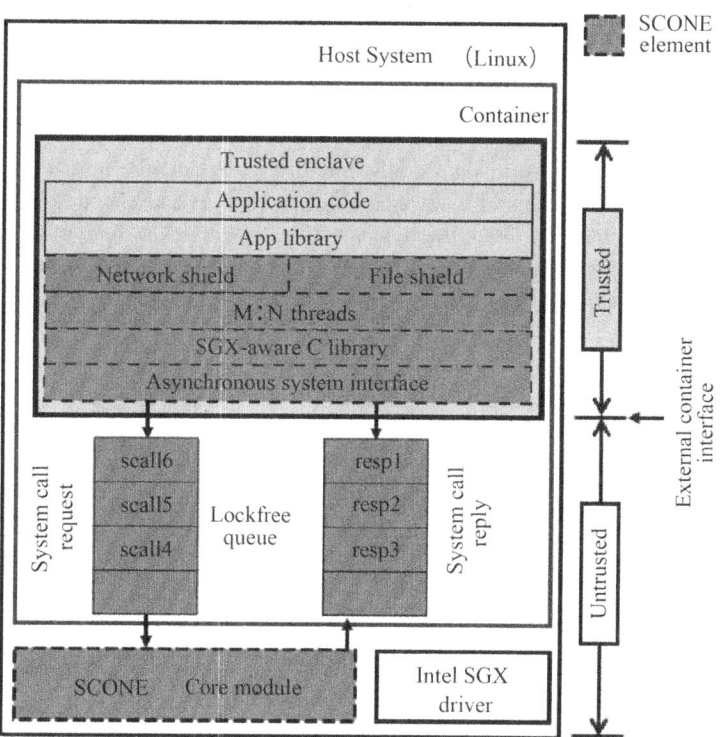

Fig. 10.8 Secure container based on hardware security technologies [3]

10.2.3 System Resilience

System resilience refers to a system's capability to defend against and recover from malicious attacks. Different from system security, which focuses only on how to prevent malicious attacks, system resilience has four main objectives:

First, the system design must consider errors caused by malicious attacks and develop countermeasures to ensure the system correctly handles system errors.

Second, the system security policies must be hierarchical, with the core functions or the minimum recovery system requiring stronger security measures.

Third, once the system is maliciously attacked, system functions must be able to recover within a specified period of time.

Fourth, the system learns from and adapts to malicious attacks so that it can cope with similar scenarios in the future.

10.3 Data Security

10.3.1 Data Encryption

10.3.1.1 Symmetric and Asymmetric Encryption

The simplest way to ensure data confidentiality and transmission security is to encrypt data. First, the sender uses algorithms to encrypt and encode data into an unidentifiable ciphertext. Then, the receiver uses algorithms to decrypt the ciphertext and restore the data to its original format. Keys are required to encrypt and decrypt data, as shown in Fig. 10.9.

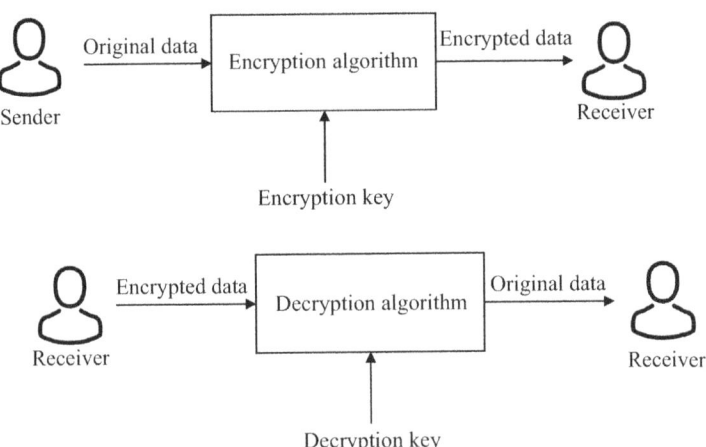

Fig. 10.9 Encryption and decryption of data

10.3 Data Security

Data encryption has been around for thousands of years. For example, there is evidence that Caesar used ciphers to send messages to his generals during his reign. It was originally more of an art form, but since then, it has evolved into an important scientific field involving things like quantum cryptography [4]. There are two main branches within the field: symmetric encryption and asymmetric encryption.

Symmetric Encryption

Symmetric encryption uses the same key to both encrypt and decrypt data. Generally, the decryption process is the reverse of the encryption process. The original symmetric encryption technology is called classical cryptography. Examples include Caesar's cipher, the rail fence cipher, and the Enigma cipher which was cracked by Turing during World War II. Modern algorithms such as the Data Encryption Standard (DES), Advanced Encryption Standard (AES), ShāngMì 4 (SM4), and the stream cipher are also forms of symmetric encryption. The symmetric encryption formula is as follows:

$$E_k(T) = C, \quad D_k(C) = T \tag{10.1}$$

In this formula,

E and D stand for encryption and decryption, respectively.
k is the key.
T is the original unencrypted message.
C is the encrypted message.

The classical cryptography focuses on the secrecy of algorithms and keys. If the algorithms used are known, then the message is said to be "leaky," because the key space size—the number of possible characters for each position in the key raised to the power of the key length—is very small. However, modern symmetric encryption technology has a large key space size which makes it impossible to crack manually. Therefore, as long as the keys remain protected, the algorithm can be disclosed. In situations where large amounts of data need to be encrypted, symmetric encryption is typically more practical than asymmetric encryption because encryption/decryption is faster and more efficient.

Asymmetric Encryption

Asymmetric encryption usually refers to the public key cryptography proposed by Diffie and Hellman in 1976 [5]. Each user has a public key and a private key. The private key is confidential. Unlike symmetric encryption, asymmetric encryption uses one key for encryption and a different key for decryption. The encryption/decryption formula is as follows:

$$E_{pk}(T) = C, \quad D_{sk}(C) = T \quad \text{or} \quad E_{sk}(T) = C, \quad D_{pk}(C) = T \tag{10.2}$$

Anyone with a public key can encrypt a message. The receiver needs to use their private key to decrypt it. During this process, no key negotiation or transfer is required, which improves data security by greatly reducing the risk of key leakage. For this reason, asymmetric encryption is widely used in modern encryption systems. Some asymmetric encryption algorithms have also been disclosed, for instance, the most commonly used Rivest–Shamir–Adleman (RSA) algorithm [6] and the ElGamal algorithm [7].

10.3.1.2 File-Level Encryption

There are different ways to ensure data confidentiality. When storing data, data encryption can be divided into two levels: file-level encryption that encrypts the data itself and storage-level encryption that encrypts the file system.

File-level data encryption can further be classified into local data encryption and remote data encryption.

Local Data Encryption

The simplest local data encryption solution is static encryption, which directly applies the encryption algorithm to the data itself. The disadvantage of this solution is that frequently accessed data (such as big data) needs to be temporarily decrypted each time, and this takes longer and compromises user experience as the volume of data increases.

It is for this reason that dynamic encryption has become more popular in recent years. Dynamic encryption, also known as real-time encryption or transparent data encryption (TDE), encrypts data during file use and stores the ciphertext on the server that users share [8]. Users who have the necessary permissions can view the plaintext, and users who do not can only view the ciphertext. Dynamic encryption does not affect the ability of authorized users to use the file, and it does not change the original flow of files in the OS, thus ensuring file security [9, 10]. The BitLocker feature that has come with Microsoft Windows since the launch of Windows Vista offers dynamic encryption in the trusted computing module.

Dynamic encryption uses the hook function of the OS to change the direction of the data flow. When a user closes a file, the system encrypts the temporary file that records plaintext and overwrites the file to the location where the user needs to store the file. When the user reads files, the system only decrypts the files being read. It creates a temporary file in the memory to store the data read from the storage media and then runs the application specified by the user to open the temporary file. This delivers real-time encryption and decryption, which means that users do not face any delays.

10.3 Data Security

Remote Data Encryption

To facilitate data use, cloud data is usually stored in blocks. The cloud program generally reads data on demand instead of by block, which means that it only reads the data that currently requires processing. Therefore, the remote data encryption technology needs to determine the location of the data.

Different encryption methods may cause data to expand or shrink after encryption, making it difficult to determine the location of data using searches based on blocks. This is a big problem for databases in particular. There are two methods we can use to help the cloud program quickly locate the data that needs to be read. We can use encryption algorithms that will not cause data deformation, or we can use partial encryption to only encrypt the data that needs protection and leave the index partially unencrypted.

10.3.1.3 Storage-Level Encryption

Storage-level encryption is usually used to encrypt entire file systems. It offers several advantages over disk-level encryption and TDE, such as easy file transfer, small encryption granularity, and file-specific encryption.

The most commonly encrypted file system is the Cryptographic File System (CFS) for Unix/Linux [11]. This system is built on the network file system (NFS) and operates at the user level. CFS creates a local or remote directory and specifies the key and encryption/decryption algorithm when creating the directory. When a user runs the mount command to access the directory folder, CFS checks whether the user name and key have been authenticated. If yes, CFS creates a mount point in the system so that the user can read and write files in the mount point. Subsequent CFSs such as NCryptfs [12], ECFS [13], Cepheus [14], and TCFS [8] have been developed using this mechanism. In particular, the NCryptfs moves from user mode to kernel mode and provides a powerful file sharing mechanism.

10.3.2 Data Integrity

Data integrity is one of the three pillars of data security. Data integrity means that data is not tampered with during transmission, storage, or processing, and if tampering does occur, it can be detected immediately. The Hamming code [15], cyclic redundancy check (CRC), error correction code (ECC), and erasure code are encoding methods that recover error data to ensure data availability. They are especially suitable for use in storage systems. Here, we introduce two methods for checking data integrity: hashing and digital signatures.

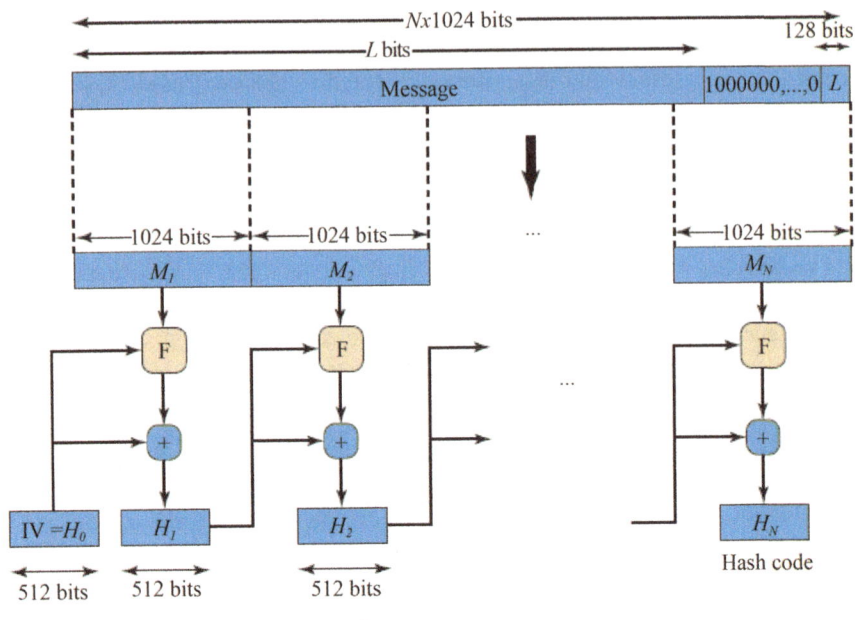

Fig. 10.10 Hashing using SHA-512 [17]

10.3.2.1 Hashing

A hash function or a message-digest algorithm compresses and maps input data of an arbitrary length to an output of a fixed, shorter length which is known as a hash code or a message digest. This process is called hashing. Hashing can be used to check data integrity because it is a one-way process that is strongly resistant to collision, meaning that the process is computationally difficult to reverse and almost never generates the same hash code for different input data. Comparing the hash codes can help us determine whether the two original data files are consistent. Most hashing methods use the Message-Digest algorithm 5 (MD5) or the Secure Hash Algorithms (SHAs) [16, 17]. Figure 10.10 shows the hashing process using SHA-512.

MD5

Since its publication in 1992, MD5 has been widely used to check the integrity of data. The specific program specifications of MD5 are defined as RFC 1321. MD5 divides the input data into 512-bit blocks and processes each block into four 32-bit message words, resulting in a 128-bit hash. The 128-bit hashes are typically represented as a sequence of 32 hexadecimal digits.

SHA

SHAs are a family of hash functions. There are four types of SHA: SHA-0, SHA-1, SHA-2 (SHA-224, SHA-256, SHA-384, and SHA-512), and, the most recent type, SHA-3, which was released in 2015. SHA is widely used to check data integrity. For example, data downloaded from the Internet typically comes with an integrity verification file that shows an SHA hash.

Hashing is also widely used in storage systems and in cloud storage systems in particular. Let us look at data deduplication as the simplest example. It uses hashing and stores one copy of the data that uses the same hash code, which greatly reduces the storage space overheads and costs of cloud storage service providers.

Merkle hash trees (MHTs) are also usually used to perform integrity checks. An MHT is a collection of hash codes, and one of the most important characteristics of it is that its elements cannot be changed. This is why MHTs can be used to check for data integrity. Examples of technologies that use MHTs include provable data possession (PDP) [18] and blockchains.

An MHT is structured in a complete, binary tree, as shown in Fig. 10.11. Every leaf node (which does not have any further children nodes) is labeled with the cryptographic hash of a data block, every node that is not a leaf is labeled with the cryptographic hash of the labels of its child nodes, and this process continues until reaching the root node. For example, the value of H_{1-2} in the figure is H_{1-2} = hash ($H1 \parallel H2$).

During an integrity check, the data block of a node and the hash code of its leaf node are used to compute the root node data block. The result is then compared to the stored root node hash code to determine whether the data is consistent. For example, if we have the root node hash code *TopHash* and want to verify the integrity of *Hash0*, we would need the value of *Hash0* and the *Hash1* (auxiliary verification hash code). We would then construct an MHT to compute the root node data

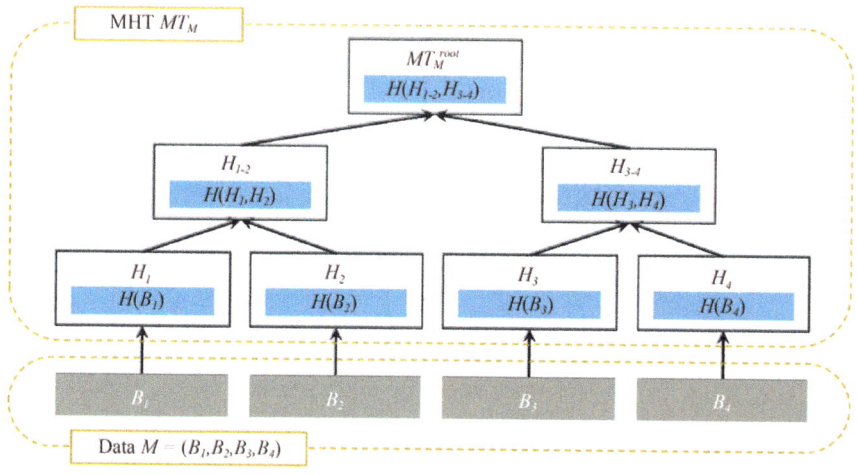

Fig. 10.11 The structure of an MHT [19]

block *TopHash'* and check whether the two are consistent to verify the integrity of *Hash0*.

MHTs are usually used to locate data quickly and to perform consistency checks. In a distributed storage system, when we check the data consistency between the primary and backup storage servers, the computing and communication overheads can be quite high if we compare all data on the two servers. By creating an MHT on both servers, we can reduce these overheads and the time this process takes because we only need to compare the data from the root node. The data on both servers is considered consistent if the root nodes are consistent. In addition, the inconsistent data can be located much more easily by comparing each data layer in the MHT.

10.3.2.2 Digital Signature

A digital signature, also known as a public key signature, consists of a message digest encrypted with the message sender's private key. It can be easily verified, and this makes it ideal for checking the integrity and source of signed data.

Recall asymmetric encryption, which uses a public key to encrypt and a private key to decrypt data, which we explained earlier in this chapter. A digital signature is the opposite. It uses a private key to encrypt and a public key to decrypt data, which means that anyone who knows the public key can verify the authenticity of the signature.

As shown in Fig. 10.12, the sender uses their private key to encrypt the timestamp and message digest and generate a digital signature, which they send together with the message. The receiver uses the sender's public key to decrypt the digital signature and obtain the message timestamp and integrity verification information.

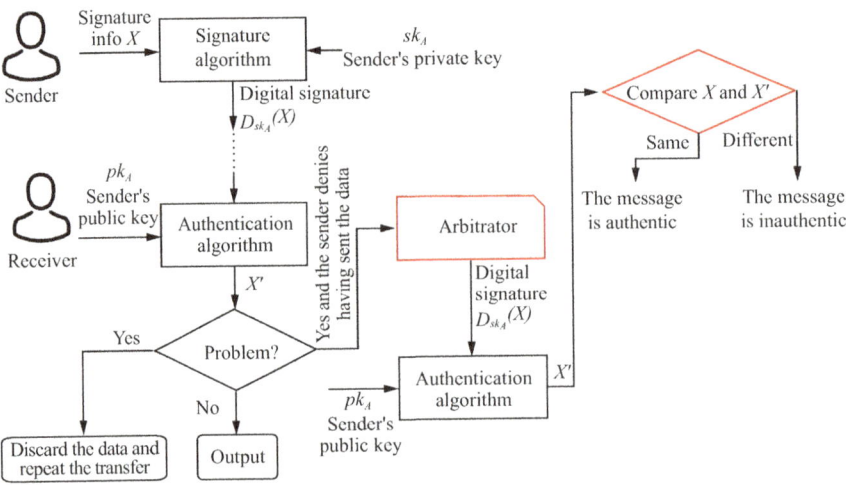

Fig. 10.12 A flow chart showing how digital signatures and their verification algorithms work together to provide non-repudiation

10.3 Data Security

This ensures that the message is not tampered with. If the signature cannot be decrypted by the sender's public key, it means that the message has been sent by someone else or that it has been tampered with. If the receiver can decrypt the signature using the sender's public key, this is undeniable proof that the message was signed by the sender. In other words, digital signatures provide non-repudiation.

Common public key algorithms, such as the RSA, ElGamal, Fiat-Shamir, Schnorr, Data Encryption Standard (DES)/Digital Signature Algorithm (DSA), and elliptic-curve cryptograph, can all be used for digital signatures. There are also many special signature algorithms, such as the blind signature, proxy signature, group signature, and threshold signature, which are used in specific cases. The consensus mechanism of blockchain uses the threshold signature.

Digital signatures play an important role in cloud storage. Take the PDP technology [18] for instance. Users may want to verify whether cloud storage service providers securely store their data given the frequent cloud storage data losses in recent years. As shown in Fig. 10.13, the user generates a verifiable digital signature label that is attached to the data uploaded to the cloud. To avoid disputes between users and cloud service suppliers, users usually entrust data integrity audits to a third-party auditor (TPA). The TPA requires the cloud to execute calculations and return aggregated data verification information, and the verifier then uses the verification information and the digital signature to run sampling verification for data integrity. Data composed of 10,000 blocks requires 460 samples to be verified in order to achieve a 99% confidence level.

This technology makes users better informed about cloud storage and reassures them that their data is safe. It also puts pressure on cloud service suppliers to improve service quality and develop new cloud storage services.

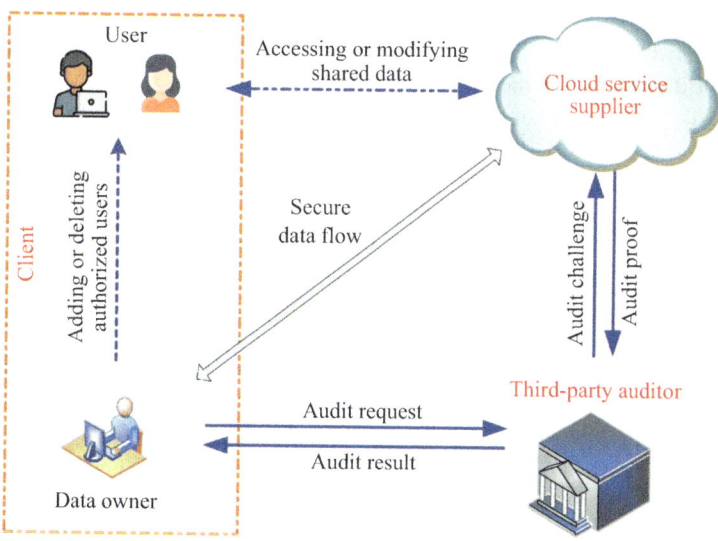

Fig. 10.13 Public audit model for cloud data integrity [20]

10.3.3 Permission Management

10.3.3.1 Access Control

Access control protects the system and resources by regulating every access request and ensuring that only authorized users have access to data in a secure system [21].

Generally, an access control policy is configured by specifying access control lists (ACLs), capability lists, or authorization tables.

If there is an arbitrator, access control typically uses methods such as discretionary access control (DAC), mandatory access control (MAC), or role-based access control (RBAC). If there is no arbitrator, the parties are in a peer-to-peer environment, and the access control is therefore called trust management. A typical example is trust management in a mobile ad hoc network.

According to both the Trusted Computer System Evaluation Criteria (TCSEC) released by the U.S. Department of Defense and the Classified Criteria for Security Protection of Computer Information System (GB 17859-1999) released by China, access control plays an important role in evaluating the security of information systems.

10.3.3.2 Encrypted Search

As cloud computing technologies continue to be developed, more and more people are choosing to upload their data to the cloud. Users usually choose to encrypt their sensitive data before uploading it to the cloud. However, this makes it hard to query and retrieve information because the encrypted data is unidentifiable ciphertext.

The most direct method is to retrieve the data from the cloud, decrypt it, and then search for the required content, but this method is time-consuming and generates a lot of traffic. To overcome this challenge, Song et al. proposed searchable encryption in 2000 [22]. Similarly to encryption, searchable encryption can be classified into symmetric searchable encryption (SSE) [22] and public encryption with keyword search (PEKS) [23]. The former is built on symmetric encryption technologies, while the latter is built on bilinear mapping technologies [24] and a series of complexity assumptions which extend the search communication from "one-to-one" to "one-to-many." Searchable encryption supports single-keyword and multi-keyword searchable encryption.

The most prominent searchable encryption is the attribute-based encryption (ABE) proposed by Sahai and Waters in 2005 [25]. It was later divided into ciphertext-policy attribute-based encryption (CP-ABE) and key-policy attribute-based encryption (KP-ABE). ABE provides fine-grained access control and search authorization while supporting encrypted search, and it ensures user identities are kept private as it offers weak anonymity.

10.3.4 Secure Destruction of Data

There are times when data needs to be cleared from storage media to prevent unauthorized access through social engineering methods to the data and maintain confidentiality, for example, if a storage device is being repurposed or disposed of. We need to take measures to ensure that the data has been completely destroyed and cannot be restored. This section describes two types of secure data destruction: local data destruction and remote data destruction.

10.3.4.1 Local Data Destruction

Local data destruction generally refers to the destruction of data stored on a local storage medium, which is most commonly a disk. According to the Information Security Technology—Security Technique Requirements for Network Storage (GB/T 37939-2019), the storage media where authentication information and sensitive data are stored need to be completely cleared. This can be achieved by methods including using the secure erase command (firmware), overwriting the data multiple times, or destroying the decryption key.

- Running the secure erase command (destruction of all data on a disk): Running this built-in command means instructing the system's firmware to overwrite the entire disk with meaningless combinations of 0s and 1s in order to erase all data from the disk in a way that makes it impossible to restore it, even with the help of data restoration software and technologies. This is also called low-level formatting. The disk is then restored to its factory settings, and the cylinders, tracks, and sectors are re-divided.
- Overwriting the data (destruction of a specific portion of data on a disk): To facilitate data recovery and accelerate deletion, an OS will often mark the metadata as having been deleted but keep the data. Therefore, to completely destroy data, you can iteratively write meaningless data to the area occupied by the deleted data to overwrite the original data and ensure the data cannot be restored.
- Destroying the decryption key (destruction of encrypted data): This method is usually used to erase encrypted data because encryption is usually very secure, and only decryption keys can decrypt and restore the original data. There are three ways to destroy a decryption key: deleting it, running the overwrite command, and running the secure erase command on the encrypted data. The chosen method will depend on the confidentiality of the encrypted data. It is worth noting that running the overwrite command will usually deploy the scrubbing algorithm [26] to iteratively write meaningless data over the original data page to erase the encrypted data and destroy the decryption key.

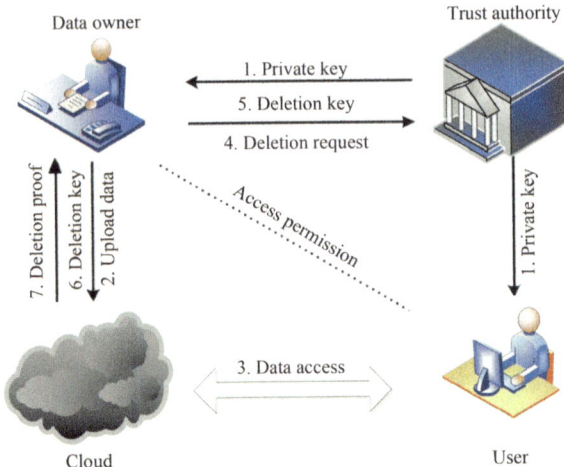

Fig. 10.14 The verification scheme for remote data destruction [28]

10.3.4.2 Remote Data Destruction

As we mentioned earlier, the development of cloud storage technologies is reshaping the storage landscape and is attracting many individuals and small businesses to upload their data to the cloud. This data may include private and confidential information. It is important to delete unneeded data from the cloud properly to ensure data security, even if it has been encrypted. However, in the era of big data, cloud companies have realized the value of data, so they may retain some user data to serve their own commercial interests. As a result, finding ways to permanently delete user data from the cloud has become a widely discussed topic.

Just as with local storage, users can encrypt their confidential data, delete it from the cloud, and then destroy the decryption keys to prevent it from being decrypted and leaked [27]. However, as we mentioned earlier, there is still the risk of the data not being permanently deleted, and this leaves users at risk. For example, cloud service suppliers may manage to obtain valuable information from the ciphertext data. Therefore, researchers have designed a verification scheme for remote data destruction that is similar to PDP [28]. It is shown in Fig. 10.14.

For access control data that was encrypted based on attributes, researchers tend to use additional data structures such as MHTs [29] and bloom filters [30] to regulate changes in access attributes. When data needs to be deleted, the data owner can modify the data structure to delete all attributes. Since no one can meet the attribute control requirements, the data remains inaccessible.

10.3.5 Secure Data Computing

10.3.5.1 Trusted Execution Environment

Cloud computing services often store data in untrusted environments, which gives attackers the opportunity to damage the security attributes of stored data and query operations or even damage and steal users' private information using other attack

10.3 Data Security

methods. To ensure data security during data processing and analysis, researchers proposed setting up a trusted execution environment (TEE) to protect code and data loaded inside the secure area.

Examples of current TEE technologies include the Arm-based TrustZone, x86-based Intel SGX, AMD Secure Encrypted Virtualization (SEV), and IBM Secure Service Container (SSC). Different TEE hardware offers different security abstraction granularities and data protection levels. For example, TrustZone uses secure physical machines, SEV uses VMs, and SGX uses code segments for security abstraction. In terms of data protection levels, SEV offers memory confidentiality but does not ensure memory integrity, TrustZone provides integrity protection but lacks confidentiality support, and SGX provides both confidentiality and integrity protection. It is clear that Intel SGX has advantages over the other TEE technologies, so now we will look at SGX in more detail.

Intel officially released SGX v1 in 2015 and SGX v2 in 2016. Figure 10.15 shows the principle of SGX. There are both trusted and untrusted parts in the system. The app runs and creates the trusted part, which is an enclave that is placed in the trusted memory and used to store key data or code. Programs running in the enclave are not affected by hardware other than the OS, basic input/output system (BIOS), and central processing unit (CPU). The edge system call interface provided by SGX needs to be used to enter or exit the enclave, and this protects the stored data and code. Typically, an enclave usually includes an enclave page cache (EPC), a trusted memory area. This area is protected by the on-chip memory encryption engine (MEE). The mapping from the virtual address to the physical address in this area is protected by the hardware address translation logic. As a result, non-enclave programs cannot access the EPC. This is how SGX protects data.

However, this security assurance comes with two overheads: when SGX processes private data, the system enters and exits the secure area constructed by SGX which causes context switching during entry and exit as well as key code and data replication between the memory of the secure area and the insecure area; when the code and data in the enclave need to be stored persistently, the MEE encrypts them, the inter-enclave local authentication or third-party remote authentication checks whether the code and data have been damaged, and this generates additional resource overheads.

SGX also has hardware limitations, such as limited trusted memory resources and insufficient data persistence support. Therefore, it is difficult to make full use of

Fig. 10.15 SGX principle

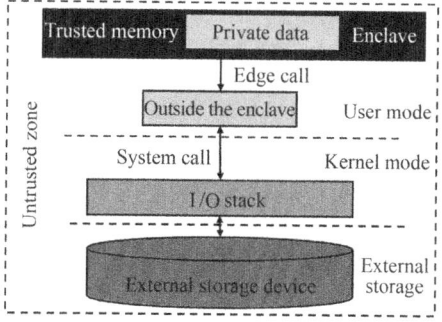

the advantages that SGX has over other TEE technologies by simply placing the storage system inside SGX without first optimizing and reconstructing the storage system. Consequently, industry experts are urgently looking for ways to ensure the confidentiality and integrity of stored data while also maintaining the performance of the storage system.

Currently, researchers primarily use the trusted memory created by SGX to ensure the confidentiality and integrity of stored data and optimize edge calls during SGX execution. They are also looking into ways to overcome the limitations of SGX, such as its limited capacity.

One well-known study is using SGX to build a trusted computing framework in the untrusted area [3, 31–48]. Take Scone [3] as an example. It uses SGX to build a secure Linux container and leverages SGX's trusted execution area enclave to protect Linux processes from external attacks. Also, Haven [31] system implements the concept of shielded execution, which uses the SGX hardware protection mechanism to protect the confidentiality and integrity of users' private data and the platform that operates the data.

Another problem with SGX is that apps in the enclave must invoke edge call functions to enter and exit the enclave, but it is not possible to invoke the system from within the enclave. The Scone system uses two lock-free multi-producer, multi-consumer queues to implement interfaces for asynchronous system invocation, which reduces the overheads associated with edge invocation. The Eleos [35] system uses a solution in which a system call is executed without the app exiting the SGX trusted memory space. The system call is executed using a remote procedure call (RPC) running in a thread outside the enclave. To address the limited capacity of the SGX EPC area [37, 42–47], the Enclage [47] system uses the SGX enclave to construct a three-tier storage hierarchy comprising EPC (trusted storage), dynamic random access memory (DRAM, untrusted storage), and external storage (persistent storage) and designs a method for secure data exchanges at the software layer. To avoid the overhead that arises from hardware paging, the Enclage system adopts a fixed trusted memory usage solution. The Aria [48, 49] solution, on the other hand, tackles the same limited capacity problem by storing key-value pair and index structures in untrusted memory and security metadata in the TEE to protect data. This uses the limited SGX resources efficiently and performs well.

10.3.5.2 Homomorphic Encryption

Homomorphic encryption is the general term for a class of encryption algorithms with special properties. It is built on a series of complexity assumptions, which means it is reliable in most cases. Different homomorphic encryption algorithms tend to have different properties. Two examples are as follows:

Additive homomorphism: $H(A) + H(B) = H(A + B)$
Multiplicative homomorphism: $H(A) \times H(B) = H(AB)$

The complexity assumptions enable the homomorphic encryption of data before computing, and this ensures data privacy and prevents data leaks. As a result, homomorphic encryption has become the foundation of many distributed security computing technologies.

Homomorphic encryption is typically quite complex and takes longer to compute than TEEs. However, it provides enhanced security as data is not decrypted during the computing process.

10.4 Security Management

The key components of the security management technology are automatic security management, self-adaptive resilience management, continuous risk assessment, and centralized policy management.

10.4.1 System Access Control (Authentication Management)

Access control, or authentication management, is the most important line of security defense. This is because the evolution of data centers has given rise to a variety of storage deployment forms (distributed, clustered, and virtualized), and the number of storage devices has consequently skyrocketed. The management and maintenance of storage devices typically require a centralized element management system (EMS), which connects to storage servers, storage network switches, and service hosts to provide full-stack management capabilities. The interconnection of the EMS with most devices in the network increases security risks, and this makes authentication management the most important part of security protection. This section describes some of the mainstream authentication management solutions.

10.4.1.1 Unified Management of Identity Authentication

Typically, user identity information is scattered across systems. This means that the maintenance personnel need to maintain multiple systems concurrently, which increases the complexity of their work exponentially. Some of the risks involved include the fact that user rights cannot be managed in a centralized manner, making least privilege impossible, disordered and shared identities such as overlapping human-machine and machine-machine accounts, and how the repeated input of usernames and passwords when frequently switching between systems complicates work and increases the risk of password leakage. As a result, a centralized and unified identity authentication management solution that integrates authentication, authorization, and accounting (AAA) and audit capabilities is required. This

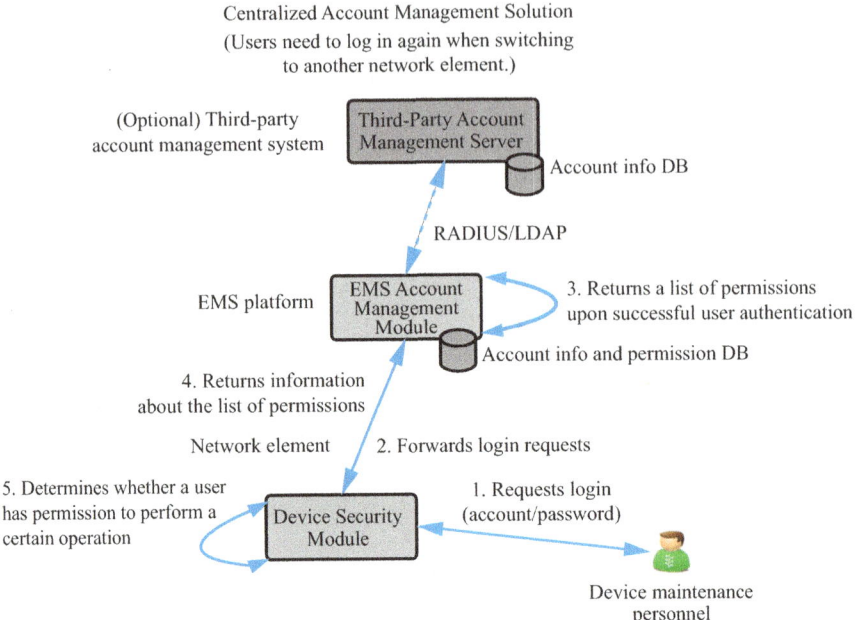

Fig. 10.16 Centralized account management architecture

solution would improve system security and management by working closely with the risk assessment module to provide authorization based on the level of risk.

Centralized account management: This solution offers unified management of servers, network devices, and app systems throughout the account lifecycle (which includes creation, deletion, and synchronization of associated accounts) as well as the management of account/password policies (shown in Fig. 10.16).

Single sign-on (SSO): This is a user-friendly identity authentication scheme that allows a user to log in to several related but independent app systems and services using a single ID. During the session, the user can access these apps and services without needing to log in again. As shown in Fig. 10.17, the authentication center is responsible for user management. The service system does not perform login authentication. Once the account has been authenticated and a temporary token encoded with the validity period and resource access information has been obtained, the Open Authorization (OAuth) protocol is used to authenticate resource access.

10.4.1.2 Two-Factor Authentication

Two-factor authentication (2FA) is an identity and access management security method that requires two authentication factors to access resources and data. It is typically used in scenarios that require strict identity authentication. For example, it might be used for high-risk storage operations, such as removing zones from

10.4 Security Management

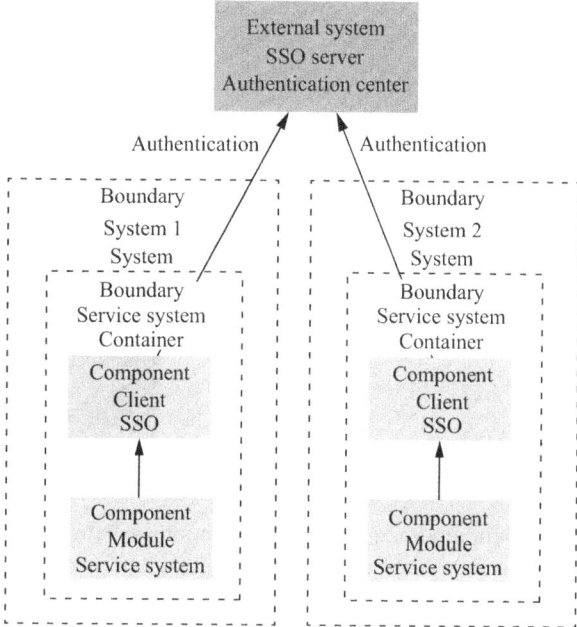

Fig. 10.17 SSO

switches, unmapping volumes from hosts, and deleting storage devices in storage maintenance.

Due to the importance of the storage EMS, the interconnection protocol it uses must support 2FA. Remote Authentication Dial-In User Service (RADIUS) is an authentication system that is defined in RFC 2865 and RFC 2866. It is the most widely used AAA. RADIUS supports multiple protocols, such as the Password Authentication Protocol (PAP), Challenge Handshake Authentication Protocol (CHAP), and Microsoft Challenge Handshake Authentication Protocol version 1 (MS-CHAPv1) and version 2 (MS-CHAPv2). Given that RADIUS uses 2FA, when a remote user enters their username and password, the password is not marked as a personal account password but rather as a PIN and token code. The token code is a six-digit number that is periodically generated by an independent hardware device.

10.4.1.3 Account Security Management

Brute-force attack prevention: The login authentication on the management interface and across trusted networks must support the brute-force attack prevention mechanism. When the number of incorrect password attempts exceeds a specified threshold (for example, three), the mechanism implements proper protection measures such as account locking, IP address locking, delayed login, verification codes, and an IP address whitelist (Fig. 10.18).

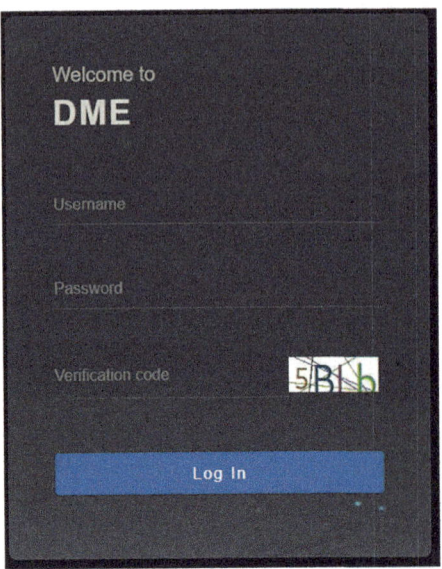

Fig. 10.18 DME login authentication interface

First login mode: Storage authentication systems typically use the first login mode to prevent attackers from obtaining default account names and passwords from public documents or historical versions. Attackers can use these accounts to access the storage system and users' data and potentially disclose confidential information.

The first login mode asks each user to set a password that meets certain complexity requirements when they first log in. Once the system starts to run, the human-machine accounts created or reset by the administrator cannot use the default passwords (such as hardcoded passwords). Instead, the administrator must set new passwords, or the system will automatically generate random passwords to meet the password complexity requirements. The passwords of the accounts created or reset by the administrator must be changed upon the first login.

Typically, the first login mode is used in two situations. First, during the installation and deployment of the storage EMS, database, OS, and EMS users are required to set passwords that meet complexity requirements to prevent them from using their default passwords. The other is when the EMS administrator creates users using the first login mode. The super administrator then creates user accounts and passwords that meet the complexity requirements. If the system detects that the user has not logged in to the system nor changed the password in the background, this mode will force the user to change the password to prevent the administrator from owning the password.

Account security management: Unauthorized users may be able to guess or deduce user accounts and passwords. In terms of security maintenance, a system must be able to manage account security and allow security administrators to adjust policies in real time to improve security.

10.4 Security Management

When the EMS provides online access for users, it monitors the login information like the users' IP addresses and compares this with the predefined network segment or IP address to determine whether the users are valid and legitimate. If not, the system generates an alarm about the unusual login and forces the user to log out. The EMS can also compare the user's login time with the routine working hours of the employee listed on the account. Again, if the login seems suspicious, it generates an alarm and takes measures to ensure security.

Generally, we advise setting a validity period for an account based on the project duration. However, if an employee resigns or a project ends earlier than expected, the administrator needs to manually deregister the account at the end of its life cycle to prevent unauthorized access to the system.

10.4.2 User Identity and Access Management

User identity and access management is the management of a user's identity and its access to data. It is the foundation of system security as it protects the system and resources from unauthorized access. The five common access control models are as follows:

DAC: The underlying philosophy of DAC is that the owner (i.e., the subject) can determine who has access to their objects. Subjects with access permissions may access specified objects as authorized. For example, if you create a file, you are the owner of this file and can grant or deny any user permission to access the file.

RBAC: In an RBAC system, the subject accesses objects based on their role. Generally, the system administrator assigns access permissions to certain roles based on the role responsibilities. If a certain role is assigned to a user, then that user has all the rights of the role. A role can come with multiple permissions and can have multiple subjects. Similarly, a single subject can be assigned multiple roles. This role-based management model supports various security policies, simplifies authorization management, and reduces management overheads.

Rule-based access control (RuBAC): This is an access control model which uses global rules, which means that the rules apply to all subjects equally across the board. Firewall is one example of RuBAC because firewalls permit or block all traffic equally.

Attribute-based access control (ABAC): This is a logical access control methodology where access permissions are managed based on evaluations of the subject's one or multiple attributes. As a result, it is more flexible than the RuBAC model.

Mandatory access control (MAC): MAC is a means of restricting access (using access rules) to objects based on the sensitivity (as represented by a label) of the information contained in the subject and object. Subjects with access permissions may access specified objects as authorized. Sensitivity labels are automatically set and maintained by the security administrator or system based on pre-determined rules (e.g., the separation of privilege model).

The privilege separation model has three default roles: System administrator, security administrator, and security auditor. This makes it suitable for use in high-security scenarios.

System administrator: This role is responsible for managing and maintaining the storage system. Its responsibilities include system monitoring and configuration, user management, resource management, data protection management, and tenant management.

Security administrator: This role is responsible for managing the security of the storage system. Its responsibilities include security configuration and role management.

Security auditor: This role is responsible for tracing and auditing the operations of the system administrators and security administrators. The storage system uses the RBAC model and the separation of privileges to ensure that users' permissions and the scope of their mandates can be controlled through roles. This is to prevent abuse of power and operational errors that may affect service system stability and data security.

Specifically, the storage system defines seven roles based on system management, resource management, data protection management, tenant management, security management, and routine maintenance requirements:

Super administrator: This role grants all privileges in a storage system. Its responsibilities include user management in scenarios such as system deployment, password retrieval, and user account unlocking.

System administrator: This role grants all privileges except for user management.

Security administrator: This role can configure system security.

Resource administrator: This role has the permissions required to manage system resources, including block storage, file storage, and network resources.

Data protection administrator: This role has local and remote data protection management privileges.

Monitoring administrator: This role has routine O&M privileges, such as information collection, performance data collection, and inspection. It does not have permissions to manage storage resources, data protection, or security configuration.

Tenant administrator: This role has all tenant management privileges.

To meet the system privilege management requirements of different industries and customers, the storage system is designed to support the customization of roles.

10.4.3 Certificate and Key Management

In 2018, expired digital certificates paralyzed the network communication services of a well-known telecom company in 11 countries. The network of one victim company was interrupted for 4 h and 25 min, during which time a total of 30.6 million users were unable to communicate. This caused a sharp drop in the value of this

10.4 Security Management

company's stocks, and 10,000 users chose to terminate their contracts in the 5 days following the event. This example illustrates the importance of maintaining digital certificates and keys properly to ensure that the systems continue to run securely and avoid huge economic losses.

10.4.3.1 Public Key Infrastructure

Definition of Digital Certificate

In order to understand public key infrastructure (PKI), we need to first understand what a digital certificate is. A digital certificate, or a public key certificate, is a form of electronic record that proves the authenticity of a device, user, or app in the digital world. It is similar to an identity card in the real world. Digital certificates typically use the latest version of the X.509 standard format [50], which is currently version 3. They contain an identity, a public key, and the digital signature value of the identity authentication authority. Digital certificates are widely used for encryption in Transport Layer Security (TLS), Secure Socket Layer (SSL), and Internet Protocol Security (IPSec) communication as well as identity authentication, authorization management, electronic signatures, and software integrity verification. Figure 10.19 shows the PKI architecture.

A certificate issuing authority is called a certificate authority or certification authority (CA). It is an authoritative, trustworthy, and impartial third-party organization that stores, signs, and issues digital certificates. The Electronic Signature Law of the People's Republic of China defines it as an electronic authentication service provider. Table 10.1 describes the fields found on an X.509 v3 digital certificate.

Now that we have understood what a digital certificate is, we can move onto learning about PKI. PKI is a support mechanism for the digital certificate, and it works like the support mechanism for our identity cards. PKI is a key management platform that follows a certain standard. It provides a management system for keys and digital certificates required by network apps to provide cryptographic services such as data encryption and digital signatures, and it ensures the security of the network environment. PKI primarily contains the following elements: a user, CA, registration authority (RA), digital certificate library (a library that stores components of digital certificates using databases, Lightweight Directory Access Protocol (LDAP), or X.500), digital certificate revocation system, and key backup and restoration system.

Fig. 10.19 PKI architecture

Table 10.1 Fields found on an X.509 V3 certificate [50]

Certificate field		Description
Certificate fields	Version	The version of the certificate, INTEGER{v1(0), v2(1), v3(2)}
	Serial number	The serial number of the certificate. It is a character string issued by a CA. The CA must ensure that the serial number is unique within the application scope
	Signature	The identifier for the cryptographic algorithm used by the CA to sign the certificate. Typically, secure certificate signature algorithms, such as the SHA256withRSA and SHA256ECDSA, are used
	Issuer DN	The distinguished name (DN) of the certificate's issuing CA. This is determined by the CA. Generally, this value comprises: Country Name (two-character abbreviation of the country name, e.g., CN) State or Province Name (e.g., Guangdong) Locality Name (e.g., Shenzhen) Organization Name (e.g., Huawei) Organizational Unit Name (e.g., Wireless Network Product Line) Common Name (e.g., Huawei Wireless Network Product CA) If a product uses a tool such as OpenSSL to issue a CA certificate, it is advised to fill in all of the above information about the Issuer DN. The Common Name field is mandatory
	Validity	The time period for which the certificate is valid NotBefore: This is the date on which the CA issues the certificate and on which it becomes valid. This date cannot be before the date on which the application was made NotAfter: This is the date on which the certificate expires. This date is calculated by adding the validity period to the NotBefore date. This date cannot be after the expiration date of the CA certificate
Certificate fields	Subject DN	The distinguished name (DN) of the certificate subject and is determined by the certificate requester. Generally, this value comprises: Common Name (it is recommended that the product code that uniquely identifies the product or non-volatile product information be included, such as the electronic serial number, IP address, MAC address, and server name) Country Name (two-character abbreviation of the country name, e.g., CN) State or Province Name (e.g., Guangdong) Locality Name (e.g., Shenzhen) Organization Name (e.g., Huawei) Organizational Unit Name (e.g., Wireless Network Product Line) If a product uses a tool such as OpenSSL to issue a device certificate, it is advised to fill in all of the above information about the Subject DN. The Common Name field is mandatory
	Subject public key	The public key owned by the certificate subject. This is usually either the RSA public key of 3072 bits or more or the ECC public key of 256 bits or more

10.4 Security Management

Certificate extensions	Authority key identifier	Optional extension
		An identifier that represents the key of the issuer. It is the unique identifier that authenticates the issuer's public key. Its value comes from the requester key identifier (Subject KeyIdentifier) in the CA certificate
		However, when multiple CA certificates bearing the same issuer DN are used, it is used to identify the CA certificate whose public key is used to authenticate the requester certificate, and its value is the hash of the public key of the issuing CA
	Subject key identifier	Optional extension
		An identifier that represents the key of the requester. It is the unique identifier of the public key contained in the current certificate. When multiple certificates with the same Subject DN are used, it is used to identify the public keys of certificates and its value is the hash of the public key of the current certificate
	Key usage	Optional extension
		A value that defines the functions and services supported by the public key of the current certificate. The value comprises a digital signature, non-repudiation, key encipherment, data encipherment, key agreement, a certificate signature, a CRL signature, encipher only, and decipher only
		The certificate signature and CRL signature are unique to CA certificates. Therefore, although the keyUsage extension field is optional, it is a crucial part of the root CA certificate, and the certificate signature and CRL signature fields are mandatory
	Extended KeyUsage	Optional extension
		A value that contains a series of object identifier (OID) values, representing the specific usage of the public key in the current certificate. Although the X.509 standard does not clearly define the extended key usage, RFC 3280 lists certain OID values related to this extension, including TLS server authentication, TLS client authentication, code signing, e-mail protection, time stamping, and Online Certificate Status Protocol (OCSP) signing

(continued)

Table 10.1 (continued)

Certificate field		Description
Certificate extensions	Basic constraints	A collection of constraints This is an essential extension for CA certificates (including the root CA certificate and level-2 CA certificate). Its value is: Subject Type = CA Path Length Constraint = None or a specific number. In most cases, the value is None, meaning a CA may have unlimited levels of subordinate CAs underneath it However, for an end user certificate, this extension is optional. Its value is: Subject Type = End Entity Path Length Constraint = None
	Subject alternative name	Optional extension A collection of alternative names (except Subject DN) for the certificate owner, including an email address, an IP address, a uniform resource identifier (URI), and the like. This extension is essential for apps that ensure e-mail security because these apps need to extract e-mail addresses from this extension and bind them to certificates
	CRL distribution points	Optional extension A collection of URLs where the base certificate revocation list (CRL) is published. The information is provided by CA

10.4 Security Management

Digital certificates for storage devices include both the device certificate and the root certificate.

A device certificate provides proof of the device's identity. Typically, a device has both a public key certificate and a private key certificate. A device certificate is a public key certificate that identifies and provides proof for the identity of a network element (NE). It is used to verify the validity of an NE during authentication. The digital certificate defined in IEEE 802.1AR is the device certificate. A device certificate has both a public key and a private key, and they are used to calculate the digital signatures or encrypt and decrypt data during NE or EMS authentication. Typically, a device certificate consists of three files: the public key, the private key, and the private key password ciphertext. Strictly speaking, the preconfigured certificate in the software package cannot be called a device certificate. It is not bound to a device, which means that the preconfigured certificates of a batch of devices would be the same, which increases the risk of private key leakage.

The root certificate, which is generally issued by a CA, is used to verify the validity of the digital signature of the other party and check whether their certificate was issued by the CA. The device only has the root certificate file, because the private key is held by the CA.

In storage, digital certificates are used to authenticate identities when an SSL connection is established, enabling secure communication within and between components and between components and EMS. The SSL connection protects data transmission at the application layer.

Why Do We Manage Digital Certificates?

There are two reasons why we need to manage digital certificates.

The first is that a digital certificate is a CA's signature for the public key of the subject. A public key and a private key are a mathematically related pair, meaning if the public key is disclosed, the private key will be at risk.

The second is that, as we mentioned earlier in this chapter, each digital certificate has a validity period and an expiration date. In addition, if a CA is acquired or disqualified, the digital certificates that it issued will become untrusted or their roots will be changed.

Digital certificate management and maintenance ensure the correct use of products. The management of digital certificates delivers faster digital certificate O&M, reduces the likelihood of errors, and minimizes the security and service stability risks that arise from improper digital certificate O&M.

Storage Device Certificate Management

The management of storage device certificates comprises the following four aspects:

The provision of CA service capabilities to support the automatic issuance of certificates during the installation and deployment of NEs.

Device interconnection with the CA using the standard certificate management protocol to facilitate the automatic issuance (or application) of digital certificates.

The unified management of certificates, including monitoring certificate validity periods, supporting certificate updates, importing revocation lists, and exporting certificate application files.

The provision of a command-line interface or a user interface (UI) via the device to support certificate maintenance, lower manual labor intensity, and improve maintenance accuracy.

In the case where the storage EMS is connected, the digital certificates of multiple devices at the data center level can be managed in a unified manner. In certain scenarios, storage device certificates can be automatically updated, and this greatly reduces certificate management costs for maintenance personnel.

10.4.3.2 The Key Distribution Mechanism

The core objective of key management is to ensure the confidentiality and integrity of keys while minimizing the security risks that arise from key leakage.

To ensure secure key management, a key hierarchy is typically used. A key hierarchy helps reduce the usage frequency of the root key and thus protects it. It also restricts the amount of data encrypted by a single key, which prevents ciphertext-only attacks and contains the impact scope of key leakage.

The hierarchical key management system typically has three layers: the root key, the key encryption key, and the work key. This hierarchy is structured like a tree, such that the lower-layer key protects (encrypts or derives) the upper-layer key.

Figure 10.20 shows the three-layer key management structure.

Work key: This is a key that is directly used by an NE or EMS to protect the confidentiality and integrity of locally stored sensitive data and data that needs to be transmitted over insecure channels. It also provides signature and cryptographic

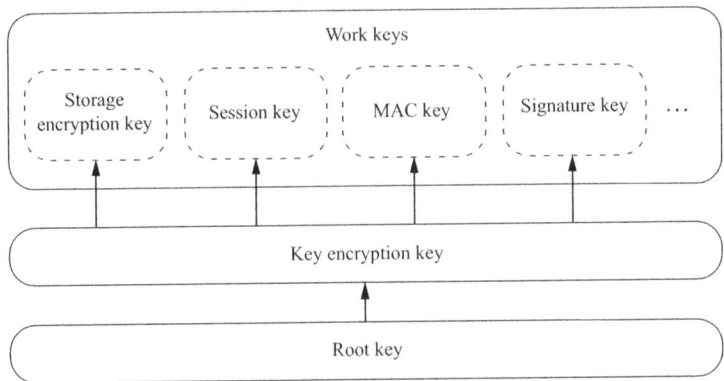

Fig. 10.20 Three-layer key management structure

10.4 Security Management

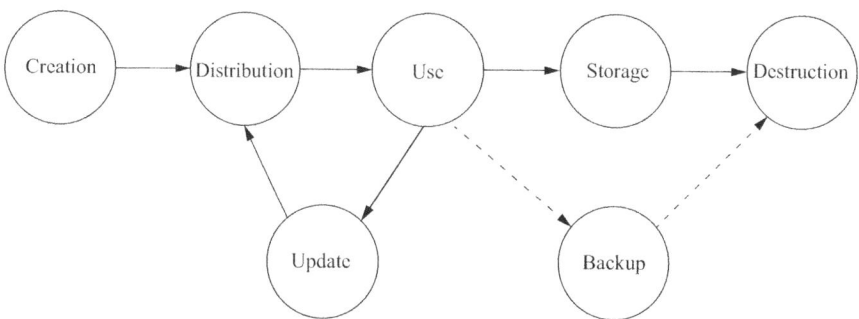

Fig. 10.21 The states that a key may transition through during its lifecycle

Table 10.2 The potential security risks in a key's lifecycle

State	Security issue caused by improper design
Creation	Keys are generated by the algorithm in a weakly random way, meaning that a key can be predicted or attackers can generate keys by themselves
Distribution	Keys are distributed in plaintext and may be intercepted by attackers
Update	Keys are never updated, so attackers can easily obtain them and access the plaintext of sensitive data
Storage	Keys are stored in plaintext in the database. As a result, attackers can easily read the keys and obtain the plaintext of sensitive data
Backup	If important keys are not backed up and they are lost, the encrypted data cannot be decrypted. This greatly reduces system reliability
Destruction	Attackers may restore keys after they have been deleted

authentication functions. Work keys can be directly used by upper-layer applications and used for keys like storage encryption key, session key, MAC key, and signature key.

Key encryption key: Each key encryption key, which is protected by the root key, provides confidentiality protection for the work key. In storage, the root key can function as the key encryption key to execute simple functions that do not require high security. In this case, there are only two layers of keys in the hierarchy instead of three.

Root key: The root key is the bottom layer of the key management hierarchy, and it is used to protect the confidentiality of all upper-layer keys (such as key encryption keys). It requires the highest security compared with the other two types of keys and is generally stored in the hardware encryption module.

A key may transition through several different states during its lifecycle, and these are shown in Fig. 10.21. A key must meet the security requirements of each state to ensure its security.

Any weaknesses in these states can lead to security risks. Table 10.2 lists the security risks that may arise from improper design at each state of the key's lifecycle.

10.4.4 Cybersecurity Management

10.4.4.1 Cybersecurity Framework

Identify—Protect—Detect—Respond—Recover (IPDRR) is a cybersecurity framework from the National Institute of Standards and Technology (NIST) of the USA. It consists of five functions:

Identify: This function assists in risk evaluation. It comprises service prioritization, risk identification, impact assessment, and resource prioritization.
Protect: This function ensures service continuity. It contains the impact of an attack on services. It mainly refers to the automatic protection measures before manual intervention.
Detect: This function identifies cybersecurity events. It monitors attacks in real time and checks whether services are running properly and whether protection measures are working as they should be.
Respond: This function responds to and handles detected cybersecurity incidents. Specific responses are selected based on the impact of the incident. This function supports incident investigation, damage assessment, evidence collection, incident reporting, and system recovery.
Recover: This function restores the system in the event of an attack and finds the root cause of the incident. It also takes preventative measures and fixes vulnerabilities.

Each function plays an important role, so this framework can help enterprises quickly build a strong and comprehensive network security system that brings these functions together.

10.4.4.2 Host-Based Intrusion Detection

The host-based intrusion detection system that is built into the storage system can be deployed hierarchically. The EMS layer runs the basic intrusion detection and can interconnect with the security operations center (SOC) to analyze the complex attack context and path. This layer uses logs, configuration information, and virus scanning information collected by NEs to evaluate and respond to the intrusion on the storage network. Figure 10.22 shows the typical intrusion detection and response deployment for telecom hosts.

10.4.4.3 Trusted Boot and Remote Attestation

Trusted boot: Hardware capabilities and startup codes can be combined to establish a root of trust for the boot platform. In a trusted boot, each stage measures the boot of the next stage in the following sequence: BIOS, bootloader, OS kernel, and system software package. This forms a complete chain of trust. The mea-

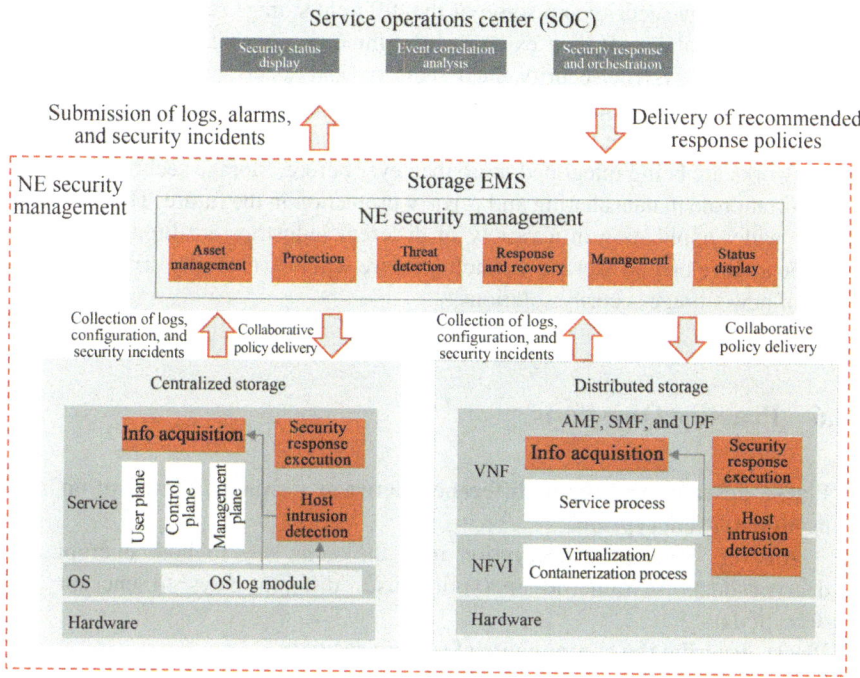

Fig. 10.22 Typical intrusion detection and response deployment for telecom hosts

surement results are then permanently saved to the trusted platform module (TPM) chip, facilitating the creation and transfer of the chain of trust and the recording of the system trust status.

Remote attestation (RA): RA consists of the RA server, RA client, and CA. The RA server is the core component in RA. It stores the reference value of each platform configuration register (PCR) as the baseline when software is released and receives the PCR value sent by the RA client. It then compares them to verify the trust status of the RA client. The RA client has a TPM chip and supports trusted boot. It responds to challenge requests from the RA server by collecting PCR values and sending them to the RA server. The CA issues attestation identity key (AIK) certificates to the RA client to prevent the RA client from being counterfeited.

10.5 Summary

Data has emerged as a new factor of production. As a result, data security has become a major concern for individuals, enterprises, and countries. This chapter provided a comprehensive analysis of storage security. It first introduced the concept of storage security and some of the different storage security systems that are currently available. It then explained the three components that make up the security system: system security, data security, and security management. This chapter also described some cutting-edge storage security applications such as encrypted search, remote data destruction, and homomorphic encryption. Storage and networks are being integrated more than ever before. Storage security will play an important role in data sharing and privacy protection in the future. The possibilities for wider application in a variety of different industries are limitless. As we continue to develop new storage technologies, we will also naturally drive the emergence of new storage security solutions.

10.6 Practice Questions

1. **Please describe the main difference between symmetric encryption and asymmetric encryption.**

   Answer: Symmetric encryption uses the same key to both encrypt and decrypt data. Asymmetric encryption uses different keys to encrypt and decrypt data.

2. **Please describe the components of system security.**

   Answer: System security comprises hardware security, container security, and system resilience.

3. **What are the two types of file-level encryption that ensure data confidentiality?**

   Answer: Data encryption at the file level can be classified into local data encryption and remote data encryption.

4. **What are the differences between hashing and digital signatures when in use?**

   Answer: Anyone can use the hash function to calculate the message digest, but you need the public key to decrypt and verify the authenticity of the digital signature.

5. **Hardware security is prone to which malicious attacks?**

   Answer: It is prone to three attacks: side-channel attacks, fault injection attacks, and physical attacks.

6. **What security isolation mechanisms can be implemented based on trusted hardware?**

   Answer: Arm TrustZone, Intel SGX, AMD SEV, and Intel TDX.

7. **What are the key elements of security management technologies?**

   Answer: The key elements are automatic security management, self-adaptive resilience management, continuous risk assessment, and centralized policy management.

8. **Please name the common access control models.**

   Answer: The common models are DAC, RBAC, RuBAC, ABAC, and MAC.

References

1. Brenner M. Biggest risks of cloud computing and how to mitigate them (2021-12-13) [2023-05-30].
2. Shu J, Shen Z, Xue W. Shield: a stackable secure storage system for file sharing in public storage. J Parallel Distrib Comput. 2014;74(9):2872–83.
3. Arnautov S, Trach B, Gregor F, et al. Scone: secure Linux containers with Intel SGX. In: Proceedings of the 12th USENIX symposium on operating systems design and implementation (OSDI). Berkeley: USENIX Association; 2016. p. 689–703.
4. Katz J, Lindell Y. Introduction to modern cryptography. Boca Raton: CRC Press; 2020.
5. Hellman M. New directions in cryptography. IEEE Trans Inf Theory. 1976;22(6):644–54.
6. Rivest RL, Shamir A, Adleman L. A method for obtaining digital signatures and public-key cryptosystems. Commun ACM. 1978;21(2):120–6.
7. Gamal TE. A public key cryptosystem and a signature scheme based on discrete logarithms. IEEE Trans Inf Theory. 1985;31:469–72.
8. Cattaneo G, Catuogno L, Del Sorbo A, et al. The design and implementation of a transparent cryptographic file system for UNIX. In: Proceedings of the 2001 USENIX annual technical conference (ATC). Berkeley: USENIX Association; 2001.
9. Xiao D. Cunchu xitong zhong de shuju anquan fangfa yu jishu [Data security methods and technologies in storage systems]. Beijing: Tsinghua University; 2008.
10. Xiao D, Shu J, Xue W, et al. Jiyu zumiyao fuwuqi de jiami wenjian xitong de sheji he shixian [Design and implementation of a group key server-based cryptographic file system]. Chin J Comput. 2008;31(4):600–10.
11. Blaze M. A cryptographic file system for UNIX. In: Proceedings of the 1st ACM conference on computer and communications security. New York: ACM; 1993. p. 9–16.
12. Wright CP, Martino MC, Zadok E. NCryptfs: a secure and convenient cryptographic file system. In: Proceedings of the 2003 USENIX annual technical conference (ATC). Berkeley: USENIX Association; 2003. p. 197–210.
13. Bindel D, Chew M, Wells C. Extended cryptographic file system (2001-01) [2023-05-30].
14. Fu KE. Group sharing and random access in cryptographic storage file systems. Cambridge: Massachusetts Institute of Technology; 1999.
15. Kumar UK, Umashankar BS. Improved hamming code for error detection and correction. In: Proceedings of the 2nd international symposium on wireless pervasive computing (ISWPC). Piscataway: IEEE; 2007. p. 498–50.
16. Gupta P, Kumar S. A comparative analysis of SHA and MD5 algorithm. Int J Comput Sci Inf Technol. 2014;5(3):4492–5.
17. Stallings W. Cryptography and network security: principles and practice. London: Pearson Education; 2020.
18. Ateniese G, Burns R, Curtmola R, et al. Provable data possession at untrusted stores. In: Proceedings of the 14th ACM conference on computer and communications security (CCS). New York: ACM; 2007. p. 598–609.
19. Koo D, Shin Y, Yun J, et al. Improving security and reliability in Merkle tree-based online data authentication with leakage resilience. Appl Sci. 2018;8(12):2532.
20. Tian H, Chen Y, Jiang H, et al. Public auditing for trusted cloud storage services. IEEE Secur Privacy. 2019;17(1):10–22.
21. Xue M. Yizhong gongxiang cunchu huanjing xia de anquan cunchu xitong [A secure storage system over shared storage environment]. Beijing: Tsinghua University; 2011.
22. Song DX, Wagner D, Perrig A. Practical techniques for searches on encrypted data. In: Proceedings of the 2000 IEEE symposium on security and privacy (S&P). Piscataway: IEEE; 2000. p. 44–55.
23. Boneh D, Di Crescenzo G, Ostrovsky R, et al. Public key encryption with keyword search. In: Proceedings of the lecture notes in computer science. Berlin: Springer; 2004. p. 506–22.

24. Boneh D, Franklin M. Identity-based encryption from the Weil pairing. Siam J Comput. 2003;32(3):586–615.
25. Sahai A, Waters B. Fuzzy identity-based encryption. In: Proceedings of the annual international conference on the theory and applications of cryptographic techniques (EUROCRYPT). Berlin: Springer; 2005. p. 457–73.
26. Wei M, Grupp LM, Spada FE, et al. Reliably erasing data from flash-based solid state drives. In: Proceedings of the 9th USENIX conference on file and storage technologies (FAST). Berkeley: USENIX Association; 2011. p. 105–17.
27. Boneh D, Lipton RJ. A revocable backup system. In: Proceedings of the USENIX security symposium (security). Berkeley: USENIX Association; 1996. p. 91–6.
28. Xue L, Yu Y, Li Y, et al. Efficient attribute-based encryption with attribute revocation for assured data deletion. Inf Sci. 2019;479:640–50.
29. Liu C, Ranjan R, Yang C, et al. MuR-DPA: top-down leveled multi-replica Merkle hash tree based secure public auditing for dynamic big data storage on cloud. IEEE Trans Comput. 2014;64(9):2609–22.
30. Yang C, Tao X, Zhao F, et al. A new outsourced data deletion scheme with public verifiability. In: Proceedings of the international conference on wireless algorithms, systems, and applications (WASA). Berlin: Springer; 2019. p. 631–8.
31. Baumann A, Peinado M, Hunt G. Shielding applications from an untrusted cloud with haven. In: Proceedings of the 11th USENIX conference on operating systems design and implementation (OSDI). Berkeley: USENIX Association; 2014. p. 267–83.
32. Shinde S, Le Tien D, Tople S, et al. Panoply: low-TCB Linux applications with SGX enclaves. In: Proceedings of the network and distributed system security symposium (NDSS). Reston: Internet Society; 2017. p. 1–15.
33. Tsai CC, Porter DE, Vij M. Graphene-SGX: a practical library OS for unmodified applications on SGX. In: Proceedings of the 2017 USENIX annual technical conference (ATC). Berkeley: USENIX Association; 2017. p. 645–58.
34. Large-scale data systems group. SGX-LKL-OE (open enclave edition); 2022 [2022-03-07].
35. Orenbach M, Lifshits P, Minkin M, et al. Eleos: exitless OS services for SGX enclaves. In: Proceedings of the 12th European conference on computer systems (EuroSys). New York: ACM; 2017. p. 238–53.
36. Weisse O, Bertacco V, Austin TM. Regaining lost cycles with hotcalls: a fast interface for SGX secure enclaves. In: Proceedings of the 44th annual international symposium on computer architecture (ISCA). New York: ACM; 2017. p. 81–93.
37. Bailleu M, Thalheim J, Bhatotia P, et al. Speicher: securing LSM-based key-value stores using shielded execution. In: Proceedings of the 17th USENIX conference on file and storage technologies (FAST). Berkeley: USENIX Association; 2019. p. 173–90.
38. Bailleu M, Giantsidi D, Gavrielatos V, et al. Avocado: a secure in-memory distributed storage system. In: Proceedings of the 2021 USENIX annual technical conference (ATC). Berkeley: USENIX Association; 2021. p. 65–79.
39. DPDK Project. DPDK (2022-01-01) [2022-03-07].
40. Kalia A, Kaminsky M, Andersen D. Datacenter RPCS can be general and fast. In: Proceedings of the 16th USENIX symposium on networked systems design and implementation. Berkeley: USENIX Association; 2019. p. 1–16.
41. Intel Corporation. Intel SGX developer SDK for Linux (2021-01-01) [2022-03-07].
42. Kim T, Park J, Woo J, et al. Shieldstore: shielded in-memory key-value storage with SGX. In: Proceedings of the 14th European conference on computer systems (EuroSys). New York: ACM; 2019. p. 1–15.
43. Zhou W, Cai Y, Peng Y, et al. VeriDB: an SGX-based verifiable database. In: Proceedings of the 2021 international conference on management of data (SIGMOD). New York: ACM; 2021. p. 2182–94.
44. Blum M, Evans W, Gemmell P, et al. Checking the correctness of memories. In: Proceedings 32nd annual symposium of foundations of computer science. Piscataway: IEEE; 1991. p. 90–9.

References

45. Tramer F, Boneh D. Slalom: fast, verifiable and private execution of neural networks in trusted hardware (2019-02-27) [2023-05-30].
46. Kim K, Kim CH, Rhee JJ, et al. Vessels: efficient and scalable deep learning prediction on trusted processors. In: Proceedings of the 11th ACM symposium on cloud computing. New York: ACM; 2020. p. 462–76.
47. Sun Y, Wang S, Li H, et al. Building enclave-native storage engines for practical encrypted databases. In: Proceedings of the VLDB endowment, vol. 14(6). New York: ACM; 2021. p. 1019–32.
48. Yang F, Chen Y, Lu Y, et al. Aria: tolerating skewed workloads in secure in-memory key-value stores. In: Proceedings of the 37th international conference on data engineering (ICDE). Piscataway: IEEE; 2021. p. 1020–31.
49. Yang F. Kexin zhixing huanjing xia de gaoxiao neicun cunchu guanjian jishu yanjiu [Research on key technologies of efficient memory storage in trusted execution environment]. Beijing: Tsinghua University; 2022.
50. X.509: Information technology—open systems interconnection—the directory: public-key and attribute certificate frameworks [2023-05-30].

Chapter 11
Data Protection

Ever-growing global data volumes have made the question, "How do we effectively protect data?" more important than ever in the field of data storage. In this chapter, we will discuss the definition, standards, metrics, and key technologies currently used for data protection.

11.1 Data Protection Background

The goal of data protection is to ensure data security [1, 2]. Article 3 of the Data Security Law of the People's Republic of China then says, "Data security refers to ensuring that data is effectively protected and lawfully used through adopting necessary measures, and to possessing the capacity to guarantee the continuous security of data." This breaks security down into two parts: data content security and data protection security. Content security relates to how we ensure the confidentiality and integrity of data content through modern cryptographic technologies, while protection security focuses on the use of modern storage technologies to ensure that data service operations are secure and that data content is not leaked, lost, or damaged and remains accessible.

Currently, both individuals and enterprises face a number of data security challenges, mainly data system hardware and software faults, human errors, cyberattacks, natural disasters, terrorist attacks, and armed conflicts. All of these can prevent normal service access or result in data leaks and damage, ultimately causing potentially immeasurable losses to individuals, enterprises, and society. The Storage Networking Industry Association (SNIA) defines data protection as "the process of safeguarding important data from corruption, compromise or loss and providing the capability to restore the data to a functional state should something happen to render the data inaccessible or unusable." Therefore, effective technical measures for data protection must be able to protect and recover the integrity, availability, and correctness of data.

11.1.1 Data Protection Standards

Many national and international standards have been set to standardize post-disaster data recovery requirements for data protection construction projects.

Currently, the data protection and disaster recovery (DR) standard most widely used internationally is the SHARE 78 standard, which was formulated by SHARE in 1992. This standard classifies data DR levels into eight tiers based on eight service principles, including application scenarios, data DR requirements, and technical mechanisms. Each tier differs in terms of DR mechanisms and data recovery times, as shown in Table 11.1.

Data protection and DR standards in China were released much later than the SHARE 78 standard, with the Information Security Technology—Disaster Recovery Specifications for Information Systems (GB/T 20988-2007) only released in 2007 [3]. These standards set basic requirements for building information system DR capabilities based on China's specific application scenarios. They were created with reference to the DR tiers defined in the SHARE 78 standard but also specify six tiers of DR capabilities and their corresponding construction requirements. The details are listed in Table 11.2.

The Chinese standards have stronger requirements for the hardware and software integrity of standby sites and standby systems than the SHARE 78 standard. Additionally, the Chinese standards effectively combine SHARE 78's fifth and sixth tiers. This is likely because a vendor with technical preferences dominated the formulation of the SHARE 78 standard, while the Chinese standards were

Table 11.1 Tiers of data DR

Tier	Differences from the previous tier
Tier 0: no off-site data	
Tier 1: data backup with no hot site	Only critical data is backed up, and the backups are stored on an off-site backup system that does not support data recovery using hot backups and usually takes a long time to recover data. Both the recovery point objective (RPO) and recovery time objective (RTO) are several days or weeks
Tier 2: data backup with a hot site	Only critical data is backed up, and the backups are housed on an off-site backup system that supports data recovery using hot backups
Tier 3: electronic vaulting	Critical data is backed up electronically to an off-site system that supports data recovery using hot backups, improving the DR speed
Tier 4: point-in-time copies	The backup software periodically backs up critical data to off-site storage electronically and specifies DR policies. System services can be quickly recovered using off-site data backups
Tier 5: transaction integrity	Data consistency between the production and DR centers is guaranteed by specific applications, and there is almost no data loss
Tier 6: zero or near-zero data loss	General backup technologies or storage technologies are used to implement real-time data backup without depending on specific applications. The RTO is in minutes or seconds
Tier 7: highly automated, business-integrated solution	A high degree of automation is incorporated into the recovery process

11.1 Data Protection Background

Table 11.2 Tiers of DR capabilities

Tier	Definition	Requirement
Tier 1	Basic support	1. Perform a full data backup once a week 2. Store backup media off-site
Tier 2	Standby site support	1. Perform a full data backup once a week 2. Store backup media off-site 3. Have some data processing and network transmission devices required for DR, or be able to allocate them to the standby site after a disaster occurs
Tier 3	Electronic vaulting and partial device support	1. Perform a full data backup once a week 2. Store backup media off-site 3. Transfer critical data in batches electronically to the standby site multiple times a day 4. Have some data processing and network transmission devices required for DR
Tier 4	Electronic vaulting and full device support	1. Perform a full data backup once a week 2. Store backup media off-site 3. Transfer critical data in batches electronically to the standby site multiple times a day 4. Have some data processing and network transmission devices required for DR that are ready or in operation
Tier 5	Real-time data transmission and full device support	1. Perform a full data backup once a week 2. Store backup media off-site 3. Replicate critical data to the standby site electronically in real time using remote data replication technology 4. Have some data processing and network transmission devices required for DR that are ready or in operation 5. Have a communication network that supports automatic or centralized switchover
Tier 6	Zero data loss and remote cluster support	1. Perform a full data backup once a week 2. Store backup media off-site 3. Support real-time remote backup to ensure zero data loss 4. Have some data processing and network transmission devices required for DR that are ready or in operation 5. Allow end users to simultaneously access both the active and standby centers over the network

developed by government authorities which were more concerned with the DR and backup needs of actual use case scenarios in China. The two standards are compared in Table 11.3.

11.1.2 The Features of Data Protection Technologies

While the above section describes the main technical means and Chinese and international standards for data protection, in this section, we will summarize the objectives, fault scenarios, RPO and RTO, and DR tier requirements for different types of data protection (see Table 11.4).

Table 11.3 Comparison between China's DR standards and the SHARE 78 standard

China's DR tiers	SHARE 78
	Tier 0: no off-site data
Tier 1: basic support	Tier 1: data backup with no hot site
Tier 2: standby site support	Tier 2: data backup with a hot site
Tier 3: electronic vaulting and partial device support	Tier 3: electronic vaulting
Tier 4: electronic vaulting and full device support	Tier 4: point-in-time copies
Tier 5: real-time data transmission and full device support	Tier 5: transaction integrity
	Tier 6: zero or near-zero data loss
Tier 6: zero data loss and remote cluster support	Tier 7: highly automated, business-integrated solution

Table 11.4 Features of data protection technologies

Item	Objective	Fault scenario	RPO and RTO	DR tier
DR	Service continuity	The service system or data center is faulty	High RPO and RTO requirements	Tiers 5 and 6
Backup	Data recovery and data copy utilization	The service system or data center is faulty	Low RPO and RTO requirements	Tiers 1–6
Archiving	Query/audit	N/A	N/A	Tiers 1–6

The objective of DR, for example, is primarily to provide system service continuity after a disaster. Backup is tasked with post-disaster data recovery and service takeover using data copies, and archiving is responsible for historical data queries and audits.

Both DR and backup are used to deal with service system or data center faults. Archiving is generally not involved in any specific fault scenarios.

Given the diverse needs of different services for continuity and data recoverability, the industry generally uses two technical metrics, RPO and RTO, during DR system construction to determine a service's DR needs [4]. RPO refers to a system's tolerance for data loss and is generally associated with the system's periodic DR and backup interval. For example, if a system's database backup interval is 5 min, the production system can use the DR and backup system to recover data saved up to 5 min prior to a disaster. This would put the system's RPO at 5 min. RTO is the duration needed for a system to resume normal services after a disaster. For example, if a system's database can recover its data and get services running within 10 min of a disaster, the RTO of the system is 10 min. DR addresses higher RPO and RTO requirements, while backup generally addresses lower RPO and RTO requirements.

DR is designed to meet high DR tiers (typically tier 5 or tier 6). When it comes to backup and archiving, organizations can choose a DR tier for data protection based on their service needs.

Next, we will further discuss the main application scenarios and technical principles of data protection technologies. First, we will look at in-system data protection technologies [5], such as mirroring, snapshots, and cloning. Then, we will dive

11.2 Data Protection Technologies

deeper into data protection scenarios, including out-of-system backup (backup for short in later sections), archiving, and DR.

11.2 Data Protection Technologies

11.2.1 Mirroring

According to the SNIA definition, mirroring is a technique that creates consistent copies of data housed in a storage system, while the storage system can access the source data and copies independently [6]. As an effective technique for data protection within a system, it is primarily used to resolve the interruption of an entire service and data loss caused by the failure of a single LUN request. Currently, mirroring can be implemented either by a storage system itself or by using a storage gateway.

When a storage system independently performs mirroring, it usually leverages a built-in mirroring feature to create a mirror copy with the same capacity as the mirror copy converted from the source LUN in the current storage system and establishes a mirroring relationship between the two copies in redundancy mode. In this solution, the two copies should be in different fault domains to ensure system services can turn to access the other copy in the event of a fault, as shown in Fig. 11.1a. When a storage gateway is used for mirroring, it typically connects a host to two backend storage systems over a network and maintains a mirror copy in each of the two systems. This allows mirroring failover to be implemented through changing the network access path without affecting the host, as shown in Fig. 11.1b.

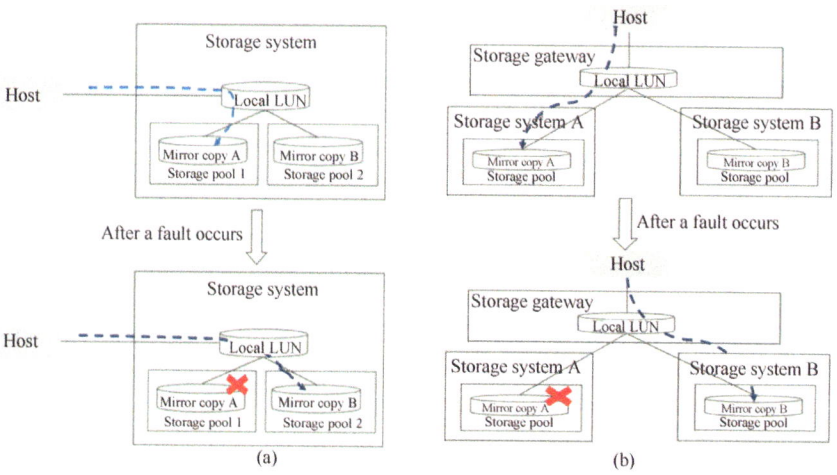

Fig. 11.1 Mirroring implementation solutions. (**a**) Built-in mirroring feature of storage devices. (**b**) Mirroring feature implemented with the storage gateway

Fig. 11.2 Mirror creation

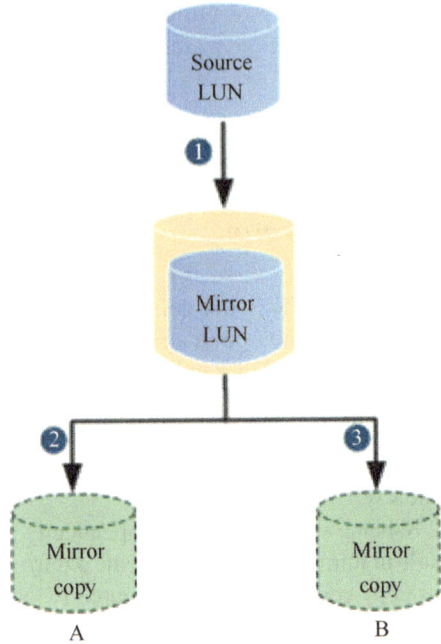

Major operations associated with LUN mirroring include creation, synchronization, and splitting.

Creation: As shown in Fig. 11.2, the typical mirroring architecture consists of a LUN and two mirror copies. Therefore, the mirror creation process can be divided into three steps. The first step is to perform a mirror creation operation on a common source LUN to convert it into a mirror LUN with the same attributes and services as the source LUN. In the second step, mirror copy A is automatically generated during the creation of the mirror LUN, and the copy inherits the storage space from the source LUN. The last step is to add a new copy (copy B) for the mirror LUN and synchronize data from copy A to copy B. After the three steps are completed, the source LUN becomes a logical unit with two mirror copies. When a host issues a read request to the source LUN, the storage system accesses the two mirror copies in round-robin mode.

Synchronization: Mirror synchronization includes initial synchronization and incremental synchronization. During the mirror creation process, initial synchronization is performed after the creation of mirror copy B to copy the data from mirror copy A to mirror copy B, as shown in Fig. 11.3. After the data synchronization is completed, the write request from the host is converted into a dual-write operation on mirror copies A and B in the storage system, thus ensuring data consistency between the two mirror copies.

Incremental synchronization refers to the process of synchronizing incremental data between two mirror copies after one of them is recovered from splitting or fail-

11.2 Data Protection Technologies

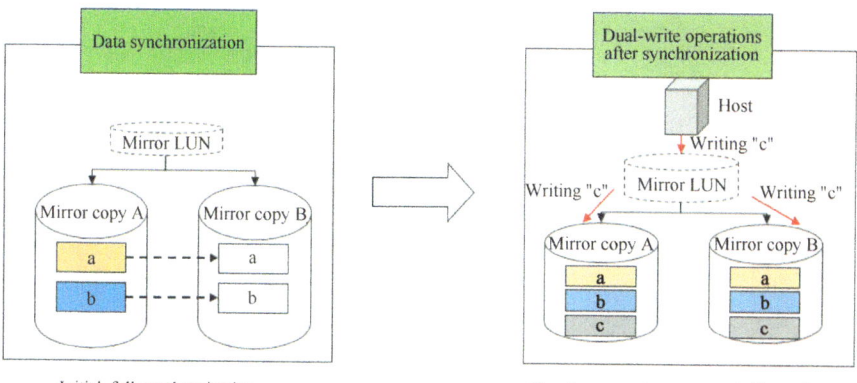

Fig. 11.3 Mirror synchronization

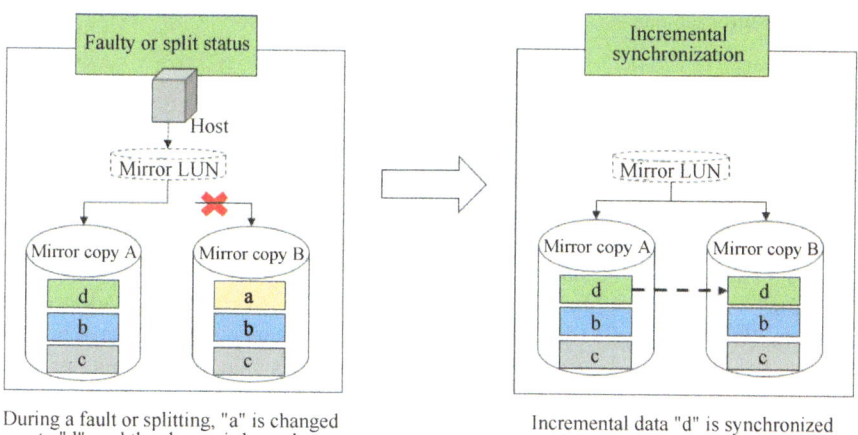

Fig. 11.4 Incremental synchronization

ures. As shown in Fig. 11.4, if mirror copy B fails or is split, the data changes made to mirror copy A during this period are logged and will then be synchronized to mirror copy B after mirror copy B is recovered to ensure data consistency between the two mirror copies.

Splitting: This operation is performed when services need to isolate a mirror copy. As shown in Fig. 11.5, splitting is performed on mirror copy B to disconnect it from the mirror LUN and suspend all data write operations on it. The split mirror copy is generally used for testing and analysis. If it is reconnected to the mirror LUN, incremental synchronization is performed to ensure data consistency between the mirror copies.

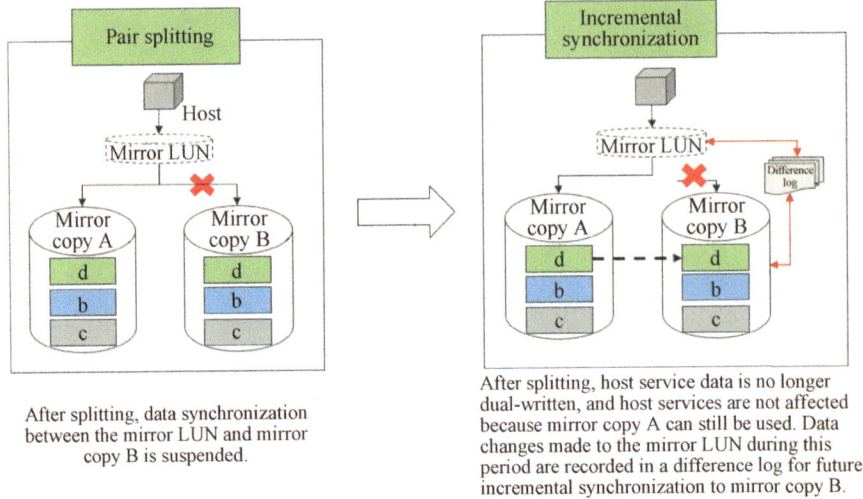

Fig. 11.5 Mirror splitting

11.2.2 Snapshots

According to SNIA, a snapshot is a fully usable copy of a defined data collection (volume), and the copy is a static image of the source data at a certain point in time. Generally, a snapshot is a virtual copy of the source data. Compared with a mirror, a snapshot stores only the status of a data collection within a certain period of time. Therefore, a snapshot occupies less space. Snapshots were originally developed for data protection. As data is becoming increasingly important, snapshots have gradually been applied to scenarios such as data analysis and application testing. Generally, snapshots are used for three purposes in data protection scenarios.

Data recovery: Snapshots can be used to quickly restore data to a specified version in the event of malicious attacks or misoperations.

Data backup: The storage system or backup and archive software suites periodically create snapshots for service data and delete earlier snapshots.

Data analysis: Read and write operations can be performed on snapshots for data testing, analysis, and mining without affecting services.

After a snapshot is created for the source volume, two identical data collections exist in the storage system. Therefore, when data is written to the source volume and snapshot volume, two solutions are available: copy-on-write (CoW) and redirect-on-write (RoW).

CoW: When a new write request is made to update the data of the source volume, the snapshot system copies the original data of the source volume to the snapshot volume and updates the data address mapping table before performing write operations on the source volume. CoW results in an extra read operation and an extra write operation in the storage system.

RoW: When a new write request is made to update the data of the source volume, the snapshot system redirects the write request to the snapshot volume and updates the data address mapping table. In such cases, the upper layer directly accesses the snapshot volume when initiating a read request for newly written data. Compared with CoW, RoW performs only one write operation in the storage system.

Different write operations result in different methods of handling read operations, snapshot rollback, and snapshot deletion.

CoW-based read operations: Since all service data is written to the source volume, all read requests for the source data are handled by the source volume. For data analysis applications, data read requests are sent to the snapshot volume. If the required data does not exist in the snapshot volume, the read request is redirected to the source volume via the address mapping table.

RoW-based read operations: Read requests made to the source volume are distributed based on the time at which the data was written and the time at which the snapshot was created. The source volume is used to respond to access requests for the data written before the snapshot is created, while the snapshot volume is used to respond to access requests for the data written after the snapshot has been created. The read requests made to the snapshot volume are also distributed based on the time at which the data was written and the time at which the snapshot was created. If the requested data is stored in the source volume, the address mapping table is used to redirect the read request.

CoW-based rollback: Because CoW preserves all data written before the snapshot is created in the snapshot volume, during a rollback, the source volume stops responding to I/O requests, and data is written from the snapshot volume back to the source volume via the address mapping table.

RoW-based rollback: Since no new data is written to the source volume after the snapshot has been created, RoW-based rollback can be implemented by disabling access to the snapshot volume.

CoW-based deletion: As the source volume has stored the latest data, snapshot deletion only requires deletion operations to be run on the snapshot volume and address mapping table.

RoW-based deletion: The source volume only stores the data written before the snapshot is created. During RoW-based snapshot deletion, the data in the snapshot volume must be copied to the source volume before the snapshot volume and the address mapping table are deleted.

11.2.3 Cloning

According to SNIA, both clones and snapshots are defined as copies of data. The difference between them is that a snapshot is just an image of the state of a storage system or the data in a storage system at one point in time, whereas a clone is a full copy of the storage system. This means that clones can exist independently from the original system, and this is why they are capable of recovering data more efficiently.

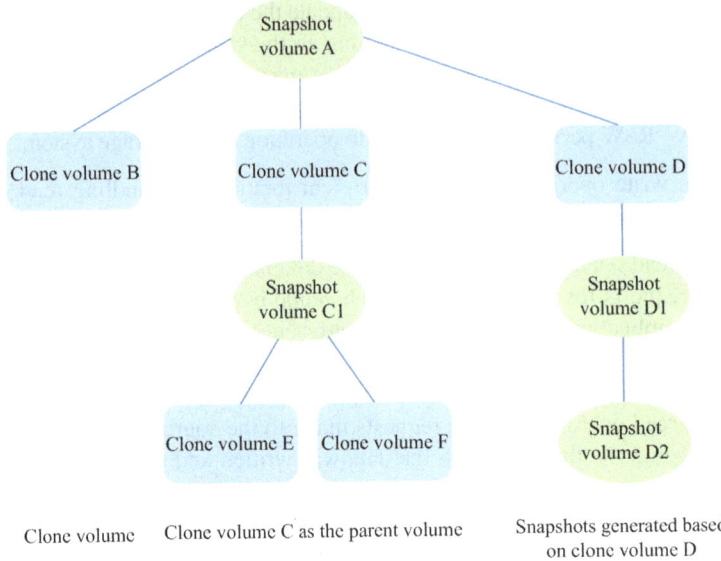

Fig. 11.6 Cloning technology

In a computer system, the purpose of cloning is to build a new system that has exactly the same functions or the same data as the source system. In that sense, clones can be considered functionally similar to writable snapshots. Cloning technology can generate multiple clone volumes based on a volume snapshot, as shown in Fig. 11.6.

Clone volumes B, C, and D can be generated based on snapshot volume A. The data on these clone volumes will be the same as the data on snapshot volume A, the original snapshot volume. Since the clone volumes are independent of snapshot volume A, subsequent read/write operations on the clone volumes will not affect the data on the snapshot volume. In addition, the clone volumes will have all the same functions as the original snapshot volume, which means that they support mirroring, snapshot, and second cloning operations on the storage system.

Data synchronization and splitting are important features that distinguish clones from writable snapshots. Data synchronization for clones is similar to data synchronization for mirrors and is therefore not discussed in this section.

Clone splitting is the process of copying snapshot data from the source LUN to a new clone LUN. After splitting, the source LUN and the clone LUN become two independent systems, and a failure in one of them will not affect the other.

11.3 Data Protection Scenarios

There are three forms of data protection: DR, backup, and archiving. They are associated but have different purposes and processes [7], as shown in Fig. 11.7.

11.3 Data Protection Scenarios

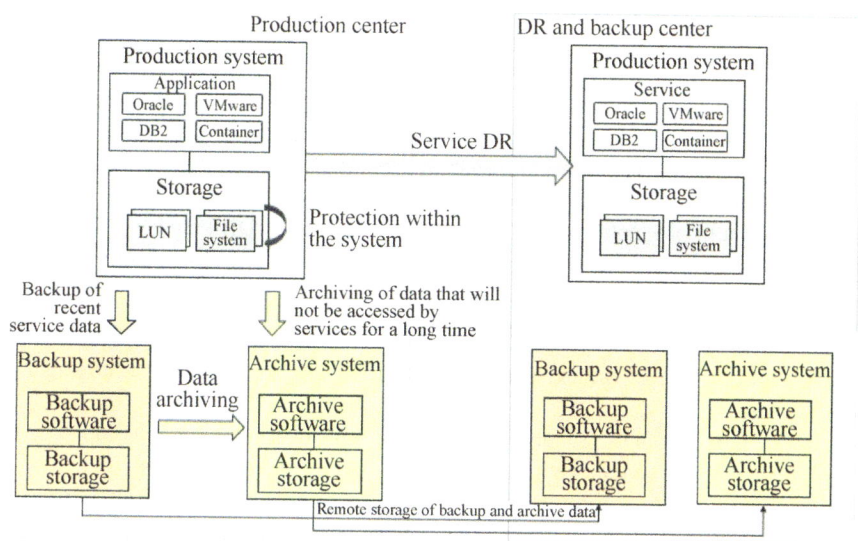

Fig. 11.7 The relationship between DR, backup, and archiving

Remote DR can meet basic data protection requirements, and the current approach to implementing remote DR is to construct a DR and backup center for an existing production center. The production center is responsible for running system services. The DR and backup center is responsible for backing up data and services and for taking over services and recovering data in the event of a disaster.

DR, backup, and archiving technologies are mainly used during the construction of the DR and backup system.

DR ensures service system continuity and data security by establishing a standby system that can quickly recover or take over services if the active system fails or a disaster occurs.

Backup is the process of periodically saving data to a backup system. If the data is corrupted, the backup data can be used to restore the production system to a historical point in time. Backup may be internal or external based on where the backup data is stored. Backup is internal if the backup data is stored locally through methods such as mirroring, snapshots, and cloning. Backup is external if the backup data is stored on an independent external storage system.

Archiving is the process of systematically storing historical system data that will not be accessed frequently.

Strictly speaking, there is no distinction between backup and archiving. Generally, backup refers to storing copies of recent system data for service system recovery, while archiving refers to the long-term (offline) storage of copies of historical data for reference and compliance purposes.

11.3.1 Backup

Essentially, backup is the process of securely storing copies of an IT system's data from different points in time at regular intervals for data recovery and utilization. Figure 11.8 shows a typical backup system.

A backup system consists of backup software and backup storage. It periodically backs up data in production environments such as Oracle, DB2, and VMware. The backup data is stored in the backup storage, and it can be used for system recovery if a fault occurs in the IT systems. Figure 11.9 shows the scenarios in which data backups can be used for enterprise IT systems.

1. Recovering data in the event that system hardware or software failures cause data loss.
2. Recovering data in the event that a virus encrypts or deletes service data.
3. Recovering data in the event that a human error results in data loss.
4. Facilitating enterprise development, testing, or analysis. An increasing number of enterprises are using backup data to develop mirror systems and utilizing mirror data for development, testing, or analysis.

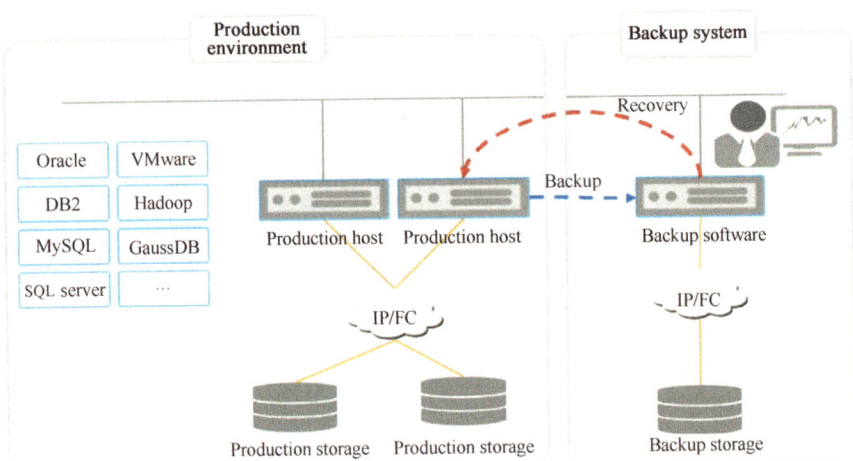

Fig. 11.8 A backup system

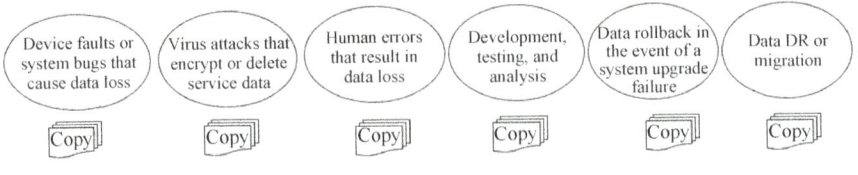

Fig. 11.9 Backup scenarios

11.3 Data Protection Scenarios

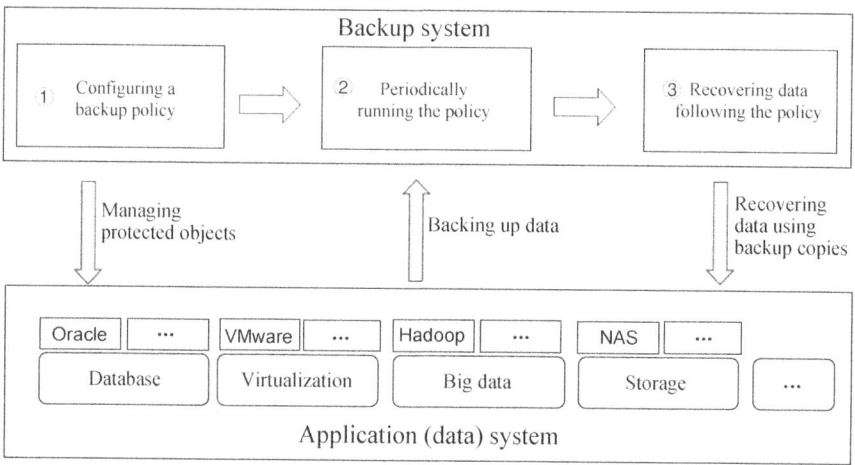

Fig. 11.10 How the backup system works

5. Rolling back data in the event that a service system upgrade fails. This requires data to be backed up before the upgrade.
6. Facilitating DR or data migration, such as for small and medium-sized enterprises that use hybrid clouds like public clouds or industry clouds as DR and backup centers to reduce enterprise IT costs.

Figure 11.10 shows how the backup system protects data for service systems. The backup process includes configuring a backup policy, periodically running the policy to write data captured from production applications to backup storage to create data copies, and choosing data copies and the recovery environment to recover data if data is lost.

The fundamentals of a backup system may seem simple, but there is a lot that needs to be taken into consideration during the design process. Here are just a few examples:

A data backup is a copy of production data. How can we ensure it is reliable and can be used to recover production data when necessary?

Enterprise data centers may use various applications that generate huge amounts of data. How can we design the backup networking architecture to accelerate data backup and improve data migration efficiency while reducing the impact on the production environment?

Periodically backing up data may create a lot of duplicate data. How can we use data reduction technologies to reduce the amount of duplicate data taking up disk space?

There are many types of application (data) systems in an enterprise data center. How can we use ecosystem compatibility technology for backup applications to obtain the application (data) system data at a specific point in time?

How can we use copy data management (CDM) technologies to make backup data quickly accessible?

Next, we will look at the different types of backup system architecture and related backup technologies that can be used to overcome these technical challenges.

11.3.1.1 The 3-2-1 System Architecture

As shown in Fig. 11.11, the 3-2-1 rule stipulates that there should be at least three copies of data, including the original data, backup data, and off-site data, and that the copies should be stored on two different types of storage media to prevent data loss caused by the same device failure. In addition, at least one copy should be stored at an off-site location, to ensure data in the production center can be recovered using off-site backups in the event of a disaster. If any copy is invalid, the other two copies can be compared to ensure the availability of recovered data, and this also helps to detect invalid copies in advance. Using at least two types of media reduces the risk of data loss due to similar reasons, because different storage media are affected differently by electromagnetism, physical vibration, temperature, and natural disasters such as floods and fires. This approach can even mitigate some of the impacts of network viruses, theft of physical media (tapes or optical disks), and limitations of storage lifespan, because the ability to handle these issues varies across different types of storage media.

Both the production center and DR center are equipped with independent backup systems. If the backup data in the production center fails or is lost, the backup data in the DR center remains available. A backup system consists of backup software, backup storage, and archive storage (or long-term storage). The backup software is responsible for the backup of production data, data copy flows, data recovery using data copies, and the management of expired data copies. Backup storage systems generally provide high capacity, large bandwidth, and a high deduplication ratio for the storage of recent backup data, such as the data copies of the past month. In contrast, archive storage is cold data storage that delivers high capacity at a low cost. It

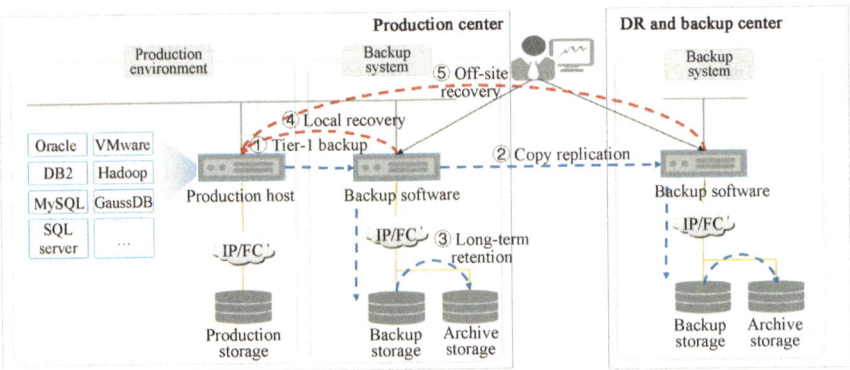

Fig. 11.11 The 3-2-1 backup system architecture

11.3 Data Protection Scenarios

is primarily used for the long-term storage of cold backup data, such as data copies that have existed for more than a month.

11.3.1.2 Networking Architecture

The data backup process involves transmitting huge amounts of data. There are two available networking architectures for the backup software to write data to the backup system (deployed on a backup server): local area network (LAN)-base and LAN-free networking.

- LAN-base networking: As shown in Fig. 11.12, a backup client and a backup server are deployed in a production center. Backup data is transmitted from the backup client to the backup server over the LAN and is then written to the backup storage system by the backup server. A downside of this approach is that it requires an excessive amount of Ethernet resources when backing up huge amounts of data, which may affect the network transmission of production service data.
- LAN-free networking: As shown in Fig. 11.13, it involves the deployment of an independent Fibre Channel (FC) backup network, which allows the backup server to write backup data flows to the backup storage media directly or via the FC switch. This approach does not occupy network bandwidth resources from the primary production service, because the data is not transmitted over the production LAN. Therefore, its impact on production services is almost zero.

As IP networks (25GE/100GE switches) advance and backup storage evolves from tape libraries and storage area network (SAN) to network-attached storage (NAS) and cloud storage, the independent FC switch in LAN-free networking can be replaced with an IP switch to form an independent LAN for backup. A comparison between the two types of architecture is shown in Table 11.5.

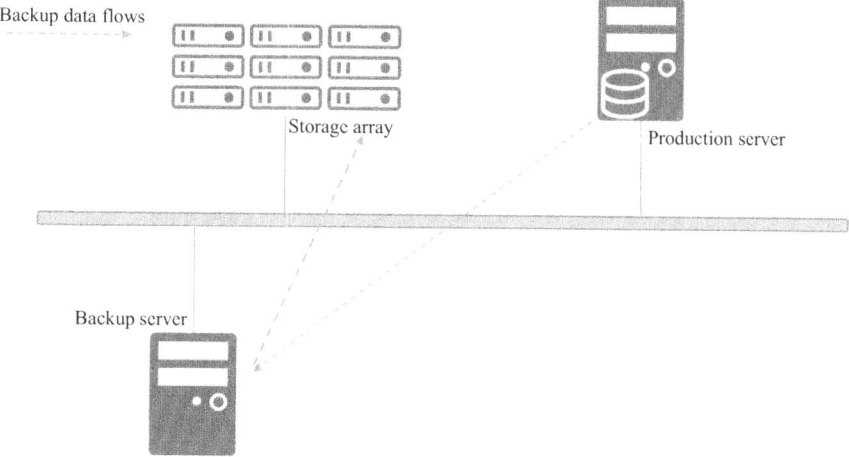

Fig. 11.12 LAN-base networking architecture

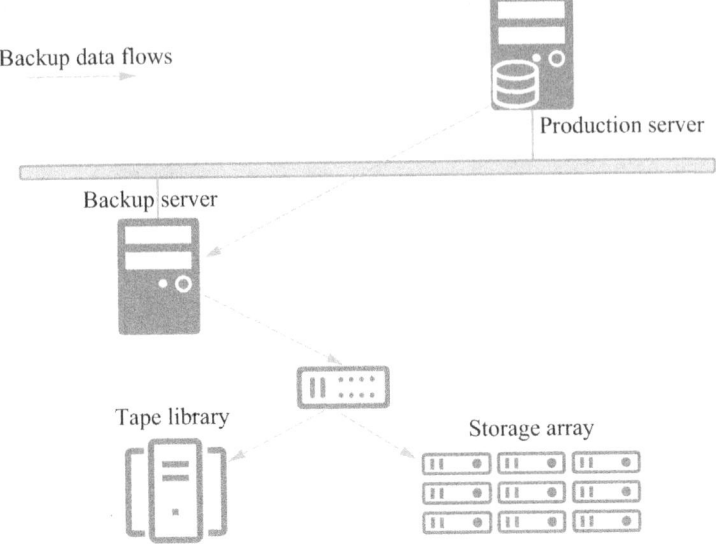

Fig. 11.13 LAN-free networking architecture

Table 11.5 Comparison between LAN-base and LAN-free networking architecture

Networking architecture	Description	Downside
LAN-base	A unified media server is deployed to store data in the backup storage, which consumes network resources	Numerous backup copies may consume too many network bandwidth resources, impacting the efficiency of service backup
LAN-free	Storage space is mapped to each host, meaning each host is a media server. This approach occupies host resources instead of network resources	Each host must have complete server functions

11.3.1.3 Converged Architecture

As mentioned above, a backup system consists of backup software and backup storage. Backup software is deployed on servers, and backup storage is a kind of dedicated external storage, typically tape libraries or SAN/NAS systems. This architecture was developed as a result of the independent division of work between backup software vendors and storage vendors. Backup software can connect to different vendors' storage devices, regardless of whether they are tape, SAN, NAS, or cloud storage, as shown in Fig. 11.14.

Given the rapid development of data protection in recent years, backup software vendors and storage vendors have jumped at the chance to launch converged architecture to further integrate backup software and backup storage. This architecture simplifies data transmission between backup software and backup storage and reduces system workloads.

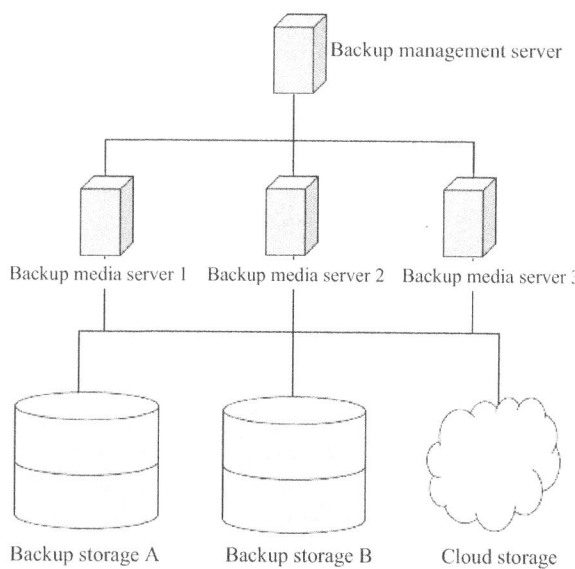

Fig. 11.14 Backup storage architecture

11.3.2 Archiving

The concept of archiving existed long before digital information systems. In the past, it mainly referred to the preservation of official documents and information for easy access. An example is the archiving and preservation of personal records in archives. In the information age, the focus of data management has shifted to the storage of electronic information, and data archiving involves moving infrequently accessed data to low-cost storage repositories [8–10]. Data archiving is a practice in which enterprises store valuable data for longer periods of time in accordance with their service needs as well as other laws and regulations so that they can quickly query and retrieve data when necessary. The architecture and functions used during data archiving are varied to meet different needs, with the most basic functional requirements being indexing and searching which ensure files remain easily accessible.

11.3.2.1 Application Scenarios

Data archiving is a common practice in multiple industries. For example, banks archive document images and transaction logs for long-time preservation. Insurance companies permanently retain copies of insurance contracts. And even enterprises typically archive emails to comply with corporate governance and legal requirements. Table 11.6 lists common data archiving scenarios in China.

The data sources and data retrieval methods used by different industries can vary wildly, so there is no unified standard for archive systems. However, regardless of

Table 11.6 Common data archiving scenarios in China

Data sources	Required retention period
Electronic imaging platforms for banks	Data is centrally archived and retained for 15 years
Electronic government archives	All data must be retained for at least 5 years, and some data must be permanently retained according to the Specification on Electronic Documents Archiving and Electronic Records Management (GB/T 18894-2016)
Public security and judicial case systems	Litigation information (including pictures and short videos) can be retained permanently, for long-term periods (60 years) or for short-term periods (30 years)
Healthcare data	The Management Regulations on Application of Electronic Medical Records (Trial) stipulates that the electronic medical records of inpatients must be stored for at least 30 years, and the electronic medical records of outpatients or emergency patients must be stored for at least 15 years
Oil and gas exploration data	According to the Regulations on Archiving and Preservation of Petroleum and Seismic Exploration Data (SY/T 5928-2009), data must be stored permanently or for a long time

the technologies they use, all archive systems generally have to consider the four factors:

1. Data security and integrity.
2. How easy it is to query and read data.
3. Legal and regulatory compliance to ensure data is not leaked or tampered with. The Data Security Law of the People's Republic of China specifies specific data protection levels and sets forth a number of requirements for data security and personal information protection.
4. Storage cost. Since archive data is used for corporate auditing and regulatory compliance purposes, it is typically not regularly accessed. Low-cost storage options are ideal for archives due to the large amount of data they must store. This has led to a boom in tiered storage, offline storage, and low-cost storage technologies (such as tape, Blu-ray, and cloud storage archiving).

11.3.2.2 Technical Principles

With the emergence of new service systems, the sources of archive data are becoming increasingly diverse and now include emails, logs, images, and databases. An archive system typically consists of archive software and archive storage, as shown in Fig. 11.15.

Archive software is mainly responsible for data processing, management, and discovery. The processing of archive data involves migrating the data to be archived from the data production system to the archive storage and compressing and encrypting it to ensure security. Archive data management includes expiration management, compliance (such as tampering prevention), and data operations audits in accordance with the data's lifecycle. Expiration management generally means simply overwriting expired data. If data is only logically deleted

11.3 Data Protection Scenarios

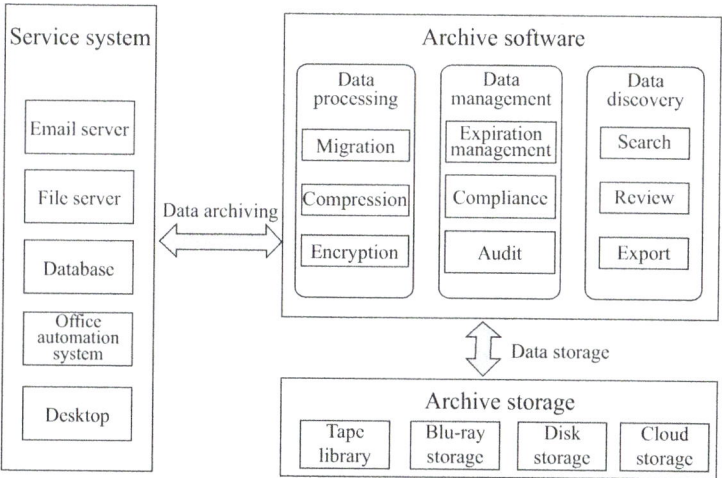

Fig. 11.15 Archiving architecture

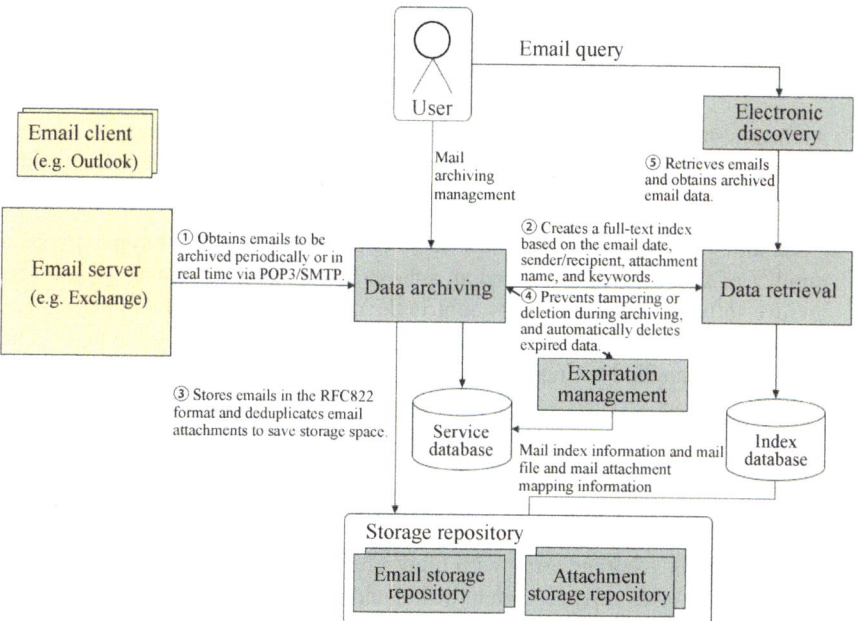

Fig. 11.16 Email archiving

from the storage media, the storage media needs to be destroyed physically. Archive data discovery requires the creation of a data index, which is then used for the retrieval, review, and export of archive data. In this section, we will use email archiving as an example to explain the data archiving process, as shown in Fig. 11.16.

The user administrator configures the archiving policy based on the archiving needs. From there, the following steps occur:

1. The archive software obtains emails to be archived either periodically or in real time via Post Office Protocol 3 (POP3) or Simple Mail Transfer Protocol (SMTP). Some mailbox systems provide email logs, and new emails can also be obtained through email logs.
2. The archive software then creates a full-text index based on the email date, sender/recipient, attachment name, and keywords. The index information is used for future archive data management and queries. This makes the data index a key technology for archive systems.
3. Emails are then stored in the RFC822 format, which is the standard format for emails. When storing mail attachments, it is important to deduplicate them to save storage space as they may be referenced multiple times in different emails.
4. During archiving, the archived emails are stored in a write-once-read-many (WORM) state to prevent data tampering or deletion. This is typically done using an object lock. Archive data must be movable to address storage device damage, DR needs, and other possible issues. Once the email archiving period expires, the archive system automatically deletes or destroys the emails.
5. During the archiving period, users can perform electronic discovery based on their service needs and search, view, or export archived emails based on the retrieval information.

Archive storage has evolved from physical and virtual tape libraries to Blu-ray storage, cloud storage (such as S3 Glacier), and more. This section will describe their basic features and differences.

Physical tape library: It is a storage device that includes one or more tape drives, a number of slots, a disk drive, and an automatic robotic arm, among others. The archive software can intelligently control the robotic arm and use it to place tapes on the disk drive, which can both read data from and write data to the tapes. The mainstream standard for tape libraries is Linear Tape Open (LTO), which is an open-format tape storage standard. Currently, it supports up to LTO-9. An LTO-8 tape, for instance, can provide a storage capacity of 12 TB, which can save approximately 30 TB of data with the tape's compression technology. It also delivers a read speed of 360 MB/s. Tape storage typically retains data for 15 years or more.

Disk storage: It is a storage array with high reliability (achieved with RAID, short for redundant array of independent disks) and excellent read/write performance. However, it is often several times more expensive than the tape library and therefore necessitates deduplication and compression technologies to reduce costs. Generally, disk arrays retain data for 5–8 years.

Blu-ray storage: It employs optical disks as persistent media and uses a blue laser to read and write the disks. The structure comprises an optical disk drive, multiple optical disk cartridges, and an automatic robotic arm. The optical disk media of Blu-ray storage only allows data to be written once. After data is written, it can no longer be modified but can be read multiple times. Blu-ray storage can keep data for more than 50 years.

Cloud storage: It is also known as object storage and is increasingly used for archiving. S3 Glacier is a typical cold cloud storage service with lower costs, and S3 data can be automatically tiered to S3 Glacier. Unlike S3 data, which is available online in real time, S3 Glacier archive data is not immediately accessible and requires a rated recovery time of 3–5 h.

Table 11.7 illustrates the differences between the main features of various storage options for data archiving.

The factors to be considered for selecting an appropriate archive storage system include data characteristics, cost (deduplication and compression, energy saving), retention period, reliability, and system maintenance.

11.3.3 DR

A data center provides computing, storage, and network services. In recent years, there has been an increasing demand for continuous data center operations, requiring systems to guarantee service continuity even during a disaster [11, 12]. Although data backup can safeguard data from damage caused by software and hardware faults or misoperations, DR measures, especially a remote DR and backup solution, are necessary for data center management in the event of a large-scale disaster. Depending on service requirements for data recoverability, service continuity, and availability, data DR is classified into active/standby DR, active-active DR, and multi-data center DR.

11.3.3.1 Active/Standby DR

In active/standby DR, a DR center is built for a production center to protect data or applications. When a disaster occurs, the DR center takes over services from the production center to ensure data recovery or service continuity. The fundamental function of this approach is to ensure that the data in the production center is recoverable, which cannot be achieved without synchronizing data between the production and DR centers. To meet different service availability requirements, two data replication techniques—synchronous remote replication and asynchronous remote replication—are used to synchronize data between the two centers.

Synchronous remote replication: After a host delivers a write I/O request, data is simultaneously written to both the primary and secondary storage systems. When the write operations are done at both sites, a write success acknowledgment is returned to the host. Therefore, the RPO of synchronous replication is 0.

Asynchronous remote replication: After a host delivers a write I/O request, data is written to the primary storage system, after which a write success acknowledg-

Table 11.7 Comparison of archive storage features

Archive storage	Capacity per disk	Bandwidth	Deduplication and compression	Power saving mode	Data retention period	Reliability	Rewritable or not
Tape library (LTO-9)	~12 TB	100+ MB/s	Compression	Power off	15+ years	No RAID	Yes
Disk storage (hard disk drive)	~16 TB	GB/s	Deduplication and compression	Disk hibernation	5–8 years	RAID technology	Yes
Blu-ray storage	~6 TB	100+ MB/s	Compression	Power off	50+ years	RAID technology	No
Cloud storage (S3 glacier)	N/A	GB/s	Deduplication and compression	Power off	N/A	RAID technology	Yes

11.3 Data Protection Scenarios

ment is returned to the host. At the same time, the storage system identifies differential data by recording I/O logs or comparing the snapshots generated during the last synchronization and this synchronization and periodically synchronizes the differential data to the secondary storage system in the background. Therefore, the RPO of asynchronous replication is the data synchronization interval, which is generally in minutes.

Regardless of the replication technology being used, active/standby DR for block storage or file system storage has the same technical focuses, which are data replication with data consistency, support for replication efficiency optimization such as link acceleration, the ability to handle various faults during the replication process, and failover upon faults.

Therefore, active/standby DR involves four phases, which are the creation of remote replication pairs, data synchronization, service switchover, and data recovery, as shown in Fig. 11.17.

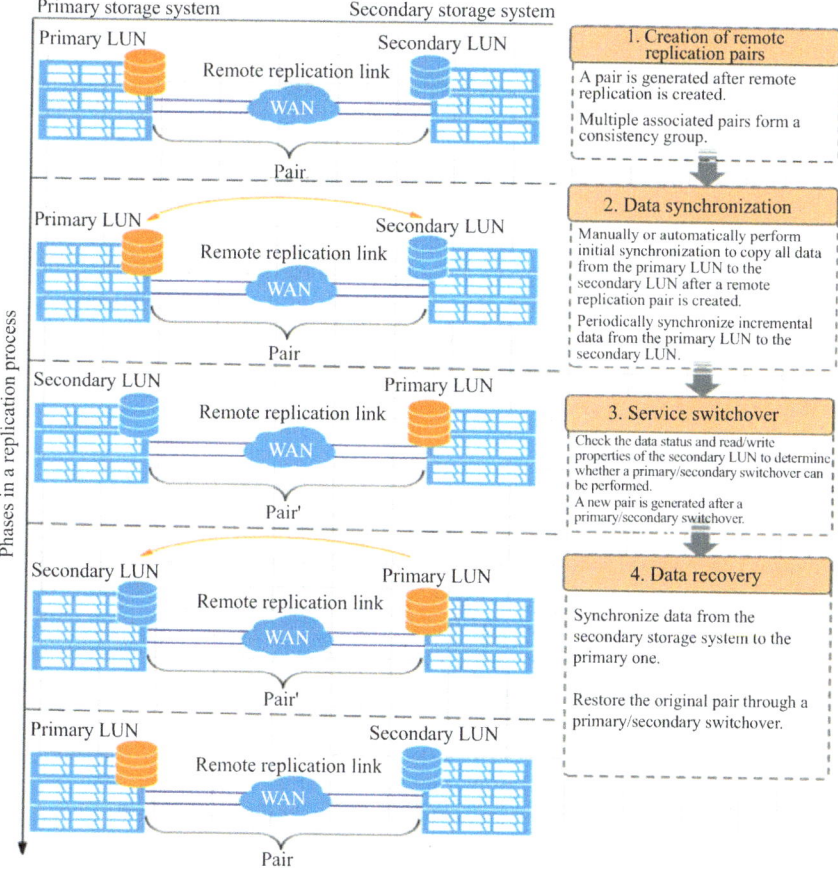

Fig. 11.17 Four phases of active/standby DR

Table 11.8 Description of each pair status during remote replication

Running status	Description
Normal	Data synchronization between the primary and secondary LUNs is complete
Synchronizing	Data is being synchronized from the primary LUN to the secondary LUN. In this state, data cannot be written into or read from the secondary LUN, and data on the secondary LUN cannot be used for service recovery if a disaster occurs. Data on the secondary LUN can be used for service recovery only when it is in a consistent state
Split	The pair relationship between the primary and secondary LUNs is manually interrupted for service needs. Data replication between the primary and secondary LUNs is suspended
Interrupted	The pair relationship between the primary and secondary LUNs is interrupted. This occurs when the links used by the remote replication are down or either LUN fails
To be recovered	If remote replication needs to be restored using a manual policy after the fault that caused a pair interruption is rectified, the pair running status changes to "to be recovered." this status reminds users of manual data synchronization between the primary and the secondary LUNs to restore the pair relationship between them
Invalid	If the properties of a pair are changed on the primary or secondary LUN (such as deletion of the pair on either LUN due to a remote replication link fault) after the pair is interrupted, its running status changes to "invalid" because the configurations of the primary and secondary sites are inconsistent

During data replication between the primary and secondary storage systems, users can check the status of replication pairs to determine whether the pairs are operating normally. A replication pair can be in one of the following states: normal, synchronizing, split, interrupted, to be recovered, and invalid. Table 11.8 describes each status.

11.3.3.2 Active-Active DR

The active/standby DR solution does prevent data loss in the event of a disaster, but it requires manual switchover and may cause service interruption. Mission-critical business requires zero interruption if a fault occurs, and that is where the active-active DR technology comes in. Active-active DR is a tier-7 DR solution with zero data loss and high DR automation. It ensures service continuity in most disasters, such as fires, floods, power failures, and network attacks, and is widely used in data centers in the financial, communications, and healthcare sectors.

As the name implies, the active-active solution requires both data centers in a service system to be capable of providing services simultaneously and that services can automatically fail over without affecting service continuity in the event of a disaster. In the active-active DR solution, the two data centers act as backups for each other. This requires the storage systems in both centers to provide two identical copies of the data in real time that support concurrent read and write operations as

11.3 Data Protection Scenarios

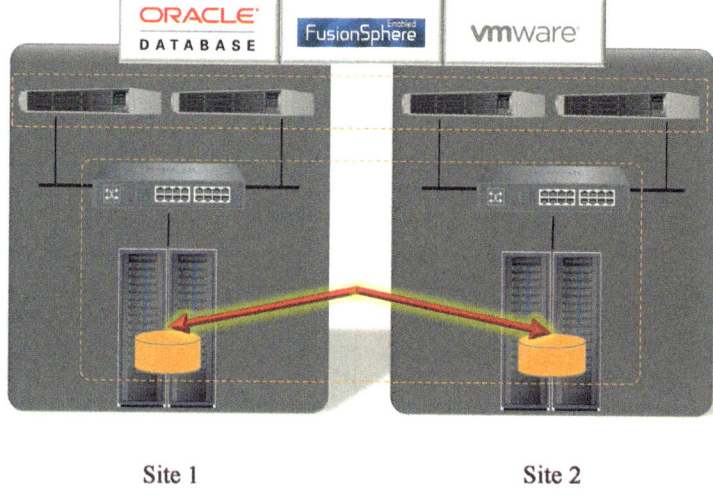

Active-active data center

Fig. 11.18 Active-active DR

shown in Fig. 11.18. The solution is also known as the active-active array solution. The active-active architecture of primary/secondary sites allows simultaneous access to both sites over all I/O paths. This enables both the primary and secondary storage arrays to handle the same service I/O request without forwarding it between systems, thus achieving service workload balancing. Additionally, a comprehensive arbitration mechanism is available to ensure seamless failover upon faults.

In terms of deployment modes, there are two available solutions: gateway-based active-active and built-in storage active-active. The gateway-based solution uses a dedicated device that performs a virtualization takeover of storage arrays to enable the active-active mode. This solution enables workload balancing and provides an arbitration mechanism to ensure service continuity in the event of a single site fault. It is important to note that more gateway devices mean higher network complexity and consequently higher purchasing and management costs. The introduction of external gateways also means more nodes for the IT system, and this leads to longer I/O paths, which increase system latency. As a result, the overall system reliability and performance are compromised. With the growing popularity of all-flash media, gateway devices are more likely to become an obstacle to the high performance and low latency of the entire system. Compared with the gateway-based solution, the built-in storage active-active solution offers a simpler network and shorter I/O paths and is therefore widely used in the industry.

The International Data Corporation (IDC) believes that an active-active array solution must meet the following requirements: Firstly, active-active mode is enabled between two independent storage systems that share no software and hardware. This is because if active-active is enabled between different engines of a single storage system, it is unable to handle problems such as suspension of the storage system, which may lead to service interruption. Secondly, both storage systems simultaneously provide upper-layer applications with read and write permissions on the same LUN or file system. This means the data copies on both storage systems are in the active state (rather than active/standby state) and the two storage systems provide mirror data volumes that are consistent in real time (RPO = 0, RTO = 0). Thirdly, there is an independent arbitration mechanism. If links between the two storage systems that provide active-active LUNs or file systems are down, real-time mirroring will be unavailable between the storage systems, and only one system is allowed to continue providing services. To ensure data consistency, an independent third-party arbitration mechanism is required to determine which storage array continues providing services. Otherwise, a split-brain may occur or services may stop. Fourthly, active-active replication links between the two storage systems support the high-performance Fibre Channel or remote direct memory access (RDMA) network to ensure uncompromised performance for mission-critical services after the active-active mode is enabled.

The key to active-active DR is to enable active-active mode and perform arbitration in fault scenarios to prevent split-brain. The following describes the active-active SAN architecture and active-active NAS architecture and their basic principles.

Active-Active SAN Architecture

As shown in Fig. 11.19, two SAN storage arrays run in active-active mode, with data being synchronized in real time. On the application host side, LUNs from both arrays are aggregated into one LUN using the multipathing software provided by the operating system or the storage vendor. Applications access the LUN using virtual disks (vDisks) of the multipathing software. To ensure the multipathing software identifies the active-active member LUNs from both arrays as the same LUN, the properties of the member LUNs that issue commands on the primary storage array are synchronized to the secondary storage array during the active-active pair configuration, and the properties of the member LUNs on the secondary storage array are also changed to be the same as that of the primary storage array. This ensures the multipathing software identifies the paths from both storage arrays as available paths. In the event of a path failure of one storage array, the multipathing software can automatically redirect all I/O requests to the path of the other storage array, thus ensuring service continuity on the host side. If it is not a path failure between the storage array and the host, but, for example, a link failure between the active-active storage arrays, the array that is out of service can report an I/O error to notify the multipathing software to switch the path.

In the active-active SAN architecture, a storage array can report the access priority of each path to the multipathing software of the host server using the Small Computer System Interface (SCSI) Asymmetric Logical Unit Access (ALUA) or NVMe Over Fabrics (NOF) Asymmetric Namespace Access (ANA) protocol. This

11.3 Data Protection Scenarios

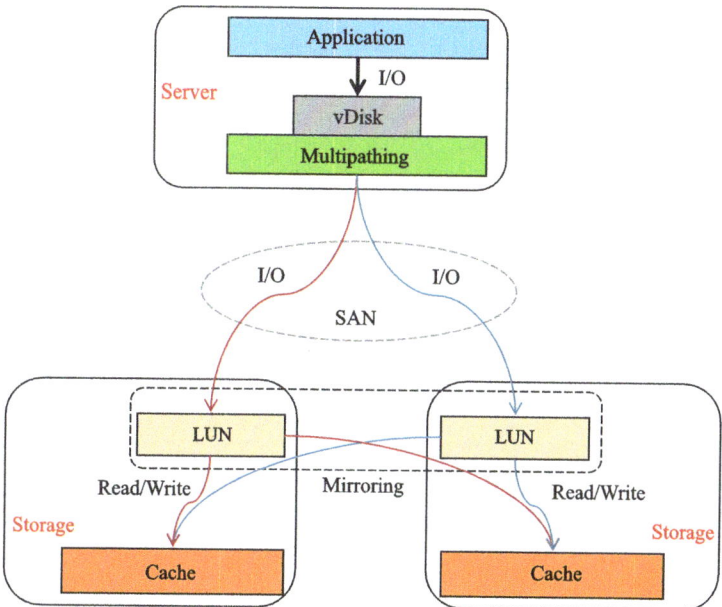

Fig. 11.19 Active-active SAN architecture

helps control the preferred path used by the host server to deliver I/O requests to the storage array. In short-distance deployment, the network latency between storage arrays is low. Therefore, all paths for active-active LUNs to connect to the host server can be configured with the same priority on the storage side by default, and the multipathing software delivers I/O requests to all paths based on the load balancing policy so that the two active-active storage arrays can evenly share half of the I/O requests to achieve optimal performance in this scenario. However, in remote deployment, when host servers at different sites access different storage arrays, the varying network distances can result in large latency differences. In this case, storage arrays at different sites can be configured to prioritize only access paths that connect to the host at the same site and to give a lower priority to access paths that connect to the host at the remote site. This ensures host servers at different sites preferentially select the paths of storage systems deployed at the same site as their own to deliver I/O requests, thus avoiding cross-network latency.

Active-Active NAS Architecture

As shown in Fig. 11.20, the active-active NAS architecture uses cross-site cluster management to enable all controllers of the storage arrays at the two sites to have read and write access to one file system simultaneously. It also provides real-time mirroring of cross-site storage data and key configuration information to implement DR failover in the event of site-level faults. The active-active NAS architecture

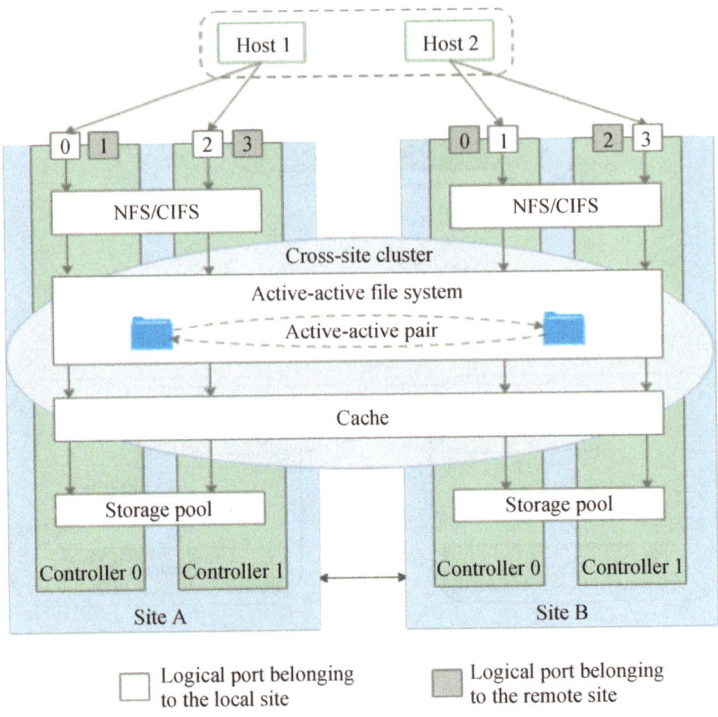

Fig. 11.20 Active-active NAS architecture

allows both sites to provide the IP address access capability simultaneously and implement IP failover to ensure uninterrupted host access.

The active-active NAS architecture has the following key features:

Front-end host access: NAS service hosts access storage systems over IP-based protocols such as Network File System (NFS) or Common Internet File System (CIFS). Therefore, server IP addresses must be configured beforehand on the logical port of storage systems. In addition, a failover group is usually configured for the port in the local storage system, and the aim is to ensure that when the physical port is faulty, the IP address can fail over between the standby ports in the failover group to ensure uninterrupted NAS access between hosts and storage devices. In active-active scenarios, IP failover between cross-site storage systems is required to ensure that when one storage site is faulty, the IP address can fail over to the active-active storage system on the other site for host access.

Global file system (GFS): A cross-site virtual cluster is created to combine controllers at both sites into a new active-active cluster. Each controller in the cluster can share the same cluster member view and service partition view through configuration synchronization. In the cluster, directories or files on the active-active file systems can be distributed to different service partitions. The quantity of service partitions in the cluster is fixed, for example, at 4096, and should be much greater than the number of nodes in the cluster. In this way, every storage controller at the two sites takes on the services of the GFS. Moreover, some service partitions can be allocated to new nodes based on changes in the number of controllers in the cluster to implement dynamic load balancing. Load balancing or random distribution can assign file system directories and files to service partitions. If the active-active sites are far away from each other, the site to which the front-end access IP address belongs can help distribute newly created directories or files to ensure that the local storage at the same site is used for the next access.

Real-time data mirroring: Based on the file ownership, the front-end system routes the I/O request to the response node, which then processes the file system semantic layer and writes the modified data to the battery-protected cache of local controllers. To ensure cross-site data reliability, the cache of one controller works with that of another at the same storage array to form a local HA mirroring relationship and works with the cache on the remote storage array to form a remote DR mirroring relationship. In this way, a total of three memory copies are maintained at the same time to ensure data reliability and consistency between sites. When it comes to flushing dirty data from the cache to disks, the data is written to the owning storage pools of the local and remote storage arrays through the volume object of the active-active file systems for persistent storage. To reduce cross-site network overhead, data sent to the remote storage array can only carry the address of the DR copy in the remote battery-protected cache. After receiving this message, the remote storage array reads data from the remote battery-protected cache through the address and directly writes the data to the storage pool.

11.3.3.3 Multi-data Center DR

To handle severe disasters, some core services require multi-data center DR solutions with higher DR levels, such as the geo-redundant three-center or even four-center solution. Having multiple copies of DR data can enhance data security and reliability so that in the event of an extreme disaster, there are still data copies available for data services. The most widely used geo-redundant three-center DR solution, or simply three-center solution, consists of three data centers: a production center, a DR center in the same city as the production center, and a remote DR center. Using the active-active feature as well as synchronous and asynchronous replication, it provides flexible and powerful data DR to implement multi-level data protection and ensure a higher level of data integrity and availability.

The three-center solution supports cascaded, parallel, and DR Star networks, which can be constructed based on multiple DR setups, such as active-active, synchronous replication, and asynchronous replication.

Cascaded three-center network: Data in production center A is replicated to DR center B in the same city and then to remote DR center C, as shown in Fig. 11.21. In addition, synchronous replication, asynchronous replication (with a short peri-

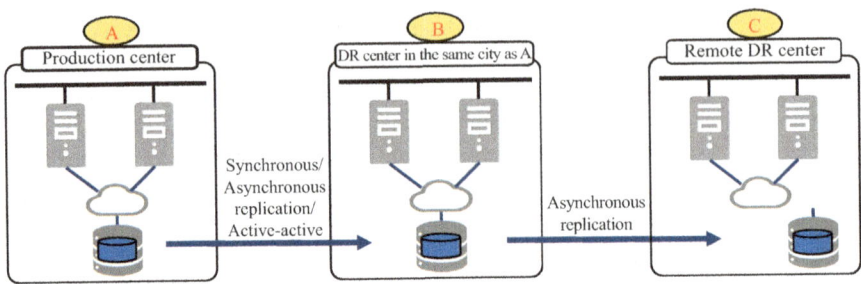

Fig. 11.21 Cascaded three-center network

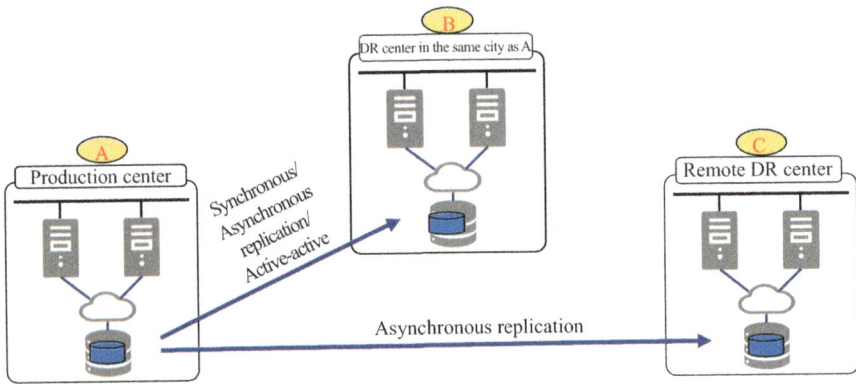

Fig. 11.22 Parallel three-center network

odic replication interval of seconds or minutes), or the active-active feature is configured between centers A and B to minimize recovery objectives (RPO = 0, RTO = 0). Asynchronous replication (with a long periodic replication interval of minutes or hours) can be configured between centers B and C to further ensure data availability.

Parallel three-center network: Data in production center A is concurrently replicated to DR center B in the same city and remote DR center C. Similarly, synchronous replication, asynchronous replication (with a short periodic replication interval of seconds or minutes), or the active-active feature is configured between centers A and B, while asynchronous replication (with a long periodic replication interval of minutes or hours) is configured between centers A and C, as shown in Fig. 11.22.

DR Star network: The DR Star solution is created by establishing physical links and a remote replication relationship between DR center B and remote DR center C in a serial or parallel three-center solution. If a fault occurs, services fail over, and incremental data replication continues. This enables reaching a normal data protection state in a shorter time. As shown in Fig. 11.23, synchronous replication and the active-active feature are common configurations between centers A

11.3 Data Protection Scenarios

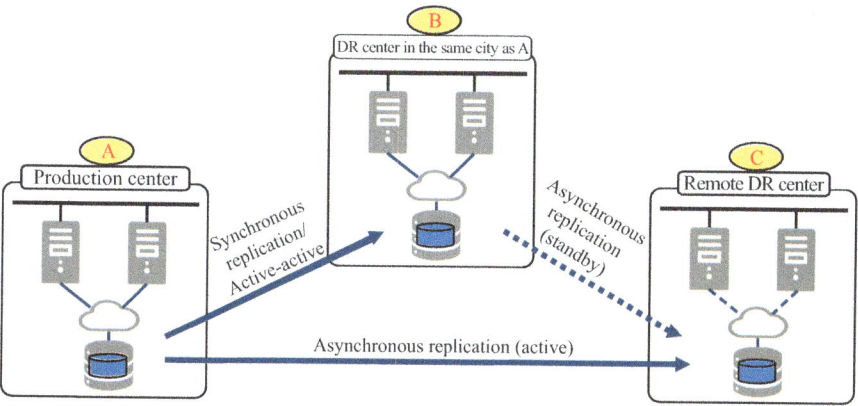

Fig. 11.23 DR Star network

and B, while asynchronous replication is configured between center A/B and center C. However, only one asynchronous replication works at a time.

Unlike the serial or parallel three-center solution, the DR Star solution provides the following benefits:

1. Standby asynchronous remote replication links and replication relationships are configured in advance to ensure they can be readily enabled in the event of a fault.
2. The DR Star solution with the active-active feature and asynchronous remote replication enables the standby asynchronous remote replication relationship to automatically take effect if a fault occurs at one of the active-active sites. This ensures service continuity, reduces manual intervention, and simplifies management.
3. The standby asynchronous remote replication relationship is usually enabled to perform incremental data replication, which provides faster and more effective DR protection for data.
4. The DR Star solution can switch between a parallel network and a cascaded network based on the storage situation and network conditions. These flexible network changes make it easier to use the DR Star solution and improve user experience.

In the preceding three network modes, the DR capabilities of the cascaded network are similar to those of the parallel network. Therefore, the following section only describes the data synchronization process in the cascaded network mode and DR Star network mode.

In a cascaded three-center network, there are two types of data synchronization, which are real-time active-active synchronization between centers A and B and periodic asynchronous replication between centers B and C, as illustrated in Fig. 11.24.

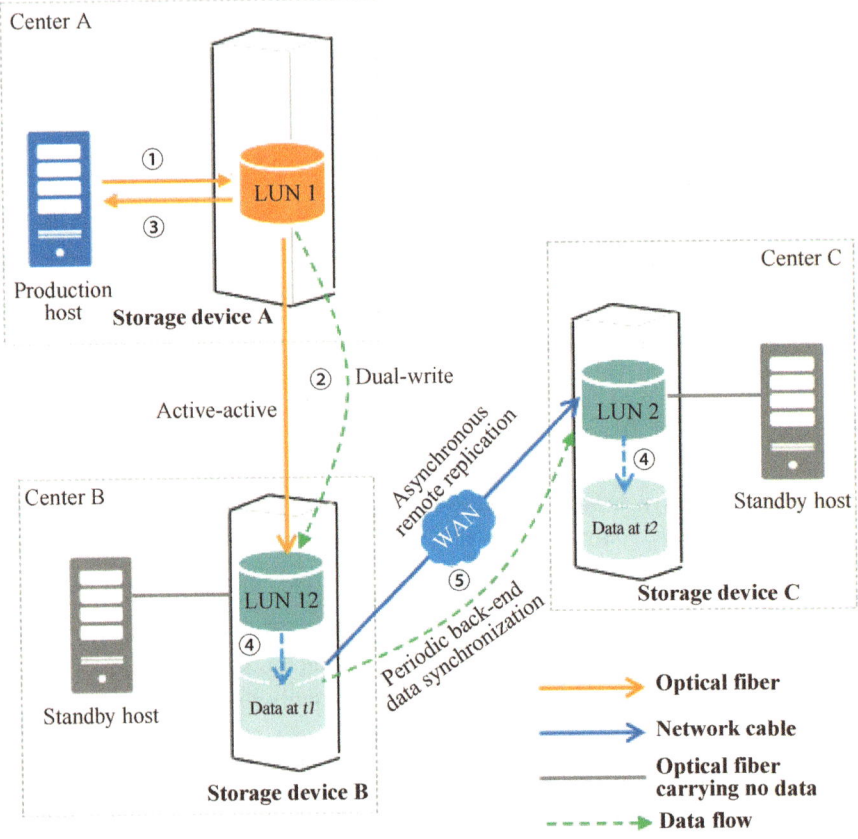

Fig. 11.24 Synchronization process of the cascaded three-center network

The process of real-time synchronization between centers A and B is as follows:

1. The production host at center A sends a write I/O request to the active-active LUN 1 on storage device A.
2. After receiving the write I/O request, LUN 1 writes data on storage A and LUN 12 on storage B at center B, which is active-active with center A.
3. LUN 1 informs the production host that the data was written successfully after receiving the write success responses from the local and remote storage systems.

The process of periodic asynchronous replication between centers B and C is as follows:

1. Asynchronous replication periodically initiates a data synchronization task and creates snapshots for primary LUN 12 at the primary (B) site and secondary LUN 2 at the secondary (C) site.
2. The snapshot data at t1 is synchronized from the primary LUN 12 to the secondary LUN 2 at the backend. When synchronization is complete, the snapshot data of the primary LUN is deleted.

11.3 Data Protection Scenarios

Because both the active-active feature and asynchronous replication are configured on LUN 12, the synchronization tasks of the two features must be mutually exclusive to prevent their data from being written to the same target LUN 12 at one time (for example, during fault recovery).

By having an asynchronous replication configuration between centers A and C, the cascaded three-center network can be transformed into the DR Star network. As illustrated in Fig. 11.25, the data from center A is asynchronously replicated to the same secondary LUN 2 at center C that the data from center B is asynchronously replicated to. Generally, only one of the asynchronous replication paths is active, either between A and C or B and C, while the other is standby. In order to continue the synchronization incrementally rather than in a full-amount way after an active-to-standby switching, the asynchronous replication on the standby path (between centers A and C in this case) is notified through the negotiation mechanism after each periodic synchronization between centers B and C is complete. In addition, the position of the difference log of incremental synchronization is updated for the asynchronous replication between centers A and C to prevent full synchronization or duplicate data synchronization after the failover.

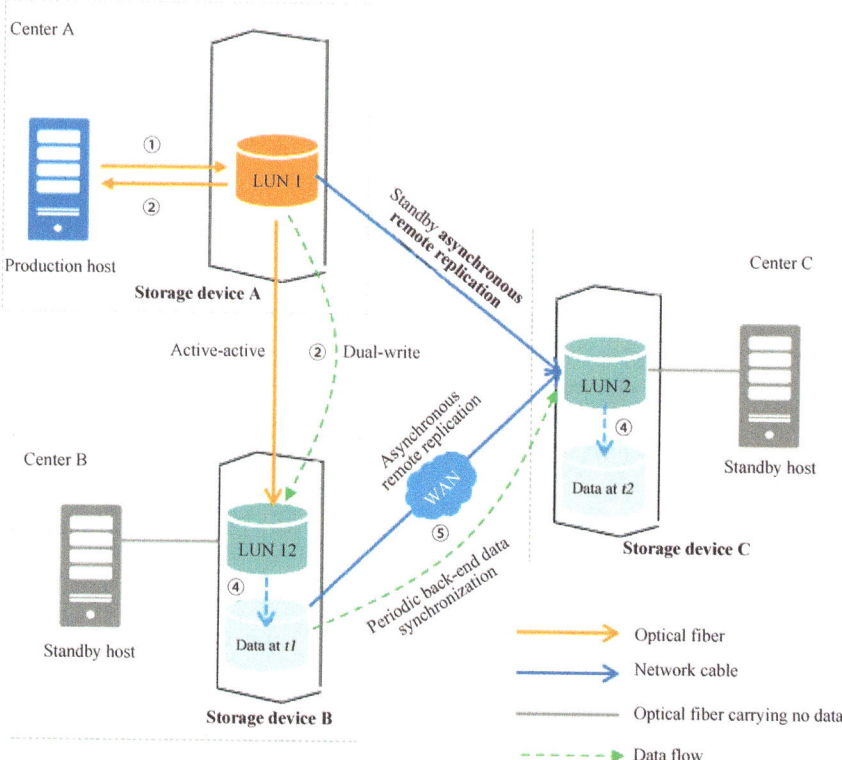

Fig. 11.25 Data synchronization process of the DR Star network

For the DR Star network, the real-time active-active synchronization between centers A and B is the same as that in the cascaded three-center solution.

The process of periodic asynchronous replication between centers B and C is as follows:

1. Asynchronous replication periodically creates snapshots for LUN 12 at the primary site (B) and secondary LUN 2 at the secondary site (C) and notifies the standby asynchronous replication at center A of aligning with the starting position of log (or snapshot) synchronization.
2. The snapshot data is synchronized at the backend at t1 from the primary LUN 12 to the secondary LUN 2. When synchronization is complete, the snapshot data of the primary LUN is deleted. In addition, the standby asynchronous replication at center A is informed of discarding all differences from before the log (or snapshot) position aligned prior to the synchronization.

When services fail over from center B to center A after center B or the link is faulty, or the active asynchronous replication between centers B and C is discontinued due to replication link faults between them, the asynchronous replication between centers A and C automatically becomes active to continue incremental synchronization from the difference log position notified when the last asynchronous replication between centers B and C is complete. This process ensures data availability at center C without manual intervention, thereby shortening the RPO.

11.4 Summary

Data is a crucial means of production in the era of big data and artificial intelligence. Loss of critical data can have serious consequences. Threats to data include system software and hardware faults, human error, cyberattacks, and natural disasters, which means data loss is inevitable. Therefore, data protection technologies rely on increasing data replication and copies to enhance tolerance for partial data loss. That is why this chapter discusses data protection from the three aspects of standards, technologies, and scenarios. With the digital transformation of enterprises around the world, the total worldwide data volume is increasing dramatically, creating new challenges in data storage and protection.

11.5 Practice Questions

1. **What are the objectives of data protection? What is the basis of data protection technology?**

    Answer: Data protection aims to ensure data security and reliability. It uses technologies such as data backup, data archiving, and disaster recovery to increase tolerance for partial data loss.

11.5 Practice Questions 345

2. **What is the role of data backup and disaster recovery in data protection? What is the relationship between data backup and disaster recovery?**

    Answer: Data backup and disaster recovery ensure data reliability and availability in data protection. Disaster recovery can quickly recover data and systems to ensure service continuity in the event of data loss or a system fault. Data backup and data archiving are the foundation of disaster recovery. Data backup is used to protect data, while data archiving is used to retain data for long periods of time.

3. **What are the data protection technologies? What are their similarities and differences?**

    Answer: Data protection technologies include mirroring, snapshot, and cloning. They are similar in that they are all used for data backup, recovery, and replication. However, they also have differences: A mirror is a copy of a complete device or system, a snapshot stores only the status of a data collection within a certain period of time and occupies less space than a mirror, and a clone is a full copy of a storage system and provides higher data recovery capabilities.

4. **What is the relationship between data protection technologies and data protection applications? Use backup as an example to describe this relationship.**

    Answer: Data protection technologies are used to meet the requirements of specific data protection applications. For example, for a backup, mirroring is used to create a full copy of data. Mirrors are usually used for system backup and migration. Snapshot technology creates a snapshot that records the data status of a storage device or virtual machine at a specific point in time for quick backup and recovery. Cloning creates an independent copy that is identical to the original object. Cloning is usually used to create a backup environment for a test environment or production system.

5. **What are the relationships between disaster recovery, backup, and archiving in data protection?**

    Answer: A backup system is for data loss caused by disasters in production systems involving system hardware and software faults, viruses, and manual operations. An archive system is for storing data that has not been accessed for a long time in production systems and backup systems. A disaster recovery system creates data copies identical to those in production systems at a remote location, and all operations on the data in production systems are synchronized to the disaster recovery system.

6. **What is data archiving? What is the role of data archiving in enterprise data management? What are the architecture and functions of data archiving?**

    Answer: Data archiving is the process of moving infrequently accessed data from primary storage to slower storage media to save primary storage space. In enterprise data management, data archiving helps reduce storage costs and improve storage efficiency while protecting data. A multi-tier storage architecture is usually used for data archiving. Data is stored in different tiers based on how frequently it is accessed and how important it is. Common tiers include hot data storage which is characterized by high performance and low latency, warm data storage which tends to deliver moderate performance at an average cost, and cold data storage which offers relatively poor performance at a low cost. The purpose of data archiving is to facilitate data movement and migration, data retrieval, data deletion, version control, and retention policies.

7. **What are the RPO and RTO in data protection? What is their role in data protection? How do we determine RPO and RTO?**

    Answer: RPO, short for recovery point objective, is the acceptable amount of time for data loss after a disaster. RTO, short for recovery time objective, is the duration of time during which a service is restored after a disaster. In data protection, RPO and RTO determine backup and recovery time points to ensure service continuity. Methods such as service requirement analysis, risk assessment, and technical feasibility analysis can be used to determine RPO and RTO.

8. **Assume that the RPO and RTO required by a company's service system are 2 and 4 h, respectively. How long does the company need to complete data recovery and get the system to run properly after a disaster?**

    Answer: RPO is the acceptable amount of time for data loss after a disaster. RTO is the maximum time required for a system or service to recover normal operations after a disaster. If the maximum time is exceeded, data may be irreparably damaged. An RPO of 2 h means that the system can recover the data from 2 h prior to the disaster. An RTO of 4 h means that the system can recover to its normal state in no more than 4 h. Therefore, the company needs to complete data recovery and get the system to run properly in 4 h.

9. **What is remote disaster recovery in data protection? What are its implementation modes? What are its advantages?**

    Answer: Remote disaster recovery is when data backup and disaster recovery facilities are deployed in different geographical locations to ensure that services can be quickly recovered if a disaster occurs. Remote disaster recovery can be implemented in active/standby, active-active, or multi-active mode. It can improve data reliability and availability, ensure service continuity, and reduce the risk of disasters.

10. **Why do we need disaster recovery technology when backup technology can periodically protect data? What are the differences between these two technologies?**

    Answer: Data backup can protect data against damage caused by software and hardware faults or misoperations. However, when a large-scale disaster (such as an earthquake or fire) occurs, backup technology cannot ensure data security. In contrast, disaster recovery can address sudden disasters by establishing multiple systems with the same functions in remote locations. If a disaster occurs in a system at one location, services on the system can fail over to a system at another location to continue running.

11. **Why do we need clones when both clones and snapshots are copies of data? How do the media access control (MAC) addresses of snapshots and clones change?**

    Answer: Clones are independent of the source data, meaning that they can be read and written independently. In addition, clone technology ensures that clones are isolated from the source data, protecting the integrity and security of the source data. When a snapshot is created, its associated MAC address does not change because it is only a pointer or metadata to the original data. In contrast, a clone is independent of the source data, so its associated MAC address changes.

References

1. Wang D, Wang L. Research on disaster recovery systems. Comput Eng. 2005;31(6):4.
2. Informatization Office of the State Council. Guidelines for disaster recovery of important information systems. Beijing: Informatization Office of the State Council; 2005.
3. National Technical Committee 260 on Cybersecurity of Standardization Administration of China. Information security technology—disaster recovery specifications for information system: GB/T 20988-2007. Beijing: Standards Press of China; 2007.
4. Garcia-Molina H, Polyzois CA. Issues in disaster recovery. In: IEEE, 35th IEEE computer society international conference on intellectual leverage. San Francisco: IEEE Computer Society; 1990. p. 573–7.
5. Xiang X. Research on key data protection technologies. Beijing: Tsinghua University; 2009.
6. Yao J, Shu J, Zheng W. Distributed storage cluster design for remote mirroring based on storage area network. J Comput Sci Technol. 2007;22(4):521–6.
7. Zheng XF, Ouyang B, Zhang DN, et al. Technical system construction of Data Backup Centre for China Seismograph Network and the data support to researches on the Wenchuan earthquake. Chin J Geophys. 2009;52(5):1412–7.
8. Whitlock MC, McPeek MA, Rausher MD, et al. Data archiving. Am Nat. 2010;175(2):145–6.
9. Piwowar HA, Vision TJ, Whitlock MC. Data archiving is a good investment. Nature. 2011;473(7347):285.
10. Hammersley M. Qualitative data archiving: some reflections on its prospects and problems. Sociology. 1997;31(1):131–42.
11. Toigo J. Disaster recovery planning: preparing for the unthinkable. ACM Digital Library; 2002.
12. Fallara P. Disaster recovery planning. IEEE Potentials. 2004;23(5):42–4.

Chapter 12
Storage Maintenance

12.1 Overview

Maintenance is mainly classified into preventive maintenance (PM) and corrective maintenance (CM) [1].

PM is a maintenance plan, which refers to performing maintenance activities at scheduled intervals or by specified standards to reduce the likelihood of device faults or function degradation. To prevent device faults, PM is implemented regularly, even if devices show no sign of faults. In this way, potential device faults can be avoided as much as possible to ensure normal service running and security.

CM refers to event-based fault rectification. If a fault occurs, fault rectification or device component replacement will be carried out to ensure normal service running and security.

PM is preventive, and CM is reactive. CM may sometimes cause device shutdown, affecting production services. Generally, modern storage needs more effective technologies to perform PM in order to prevent CM from disrupting services.

12.2 Preventive Maintenance

Data is critical to enterprise informatization. The increasing number and types of devices complicate maintenance. Enterprises will suffer huge losses once device faults occur.

Traditional passive maintenance based on manual experience faces the following challenges:

Difficulty in meeting service-level agreement (SLA): The response is passive when device faults occur and the subsequent fault locating process is complex.

High operating costs: A greater device quantity and hybrid deployment of multiple services are accompanied by more complex systems. This requires more professional maintenance personnel, resulting in higher annual management costs.

Inappropriate resource utilization: Increasing services requires more storage space. Subsequently, the planning of storage space allocation becomes more difficult due to growing storage capacity fragments, and as a result, cost control is a headache. Threshold and defined filtering rules of traditional resource monitoring technologies are too general to perform refined management. This results in low resource utilization.

Long service rollout time: Comprehensive service planning and evaluation are required before the new service rollout. As the number of devices and services increases, manual planning and evaluation take several weeks or even months, holding back service rollout.

Modern storage uses big data analytics and artificial intelligence (AI) technologies to carry out PM. In this regard, faults can be identified in advance to mitigate or avoid their impact on services. In addition, future performance trends and capacity needs can be accurately predicted and proactively planned ahead of time. PM of storage can focus on disks from the following four aspects:

Disk health prediction [2]: Adopts AI technologies to perform dynamic analysis on disk data indicator changes and predict disk failures based on its performance and workload characteristics. In this way, proactive fault prevention is achieved, significantly improving system reliability.

Capacity trend prediction: Collects historical performance and capacity data of storage systems, then employs machine learning training, and selects the optimal prediction models to predict capacity needs of services in order to enable storage systems to better meet service requirements.

Performance anomaly detection: Learns service feature changes from historical service performance data to identify performance anomalies and detect device performance risks in a timely manner.

Performance fluctuation analysis: Identifies peak and off-peak hours of services according to historical performance fluctuation and further provides the reference for corresponding maintenance plans to prevent device performance insufficiency caused by operations during peak hours.

12.2.1 Disk Health Prediction

Disks are the basic components and the largest wear parts of storage arrays. Despite the wide use of various redundancy technologies, storage arrays can continue to work only when faults occur on a limited number of disks. For example, a redundant array of independent disks 5 (RAID 5) only tolerates single-disk failure. If two disks fail simultaneously, the storage system will stop providing services to ensure data reliability.

With collected S.M.A.R.T. information, I/O link information, and reliability indicators of disks, the disk failure prediction model can accurately predict the

12.2 Preventive Maintenance

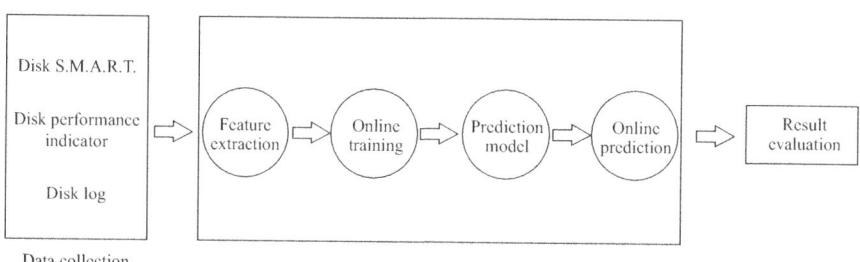

Fig. 12.1 Disk health prediction process

solid-state drive (SSD) lifespans and disk failure risks, achieving fault prevention for higher system reliability.

As shown in Fig. 12.1, the key process of disk health prediction is as follows:

12.2.1.1 Data Collection

Data sources include disk S.M.A.R.T. information, disk performance indicators, and disk logs.

Disk S.M.A.R.T. information refers to the log page information provided by SSD interfaces. The page records details regarding the current status, performance indicators, and read/write error information of disks. Disk performance indicators include the average I/O size per minute, I/O operations per second (IOPS), bandwidth, daily processed bytes, latency, and average service time. Disk logs include I/O error codes and disk lifespan.

12.2.1.2 Analytics Platform

The analytics platform can carry out feature extraction, establish a machine learning model library through online training, and then perform online prediction.

Feature extraction uses machine learning algorithms with mass sample data to perform automatic feature transformation and extraction on disk indicator data. Based on multiple model algorithms, online training conducts mass training on data extracted by features to optimize model algorithms. Online prediction adopts optimized training models to predict disk failures.

12.2.1.3 Result Evaluation

Disk health prediction can be abstracted as a classification problem. Therefore, false discovery rate (FDR), which is the proportion of correctly predicted faults to the total number of faults, and false alarm rate (FAR), which is the proportion of samples that are incorrectly reported as faults to the total number of samples, can be adopted to provide overall evaluation over the prediction effect.

$$\text{FDR} = \frac{\text{TP}}{\text{TP} + \text{FN}} \quad (12.1)$$

$$\text{FAR} = \frac{\text{FP}}{\text{FP} + \text{TN}} \quad (12.2)$$

TP is the total number of samples that are correctly predicted to be faulty, and FN is the total number of samples that are not correctly predicted to be faulty. FP indicates the number of samples that are actually normal but detected as faulty, in other words, false positive quantity.

12.2.2 Capacity Trend Prediction

Enterprises have been looking for capacity management solutions to strike a balance between low cost and high performance, but due to various factors (such as holidays and temporary events), the needed capacity for service growth cannot be accurately predicted, resulting in temporary emergency capacity expansion. Moreover, manual capacity prediction fails to provide accurate mid- and long-term service requirement planning, which has plagued maintenance personnel.

The change in system capacity is subject to various factors. A single prediction algorithm cannot ensure the accuracy of the prediction result. Multiple prediction models can be used for online prediction to output corresponding prediction results. As shown in Fig. 12.2, data in the historical data warehouse can be used for training to optimize model parameters and obtain measurement indicator statistics of each model. Then, the measurement indicator statistics of the online training are compared with the online prediction results to select the optimal prediction result.

The key process is as follows:

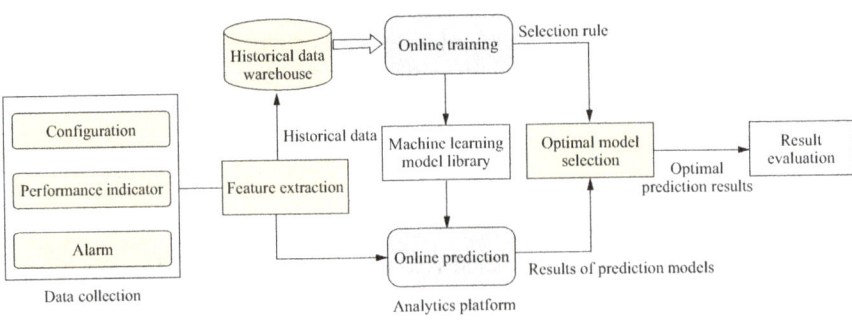

Fig. 12.2 Capacity trend prediction process

12.2 Preventive Maintenance

12.2.2.1 Data Collection

Data collection refers to collecting the current configurations, performance indicators, and alarm information of storage devices to reduce interference of multiple factors on machine learning training results. It collects historical device capacity data over the latest year and stores it in the historical data warehouse.

12.2.2.2 Analytics Platform

Feature extraction uses machine learning algorithms to perform automatic feature transformation and extraction on capacity data.

Online training based on mass samples generates the prediction measurement indicator statistics of each model and selection rules for optimal models. Furthermore, it can improve model algorithms with the current and historical capacity data.

Online prediction adopts optimized models in the real environment to output the prediction results of multiple models and mean absolute percentage error (MAPE) values of measurement indicators.

$$\text{MAPE} = \frac{1}{m}\sum_{i=1}^{m}\left(y_i - \widehat{y_i}\right)^2 \tag{12.3}$$

In the preceding formula, y_i indicates the actual capacity, $\widehat{y_i}$ indicates the predicted capacity, and m indicates the number of samples.

Optimal model selection refers to weighing the model statistics of online training and online prediction results to obtain the optimal prediction result.

12.2.3 Performance Anomaly Detection

The biggest concern of enterprises is smooth service running. However, because performance issues are complicated and difficult to identify and solve early, such issues get worse and eventually affect services, which causes loss to enterprises.

For service latency issues, performance anomaly detection learns service features from historical performance data and combines industry and expert experiences to obtain device performance profiles and perform real-time abnormal data detection, which can further provide accurate fault locating and corresponding rectification suggestions.

Performance anomaly detection boasts speed and accuracy. It provides minute-level anomaly detection and 24/7 real-time protection. On top of that, it troubleshoots service latency issues and provides corresponding rectification suggestions based on I/O changes, service configurations, hardware bottlenecks, and hardware faults.

The following indicators can be used to evaluate the anomaly detection effect, including precision and recall.

$$\text{Precision} = \frac{TP}{TP + FP} \tag{12.4}$$

$$\text{Recall} = \frac{TP}{TP + FN} \tag{12.5}$$

FN indicates the number of samples whose actual performance is abnormal but whose detected performance is normal, namely, the number of unreported samples with performance anomaly.

12.2.4 *Performance Fluctuation Analysis*

It is well known that periodic services (scheduled snapshots) or temporary service changes (online upgrade, capacity expansion, and spare part replacement) should be performed when the service load is light to prevent affecting online services. Experienced maintenance personnel selected the proper time window for service change according to performance indicators of a past period of time. This traditional method depends on the subjective experience of maintenance personnel, which is not refined, accurate, and scientific. Furthermore, it is prone to accidental factors, making it difficult to obtain the periodic performance fluctuation rules.

Based on historical device performance data, performance fluctuation analysis adopts big data and AI technologies to analyze the service period rules in terms of workloads, IOPS, bandwidth, and latency, which are further described in Table 12.1. The graphical user interface (GUI) provides guidance for maintenance personnel to select a proper time window for periodic or temporary service changes, ensuring an uninterrupted user experience during peak hours. Compared with the traditional method, performance fluctuation analysis has the following advantages:

Scientific and accurate: The fluctuation rules are identified more accurately given that performance data of hardware such as controllers, disks, and ports is taken

Table 12.1 Performance fluctuation analysis indicators

Indicator	Description
IOPS	The average processed requests per second by a device for systems' I/O processing capability measurement
Bandwidth	Average processed data per second by a device for system performance measurement
Load	Service workloads of a device for device performance pressure measurement
Read/write latency	Average time required for the device to process a read/write request

into consideration. Furthermore, the interference brought by causal factors is ameliorated since the historical performance data is also covered in the overall analysis.

Easy to understand: Performance fluctuation analysis uses visualization technologies including heat maps to display analysis results. For example, workload levels can be color-coded, helping maintenance personnel manage service changes.

12.3 Corrective Maintenance

Common CM methods for storage systems include proactive troubleshooting, upgrade [3], and capacity expansion.

Proactive troubleshooting refers to a comprehensive process of automatic monitoring of device health (including alarms and events) as well as subsequent handling.

Upgrade refers to a set of processes carried out during software version changes and includes solution formulation, upgrade preparation, upgrade implementation, and post-upgrade verification.

Capacity expansion is the process of adding controller nodes, disk enclosures, and disks to the existing system when the current system cannot meet service capacity or performance requirements. It includes solution design, capacity expansion preparation, expansion implementation, and post-expansion verification.

12.3.1 Proactive Troubleshooting

Traditional services rely on technical personnel to provide support. However, technical personnel may be unable to identify and handle faults in a timely manner and may fail to transfer the relevant fault information to the relevant party.

Proactive troubleshooting is to establish a secure and controllable network connection between devices and the technical support center of the vendor or agent. Through this connection, the maintenance system regularly collects and sends information to the technical support center.

The proactive troubleshooting system provides the following capabilities:

Proactive alarm monitoring: This capability supports 24/7 monitoring of the health status of customer devices. Whenever a device generates an alarm, the system automatically creates a service ticket and sends it to the corresponding engineer for handling, enabling proactive identification and rectification of issues.

Intelligent alarm analysis: This capability leverages AI technology and an alarm feature model library containing the alarm features of numerous devices to automatically mask irrelevant alarms, suppress repeated reporting of alarms of the same type, and filter redundant alarms, thus improving the accuracy and efficiency of alarm handling.

Service ticket process tracking: The tracking system displays the real-time progress of the service tickets created from device alarms. This enables customers and maintenance personnel to learn about the troubleshooting progress of faults in a timely manner.

12.3.2 Upgrade

Storage system upgrades can be conducted either online or offline. Offline upgrade is fast, but requires suspending host services. It is mainly used in device commissioning or scenarios where upper-layer services are relatively less important. In contrast, online upgrade does not need to stop host services. During an online upgrade, controllers can be upgraded simultaneously or in different batches. Upgrading in batches is to sequentially upgrade the controllers, during which the services of controllers being upgraded are taken over by those yet to be upgraded. In contrast, simultaneous online upgrade enables service software on controllers to be simultaneously restarted within seconds. This greatly improves the upgrade efficiency and reduces the dependency on host software.

Table 12.2 lists a comparison between offline upgrade and online upgrade.

12.3.2.1 Online Upgrade in Batches

Figure 12.3 shows the process of online upgrade in batches.

1. Before the upgrade of controller B, the link between P1 on controller B and P1 on the host bus adapter (HBA) is disconnected.
2. P1 on the HBA detects the link interruption and returns the link fault information to the multipathing software if the interruption is not recovered within the timeout period (may exceed 30 s by default) set by the HBA driver.
3. The multipathing software switches the P1-P1 logical link to the P0-P0 physical link.
4. After the P1-P1 link is disconnected, App 2 attempts to deliver I/Os again within a certain timeout period. (Services will be interrupted if the timeout period is shorter than that of the HBA driver.)
5. The system starts to upgrade controller B.

Table 12.2 Comparison between offline upgrade and online upgrade

Upgrade mode	Host service suspension before the upgrade	Controller restart during the upgrade
Offline upgrade	Yes	Yes
Online upgrade in batches	No	Yes
Simultaneous online upgrade	No	No

12.3 Corrective Maintenance

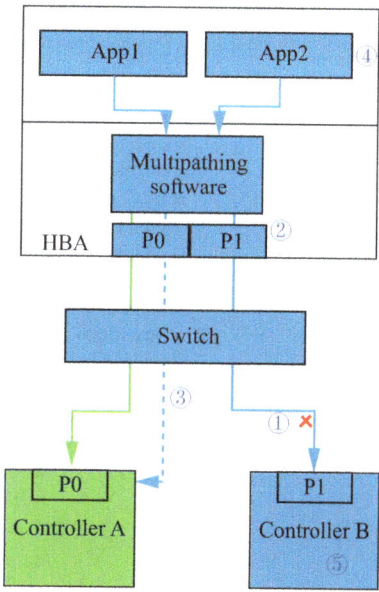

Fig. 12.3 Online upgrade in batches process

The preceding steps show the online upgrade process when the host accesses the SAN storage through Fibre Channel (FC) ports. Step (2) depends on the host HBA, step (3) depends on the host multipathing software, and step (4) depends on application software such as databases and backup software. Any changes to the behavior of the dependent objects may result in serious consequences such as service interruption and database damage.

Other scenarios using iSCSI and NFS face similar challenges and more complex dependencies. For example, in iSCSI scenarios, step (2) depends on the version and configuration of the iSCSI initiator; in NFS scenarios, step (2) depends on the fault tolerance processing of different NFS versions.

Customers and storage vendors have been plagued by host compatibility issues during online upgrade. Thanks to technological advancement, different solutions have been developed to reduce the dependency on host software during online upgrade in batches, as shown in Table 12.3.

12.3.2.2 Simultaneous Online Upgrade

Figure 12.4 shows the simultaneous online upgrade process.

1. The connection session with the host is saved to the connection keepalive process.
2. Service processes of all controllers are upgraded at the same time. During the upgrade, I/Os received by the connection keepalive processes are placed in queues and will be delivered after the upgrade is complete.

Table 12.3 Different solutions for performing online upgrade in batches

Solution	Description	Constraint
Active switchover	The multipathing software performs a switchover to the path connected to controller B before controller A is upgraded	Dependent on the multipathing software
IP address failover	A logical IP address fails over from a faulty port to another available port. This solution is easy to implement and widely used in distributed and unstructured scenarios where host services are insensitive to latency	1. Not applicable to FC networks 2. The handles and sessions saved on the original controller for communication with the host are still lost and need to be re-established. Dependency on hosts still exists 3. Redundant paths are required
FC port failover	This solution is similar to IP address failover and is mainly used in FC scenarios	1. Dependent on FC switches 2. The handles and sessions saved on the original controller for communication with the host still need to be re-established. Dependency on hosts still exists 3. Redundant paths are required
Front-end interconnect I/O module	The front-end interconnect I/O modules forward I/Os to controllers that are not upgraded	The cost of interconnect I/O modules is high

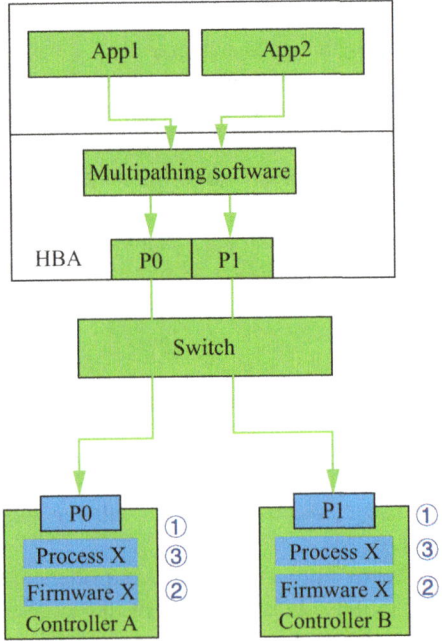

Fig. 12.4 Simultaneous online upgrade process

12.3 Corrective Maintenance

3. The service processes start quickly (generally within 10 s) after the upgrade is complete.

The connection keepalive technology can work either with or without using front-end interconnect I/O modules. When front-end interconnect I/O modules are used, links between the host and the storage system are established over the I/O modules. After receiving service requests from the host, the I/O modules distribute the requests to the controllers. During the restart of service processes on the controllers, the I/O modules save the I/Os received from the host and distribute them to the controllers after service processes run properly. The restart of service processes takes only a few seconds, so I/Os issued by the host do not time out, ensuring zero impact on host services. When front-end interconnect I/O modules are not used, the daemon process keeps the links between the controllers and hosts connected during service process startup. Similarly, this procedure takes only a few seconds, so I/Os delivered by hosts do not time out, eliminating the host compatibility and dependency issues mentioned in the previous section.

12.3.2.3 Large-Scale Node Upgrade Technology

In scenarios where a storage system has a large number of nodes, for example, a distributed storage system, long upgrade duration is a pain point that needs to be resolved. Common technical solutions for this issue include component-based upgrade [4], gray upgrade, and parallel upgrade in a storage pool.

Component-Based Upgrade

Component-based upgrade divides all software in a storage system into multiple components that can be independently tested, delivered, deployed, and upgraded. This allows each upgrade to be performed only on the necessary components, reducing both the upgrade duration and the related impact on services.

Component-based upgrade has high requirements on vendors' software capabilities. Major concerns when conducting a component-based upgrade include appropriately dividing software into different components, ensuring independence between the components, and covering testing for different software version combinations.

Gray Upgrade

Gray upgrade only upgrades certain nodes in the system at a time. This not only divides the upgrade duration into multiple time windows, but also reduces the impact caused by the upgrade. Gray upgrade is widely used for Internet- and cloud-based software.

Similar to component-based upgrade, gray upgrade has high requirements on compatibility between different software versions, since these versions may coexist in the same system for a long time.

Parallel Upgrade in a Storage Pool

Generally, distributed storage is upgraded by a node in serial mode. Due to its complex logic and the large number of nodes, the upgrade takes a long time to complete even when the storage pool management software is divided into components.

Two technologies can be used to implement parallel upgrade in a storage pool to reduce the upgrade duration. For storage pools with a large number of nodes, the cabinet-level security policy is recommended for performing parallel node upgrades in the cabinet. Storage pools with both a large number of nodes and server-level security can be divided into different disk pools when being created, and then, parallel node upgrades can be performed in the different disk pools.

12.3.3 Capacity Expansion

The acceleration of enterprise informatization and continuous expansion of services have led to the rapid growth of service data, resulting in the default capacity of storage systems often being unable to satisfy current service requirements. In such cases, the maintenance personnel need to conduct capacity expansion.

12.3.3.1 Capacity Expansion Type

Capacity expansion is classified into scale-up and scale-out.

Scale-Up

Scale-up, as shown in Fig. 12.5, is the process of adding storage components, such as disk enclosures and disks, to expand the capacity of a storage system in order to satisfy users' increasing service requirements.

Scale-Out

Scale-out is the process of adding controllers to satisfy users' service requirements for both higher performance and capacity, as shown in Fig. 12.6.

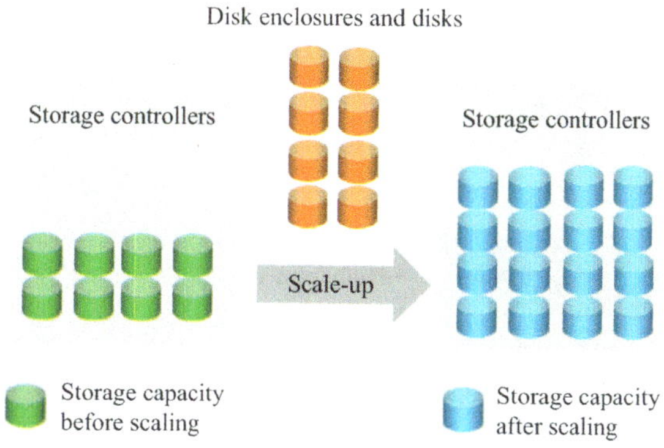

Fig. 12.5 Scale-up

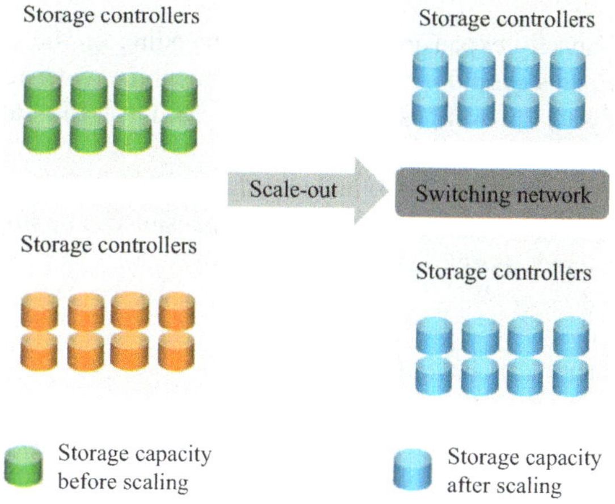

Fig. 12.6 Scale-out

12.3.3.2 Capacity Expansion Process

The capacity expansion process includes preparation, implementation, and post-expansion check.

Capacity Expansion Preparation

Preparations include live network information collection, capacity expansion planning, and pre-expansion check.

The purpose of collecting live network information is to provide reference information for capacity expansion plans and hardware installation. The collected information includes storage system information and host application information.

Table 12.4 Characteristics and application scenarios of capacity expansion methods

Type	Method	Characteristic	Application scenario
Scale-up	Adding disks	• No need to stop services • Easy to perform • Fast • Cost-effective	• There are vacant disk slots for capacity expansion to meet service needs
	Adding disk enclosures	• No need to stop services • High capacity	• No vacant disk slots • Vacant disk slots are insufficient to meet service needs
Scale-out	Adding controllers	• No need to stop services • High capacity • System performance boost	• No vacant disk slots • Vacant disk slots are insufficient to meet service needs • The system performance falls short of service performance requirements

Capacity expansion plans vary depending on the capacity expansion methods. The plans should be feasible and efficient and cater to the compatibility of added hardware as well as the operating status of the storage system, as shown in Table 12.4.

Pre-expansion check includes the checking of storage system health status and application server status. Capacity expansion can be smoothly performed only when the storage system is working correctly.

Capacity Expansion Implementation

Expanding capacity by adding disks is easy, fast, and cost-effective, suitable for scenarios requiring relatively small expansions in capacity. The steps include installing new disks and confirming the status of the new disks.

Expanding capacity by adding disk enclosures is applicable to scenarios requiring relatively high capacity. The steps include installing disk enclosures, connecting cables, powering on the added disk enclosures, and confirming the status of the added disk enclosures.

System performance can be enhanced by adding more controllers if it fails to meet growing service needs. Expanding capacity by adding controllers includes installing new controller nodes, connecting cables, connecting switches, powering on the added controllers and disk enclosures, and confirming the status of the added disk enclosures.

Post-expansion Check

Post-expansion check includes storage system status check and service verification to ensure that services are running properly.

12.4 Practice Questions

1. **What is the essential difference between PM and corrective maintenance CM?**

   Answer: PM is preventative, while CM is reactive. Effective PM measures can reduce or avoid corrective maintenance impacts on services.

2. **What are the similarities and differences between performance anomaly detection and performance fluctuation analysis? Try to analyze them in terms of application scenarios and technical solutions.**

   Answer: (1) Performance anomaly detection is applicable to scenarios where performance anomalies have occurred, and technical means are used to detect performance anomalies in the time and space dimensions. Performance fluctuation analysis is applicable to scenarios where there is performance fluctuation (no performance anomaly occurs in most scenarios), and technical means are used to provide a visualized display of performance fluctuation rules to help maintenance personnel identify service peak and off-peak hours. (2) In terms of technical solutions, performance anomaly detection focuses on learning service feature changes. It identifies performance anomalies by comparing the current performance data with historical performance data. Performance fluctuation analysis focuses on identifying the service workloads in a certain period. It identifies the performance fluctuation by comparing the current performance data with historical performance data.

3. **Which PM measures can help maintenance personnel improve O&M efficiency and quality in service change scenarios including new service rollout and service migration?**

   Answer: (1) Capacity trend prediction helps maintenance personnel predict the capacity growth trend of devices so that appropriate capacity planning can be performed to prevent future capacity insufficiency. (2) Thanks to performance fluctuation analysis, which helps identify off-peak and peak hours of services, service change can be carried out during off-peak hours to effectively prevent service impact caused by service changes during peak hours.

4. **What are the differences in key technologies between online upgrade in batches and simultaneous online upgrade?**

   Answer: The key technical challenge of online upgrade in batches is reducing the impact of path switchover on host services. Related key technologies include active multipathing switchover, IP address failover, and FC port failover. The key technical challenge of simultaneous online upgrade is to complete the online upgrade without triggering path switchover. Related key technologies include front-end interconnect I/O modules, front-end link keepalive, and quick startup of service processes.

5. **What technologies are available for large-scale node upgrade? What are their corresponding application scenarios?**

   Answer: The technologies include component-based upgrade, gray upgrade, and parallel upgrade. Component-based upgrade applies to projects that use component-based or microservice-based software. Gray upgrade applies to projects comprising numerous nodes and requiring high reliability. Parallel upgrade within a storage pool applies to projects involving large storage pools and requiring fast upgrades. The three technologies are complementary and can be used in the same project.

References

1. Wang Y, Deng C, Wu J, et al. A corrective maintenance scheme for engineering equipment. Eng Fail Anal. 2014;36:269–83.
2. Narayanan I, Wang D, Jeon M, et al. SSD failures in datacenters: what? When? and why? In: Association for computing machinery. Proceedings of the 9th ACM international on systems and storage conference (SYSTOR'16). New York: Association for Computing Machinery; 2016. p. 1–11.
3. Floyd P, Hawkins M. High availability: design, techniques, and processes. London: Prentice Hall Professional; 2001.
4. Richardson C. Microservice patterns: with examples in Java. Beijing: China Machine Press; 2022.

Chapter 13
Storage Solutions

13.1 e-Government Converged Storage Resource Pool Solution

With the ongoing enhancement of global social governance capabilities, government functions have shifted from administration-oriented to service-oriented. Digital transformation is revolutionizing governance instruments through the application of advanced information technologies. On one hand, information systems must be built in an intensive and centralized manner. For example, the e-government cloud platform must be able to carry the services and data of various government sectors, enabling seamless data exchange, sharing, and efficient management of public services. On the other hand, emerging ICT technologies such as IoT, video cloud, and big data enable new capabilities like AI analytics and emergency dispatch to facilitate security supervision and city governance.

13.1.1 Scenario Requirements

Government sectors cater to the needs of numerous industries and face the demands of diverse services. For example, core systems such as taxation and human resources and social security (HRSS) require ultra-high performance and low latency to effectively handle service surges. Meanwhile, AI and big data analytics systems require large capacity to process massive amounts of data. To address these requirements, it is imperative to establish a data storage foundation that breaks data silos. This storage foundation should be able to tailor service levels according to the specific requirements of each application, ensuring that the appropriate resources are allocated accordingly. Additionally, the foundation should incorporate multiple types of media and constantly enhance its capabilities to deliver high performance, reliability, and elastic scalability. This will enable it to precisely fulfill service requirements and facilitate resource allocation on demand.

13.1.2 Converged Storage Resource Pool Solution

The industry generally utilizes the converged storage resource pool solution, as illustrated in Fig. 13.1, to cater to the diverse demands of government services. This solution deploys an ultra-high-performance converged storage system in a metropolitan data center, serving as the data storage foundation. The storage system supports the interworking of multiple storage protocols and enables data sharing across multiple clouds, empowering a wide range of industry applications within the city.

The converged storage resource pool solution consists of multiple storage types, including all-flash storage, distributed storage, and backup storage, to address the demands of various service systems.

All-flash storage offers outstanding performance for service systems. The highly reliable design of all-flash storage protects services from the failure of any controller. The gateway-free metropolitan active-active and remote backup capabilities ensure high service availability and facilitate remote data recovery. These robust protection measures guarantee uninterrupted service continuity in fault scenarios, effectively meeting the demanding requirements of core taxation, HRSS, and finance database services in terms of performance and reliability.

The distributed storage integrates multiple storage protocols, including block, file, object, and HDFS, to cater to the varied needs of HPC, big data, virtualization, and cloud platforms. The decoupled storage-compute big data solution allows for on-demand expansion of computing and storage resources, resulting in enhanced resource utilization. In addition, the elastic EC technology enables a remarkable disk utilization rate of over 90%, greatly improving usable storage capacity and

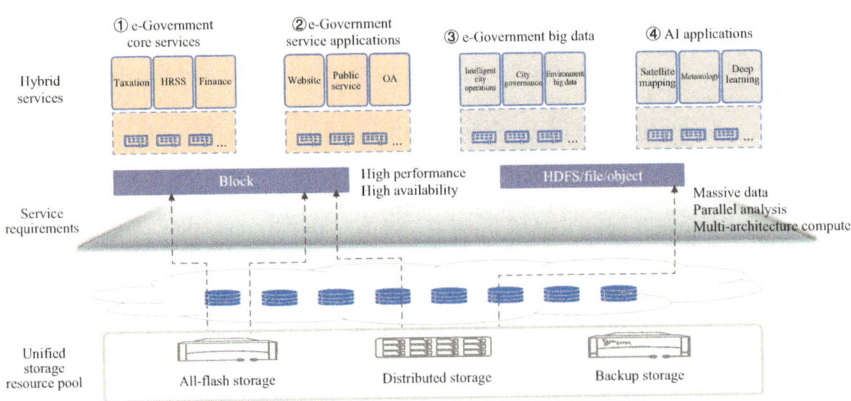

Fig. 13.1 Logical architecture of the converged storage resource pool solution

optimizing the cost per bit of data. The distributed storage also provides high access bandwidth for HPC applications such as meteorology, surveying, and mapping. Its elastic scalability allows resource pools to effortlessly expand in response to the growing demands of services.

The backup storage is specially designed to securely store backup data. In the event of accidental deletion or malicious tampering of production data, the backup data serves as a reliable source for data recovery. The backup storage necessitates high backup and recovery bandwidth to efficiently support backup tasks, as well as strong deduplication and compression capabilities to lower backup costs. Furthermore, the storage must take end-to-end data encryption and ransomware protection measures to guarantee the best possible security of backup data.

The converged storage resource pool offers a holistic suite of DR and backup services, including cloud server backup, DR, and HA, as well as cloud disk backup and HA. These services are designed to be self-service, allowing for easy application and deployment by users. By harnessing its diverse storage capabilities, the converged storage resource pool guarantees uninterrupted service continuity and eliminates any risk of data loss, laying a solid foundation for agile service innovations in government sectors.

13.2 Disaster Recovery Solution for the Financial Industry

With the financial industry experiencing exponential growth in customer volume, service volume, and data volume, the need for a robust DR system is becoming increasingly urgent. A DR system is crucial to ensuring the high reliability and availability of financial data while also providing the ability to tolerate potential data corruption caused by unforeseen disasters. Among the available options, the geo-redundant 3DC DR solution, comprising a metropolitan DR center and a remote DR center, has emerged as the favored choice within the financial industry [1].

13.2.1 Scenario Requirements

The financial industry's business systems can be categorized into three sections: the front-end, middle-end, and back-end. The middle-end carries major financial systems, such as class A/A+ key service systems for transaction settlements, which offer extensive support for the public, retail, and interbank businesses. Furthermore, the middle-end runs numerous class B service systems.

In recent years, financial regulators have been implementing increasingly strict requirements on the data security, stability, and reliability of the entire industry. Financial systems should ensure 24/7 uninterrupted service continuity and be adaptive to business changes by leveraging elastic scaling and quick response capabilities. The three sections of the financial industry work in synergy, with a growing focus on the development of the middle end.

In this context, there is an urgent need to upgrade the industry's overall infrastructure. As a vital component of this infrastructure, the storage technologies and solutions for each module should be prioritized to effectively support the transformation of the financial industry, as depicted in Fig. 13.2.

The core banking system is the pivotal transaction and settlement system in the financial industry, demanding the highest levels of storage reliability and security, alongside low latency. In this scenario, the all-flash storage solution offers comprehensive assurance measures across modules, systems, and the overall solution to guarantee the stability, speed, and cost-effectiveness of banking services. Moreover, the end-to-end NVMe technology of this solution ensures supreme performance of core services.

The development of a converged storage solution for massive data aims to help financial institutions efficiently store massive amounts of data, such as check images and videos. This solution builds a holistic converged data storage resource pool, enabling the integration of file, HDFS, block, and object storage protocols. Furthermore, the pool supports seamless data interoperability and multi-terminal access while also offering scalability up to PB-level storage capacity.

In emerging scenarios like mobile Internet, the converged resource pool for Internet finance virtualizes storage resources into pools, enabling on-demand scaling and full utilization of resources. This advanced technology facilitates elastic VM provisioning, drastically reducing the provisioning time from weeks to minutes in order to support the rapid rollout of innovative services. These advancements would be impossible without a highly elastic, reliable, and easy-to-manage storage foundation that effectively handles service surges and significantly improves efficiency.

The geo-redundant 3DC DR solution, consisting of a metropolitan DR center and a remote DR center, establishes a solid framework for storage system construction in the financial industry, catering to the industry's stringent demands on data backup and protection. This solution boasts high availability and robust DR capabilities in supporting public, retail, and interbank businesses, helping them establish reliable middle-end and back-end services. This solution is primarily applied by

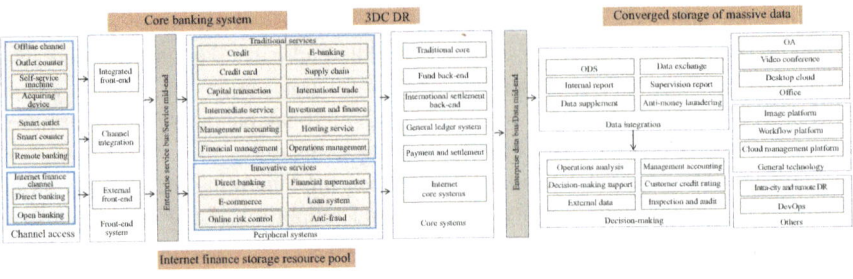

Fig. 13.2 Financial industry solution panorama

core banking systems, asset management, and general ledger, as well as their associated front-end and back-end systems.

13.2.2 DR Construction Requirements

Financial services require a holistic data storage solution that supports the entire lifecycle of data, including data generation, storage, analysis, backup, and archiving. Whether dealing with traditional data centers or cloud environments, ensuring uninterrupted service continuity has always been a key focus of the industry. Service downtime poses risks of huge economic losses and damaging reputations for banks. Therefore, the construction of a DR system is the top priority in infrastructure development across the entire industry. The primary concerns of DR construction in the financial industry are service continuity and data reliability. Ensuring service continuity means guaranteeing the normal operation of financial information systems in the event of disasters, such as floods and earthquakes. Data reliability measures are taken to fulfill the requirements related to operational risk and systematic prevention outlined in the Basel Accords.

Take China's national standard as an example. According to the Information Security Technology—Disaster Recovery Specifications for Information System (GB/T 20988-2007), the DR system of the financial industry must comply with the level-6 standard (the highest level), ensuring minimal to zero data loss. Active-active synchronous replication enables minute-level recovery and zero data loss. Similarly, the active/standby DR and 3DC DR methods adhere to level-3, level-4, and level-5 standards, while local and remote backup methods meet level-1 and level-2 standards. According to best practices in the financial industry, Fig. 13.3 illustrates the DR requirements of specific banking service systems.

13.2.3 Geo-redundant 3DC DR Solution

The 3DC solution integrates the independent local and remote DR methods, effectively overcoming their respective limitations and capitalizing on their strengths. This solution is not only suitable for meeting large-scale DR requirements, but also enables quick response to small-scale regional disasters and manual operation risks. It significantly improves RPO and RTO while ensuring zero service data loss.

The 3DC solution establishes three data centers to guarantee service continuity, even if any two data centers are down, ensuring high availability of service systems. The three data centers consist of a production center, a metropolitan DR center, and a remote DR center. In the event of a disaster at the production center, services are quickly switched to the metropolitan DR center. If both the production center and metropolitan DR center are down, data copies stored at the remote DR center are used to recover services, providing maximum service continuity.

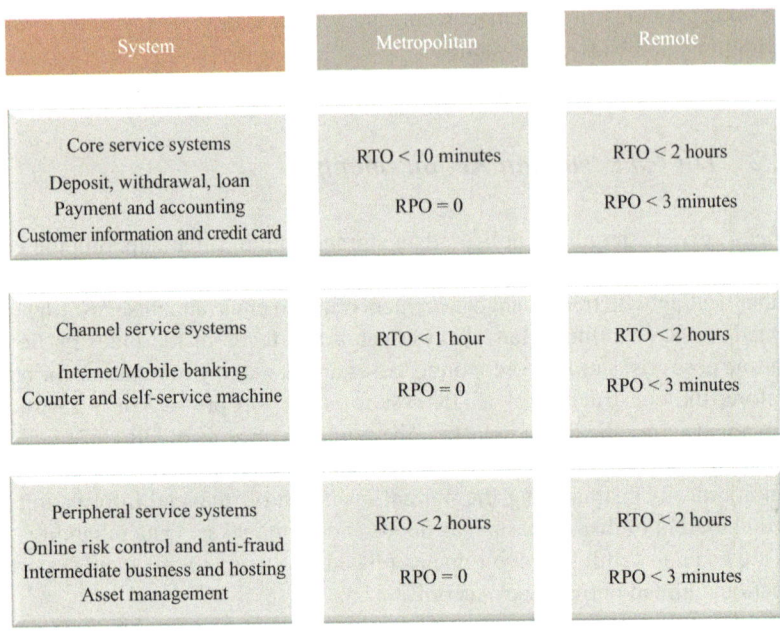

Fig. 13.3 DR requirements of banking service systems

In the financial industry, there are multiple models available for building DR systems. These include local active-active + metropolitan/remote DR, metropolitan active-active + remote DR, and the 3DC DR in a ring topology. The solution shortens the fault recovery time from hours to minutes. It is also gateway-free and supports interoperability across high-end, mid-range, and entry-level storage models, reducing costs by 30%. This solution offers application-level active-active deployment to ensure zero service interruption and zero data loss, thus achieving service reliability of up to 99.99999%.

Figure 13.4 depicts the overall solution. The local production center and metropolitan DR center operate independently but collaborate with each other. In the event that the link between the production and metropolitan DR centers is interrupted, or the storage system at either center is faulty, asynchronous data replication to the remote DR center remains unaffected. Once the affected link or storage system recovers, the production center synchronizes incremental data to the metropolitan DR center, thereby restoring consistency across all three data centers. Likewise, if the asynchronous replication link to the remote DR center is down or a device at the remote DR center is faulty, metropolitan synchronous replication is not affected.

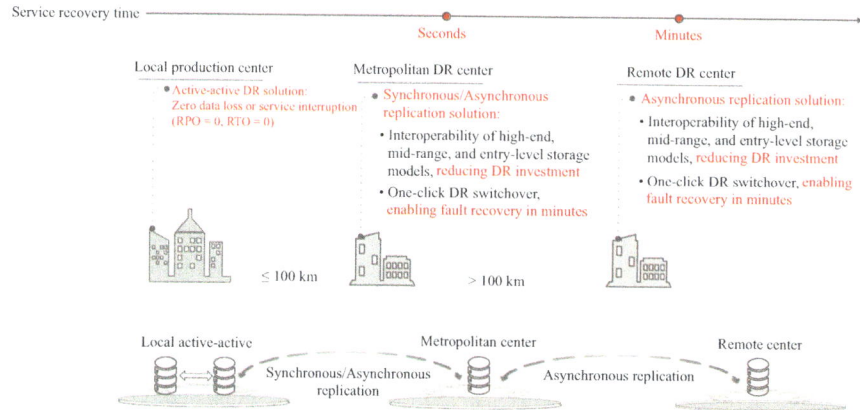

Fig. 13.4 Geo-redundant 3DC DR solution

13.3 Storage Solution for the Healthcare Industry

As the development of smart healthcare systems gains momentum, it is imperative for the picture archiving and communication system (PACS) to not only offer sufficient image storage capacity, but also enable efficient and prompt responses to service needs, as well as ensure the reliable and stable operation of the entire system. Furthermore, with the advancements in imaging devices generating an exponential increase in image data, cost is becoming a significant factor to consider when selecting PACS solutions. Taking all these factors into account, a PACS solution that incorporates online, near-line, and offline tiered storage modes is the optimal choice to effectively fulfill customer requirements.

13.3.1 Scenario Requirements

The development of hospital informatization can be traced back to the 1980s. Originally, the hospital information system (HIS) was designed for individual basic services, and later, it gradually evolved to support electronic medical records and digital images based on service applications. This improvement enabled the reorganization of fragmented data, thus providing patients with better services. Today, the HIS has developed into a comprehensive platform that integrates management and service functionalities. Typical systems include the electronic medical record (EMR), laboratory information management system (LIMS), radiology information system (RIS), and PACS.

As global society wrestles with the challenges of aging populations and urbanization, the demand for healthcare services continues to escalate. However, existing healthcare service systems cannot meet these increasing demands due to inadequate deployment, insufficient resources, and improper allocation of resources. Therefore, emerging services like telemedicine and Internet hospitals are set to play an important role in addressing these challenges and accelerating the transformation of healthcare information systems.

By integrating new technologies such as 5G, AI, IoT, and cloud computing into medical science, hospital information systems are upgrading toward smart healthcare that incorporates service, management, and healthcare capabilities. This is empowering the development of smart applications like medical research big data, AI-assisted diagnosis and treatment, gene sequencing, medical brain, and Internet hospitals.

Healthcare is the core service of hospitals and relies on support from the HIS, EMR, LIS, RIS, PACS, and integration platforms. Among these, the HIS serves as the core operation system that must run 24/7 to ensure the stable and reliable operation of critical services like charging and registration. To fulfill these demands, the industry commonly employs all-flash storage and the active-active DR solution to guarantee service continuity.

PACS, another vital system for hospital operations, stores key treatment data, including MRI, CT, ultrasound, and X-ray data. In the upcoming section, we will have a comprehensive exploration of the PACS storage solution.

13.3.2 PACS Storage Solution

PACS is one of the most prevalent service systems in hospitals. Through the establishment of PACS, hospitals are striving to facilitate the efficient sharing of medical information and real-time coordination of high-quality medical resources, in order to offer patients high-quality medical services.

As hospitals expand their service scale and employ advanced technologies like 3D imaging and AI-assisted diagnosis, the demand for faster image retrieval speed is growing. In terms of data volume, the PACS of a top university-affiliated hospital generates hundreds of GB of data every day, with an annual increment reaching hundreds of TB. Generally, image data generated within a year is more frequently accessed and thus requires quick access speeds to guarantee a favorable retrieval experience. Image data generated more than a year ago is seldom accessed and can therefore be archived for future queries. To fulfill these demands, the PACS must provide the following capabilities:

1. High performance, enabling fast online retrieval of massive image data.
2. PB-level capacity and high scalability: A top university-affiliated hospital generates millions of files each year, totaling hundreds of TB, with a data growth rate exceeding 40%/year. Moreover, these files must be retained for a long term (15–30 years) according to applicable laws and regulations.
3. Simplified management: The coexistence of structured data such as medical records and unstructured data such as images poses difficulties regarding

information retrieval. Therefore, it is imperative for the PACS to streamline the management of structured and unstructured data in order to enhance the work efficiency of medical staff.
4. Comprehensive DR mechanism: In the event of a single point of failure caused by a virus attack or human error, the PACS must be able to guarantee the integrity of critical image data.

The massive distributed file storage solution is the highly preferred option in the industry for establishing the PACS. It offers high-performance nodes for storing near-line files and capacity nodes for storing files that will be retained for more than 15–30 years.

The storage tiering capability dynamically migrates image files between the high-performance and capacity nodes of the massive distributed file storage according to their creation time, modification time, and read/write frequency, offering high-performance and large-capacity storage services to support the online file retrieval on the PACS platform.

The remote replication feature of the massive distributed file storage provides DR capabilities for data centers. In the event that a data center is down, a secondary data center automatically takes over services, ensuring uninterrupted operation of the PACS services.

13.4 Storage Solution for the Education Industry

The advancement of education informatization has become a crucial factor in assessing a country's overall strength. The construction of cutting-edge information infrastructure in primary education sectors and universities is pivotal to the evolution and development of the education industry and facilitates the implementation of modern education management approaches. In addition, such infrastructure holds far-reaching significance in terms of promoting education equity, enhancing education and research capabilities, nurturing innovative talents, and ultimately improving people's satisfaction.

According to government policies, the development of new education infrastructure is crucial to new national infrastructure and will act as a catalyst for educational transformation in the era of informatization. This development also represents a strategic measure to accelerate the modernization of education and build a nation that possesses strong educational capabilities. The new education infrastructure focuses on the sharing and intensive construction of cutting-edge smart campus infrastructure and high-performance scientific research equipment in universities. This intensive, efficient, secure, and reliable infrastructure lays the solid groundwork for implementing education modernization.

13.4.1 Scenario Requirements

Education services are currently undergoing close integration with new intelligent technologies, enhancing intelligent, integrated, and personalized functionalities. In addition, education management systems are becoming more comprehensive, while unified identity authentication and public databases are promoting data sharing. IT infrastructure is gradually becoming more efficient, intelligent, and reliable, harnessing diverse information technologies to fully support teaching, scientific research, and management activities.

Based on the national plan for building next-generation exascale computing infrastructure, universities assume critical responsibilities in terms of national scientific research and interdisciplinary innovation. Their efforts are driving the advancement of high-performance computing (HPC) and data analytics platforms within the education and scientific research field.

13.4.2 High-Performance Data Analytics for Education and Scientific Research

HPC is evolving toward high-performance data analytics (HPDA) in order to build data-centric supercomputing systems for scientific computing workflows. HPDA handles data-intensive workloads, including big data and AI workloads. By integrating data analytics with HPC, HPDA harnesses the parallel processing capabilities of supercomputers to deliver over trillion-level floating-point operations per second (FLOPS) for data analytics software. This approach enables quick retrieval and analysis of large data sets, facilitating the formulation of informed conclusions.

Multi-disciplinary development in universities necessitates the use of big data, AI, and numerical computation technologies to foster research and innovation across diverse disciplines like life sciences, geosciences, analytical finance, engineering simulation, and information engineering. As a result, HPC platforms in universities are shifting from separate systems to centralized university-level public platforms.

Disciplinary research conducted in universities has the following characteristics:

Large data volume: The volume of data generated by high-performance scientific research in universities has now reached the PB scale. Specifically, researches in meteorological prediction, gene sequencing, high-energy physics, and astro observation have each amassed data volumes of up to 10 PB. Furthermore, this data volume is growing at an annual rate of 27.2%. Consequently, there is a pressing need for extensive data analytics and storage capabilities.

Complex I/O model: The scientific research platforms in universities carry a wide range of research services and applications, which typically vary in file size (ranging from KB to GB), I/O type (random or sequential), and workload model (metadata-intensive or data-intensive). Therefore, the corresponding storage system must be able to handle these hybrid workloads while still delivering the optimal performance for each application.

Multiple protocols involved: The integration of numerical computation, big data, and AI with disciplinary research requires data access across various protocols like file, object, and HDFS. Therefore, the storage system must support multi-protocol data access to satisfy the needs of different applications.

The prevailing HPDA solutions for education and scientific research in the industry rely on distributed parallel file storage as their core, which coordinates with peripheral resources such as compute, network, and scheduling software to achieve joint tuning and thus fulfill the demands of high-performance data analytics.

The mainstream distributed parallel file storage systems for HPDA have the following advantages:

High-density hardware for massive amounts of data: The storage systems are commonly delivered in the form of high-density, integrated cabinets, offering distributed expansion capabilities to potentially support the storage of petabytes of scientific research data.

Superb performance for hybrid loads: The storage systems excel in handling both bandwidth-intensive and IOPS-intensive workloads, offering unmatched read/write concurrency and bandwidth for multi-disciplinary development. In addition, by deploying distributed parallel clients on compute nodes, the storage systems overcome the limitations of traditional NFS clients in terms of single-stream and single-client performance, facilitating parallel I/O access (MPI-I/O) for HPC.

Multi-protocol interworking for enhanced efficiency: Data can be read using multiple protocols, regardless of the protocol used for writing. This eliminates the need for data migration and significantly enhances the collaborative analysis efficiency of big data and AI applications.

13.5 Practice Questions

1. **What are the characteristics of a decoupled storage-compute big data solution? What are the core big data storage technologies that support the decoupled storage-compute architecture?**

    Answer: Decoupled storage-compute big data solutions have (1) flexible and on-demand expansion capabilities for computing and storage resources, which enable high resource utilization; (2) global data resource pooling, which improves data analysis efficiency; and (3) independent deployment and maintenance for software and hardware components, which ensures high availability.

    The core technologies in big data storage that support the architecture include large-ratio erasure coding, automatic data tiering, dynamic cluster expansion, and HDFS semantic APIs.

2. **What are the core components of an all-in-one backup solution? What key technologies does an all-in-one backup solution use to enhance backup efficiency?**

    Answer: The core components of an all-in-one backup solution include backup software, data protection agents, backup storage media, and backup networks. The solution improves backup efficiency in two ways. (1) It uses native copies to back up production data to the backup media without requiring any changes to the data format. This eliminates the overhead associated with converting data formats during backup and recovery. (2) It unifies the management and O&M of various backup components to simplify the backup workflow.

3. **What is the geo-redundant 3DC DR solution? What key storage technologies does it use?**

    Answer: The geo-redundant 3DC DR solution is an HA DR solution that adds a remote DR center to the metropolitan DR center. This solution comprises a production center, a metropolitan DR center, and a remote DR center. It ensures core service continuity in the event that any two data centers go down. It is applicable to scenarios demanding high reliability, such as core financial operations.

    The geo-redundant 3DC DR solution provides multiple networking models to meet different requirements. Typically, it employs active-active or synchronous/asynchronous remote replication between metropolitan DR centers and asynchronous remote replication between remote DR centers.

4. **How does the geo-redundant 3DC DR solution address reliability needs across different service scenarios within the financial industry?**

    Answer: (1) For core financial service systems, the metropolitan active-active technology is used to implement lossless data recovery within seconds, ensuring 0 RPO and an RTO under 10 min. Asynchronous remote replication enables the remote center to provide an RTO under 2 h and an RPO under 3 min. (2) For channel and peripheral service systems, metropolitan active-active or synchronous remote replication is used to ensure 0 RPO and an RTO under 1–2 h. Asynchronous remote replication enables the remote center to provide an RTO under 2 h and an RPO under 3 min.

5. **What are the characteristics of a PACS in the healthcare industry? Which storage system is typically used to construct a PACS in the industry? Why?**

    Answer: PACS systems are large-capacity systems, designed to handle multiple data types, fast data growth, high access performance, and long data retention periods. High-performance distributed file storage solutions are used as the data foundation for PACS systems. The distributed architecture meets the scalability requirements that come from increasing volumes of data and services. The distributed file system provides high-performance and high-concurrency access. Tiered storage and automatic data migration capabilities deliver high performance and cost-effectiveness.

6. **Which storage architecture is typically used to construct the HPDA systems of high-performance computing? Why?**

    Answer: Typically, a distributed parallel file storage system is used as the data foundation of HPDA systems. The high-density and distributed architecture is scalable to hundreds of PBs of capacity which meets the demands of scientific research data growth. High-performance parallel file systems deliver maximum performance under hybrid loads. Multi-protocol interoperability for unstructured data also facilitates data sharing and multi-mode access throughout the data analytics process.

Reference

1. Shu J. Network storage column. Network storage area disaster recovery (Vol II). China Educ Netw. 2007;6:67–9.

Chapter 14
Storage Technology Trends and Development

With the rapid development of technologies such as big data, cloud computing, artificial intelligence (AI), and blockchain, users are developing increasingly stronger demands for high-performance, high-capacity, and high-reliability data storage. This shows how it is crucial to explore new storage technologies to fulfill these demands. This chapter focuses on the trends of storage technologies in new storage models, non-volatile storage systems, and application storage optimization and introduces cutting-edge technologies such as flash storage system, near-data computing, persistent memory, in-network storage, intelligent storage, edge storage, blockchain storage, separated data center architecture, and high-density new storage.

14.1 Flash Storage System

Increasing storage density and declining prices are sending flash memory to desktop computers, servers, and data centers. Compared with traditional disks, flash memory has higher performance, lower energy consumption, and smaller size. Such advantages mean both opportunities and challenges for building efficient storage systems. However, flash memory is also facing emerging challenges. The form of flash memory, in practical applications, has evolved from early solid-state disks (SSDs) and flash cards to flash arrays and flash cluster systems. Researchers are committed to further innovating existing storage structures, system software, and distributed protocols [1]. Examples of related research include bare flash-based storage systems without a flash translation layer (FTL) on flash devices, optimized data migration on flash clusters to extend the lifespan of SSDs, and software-defined storage that combines application-specific load characteristics.

The research into flash cluster systems has two representative works: FAWN [2] and Gordon [3]. Fast array of wimpy nodes (FAWN) is a scalable, low-energy, and high-performance cluster system proposed by Carnegie Mellon University based on flash media. Unlike the design of flash cards and flash arrays that focuses only on the performance and reliability of subsystems, FAWN considers the matching of flash memory and processors from the perspective of the overall cluster design to reduce total energy consumption. It uses low-frequency and low-energy CPUs in collaboration with flash storage to improve the utilization of each system component involved in data-intensive computing. The key-value store system based on FAWN has an energy cost of 1 J for 364 queries, much lower than 1 J for 1.96 queries in normal desktop systems, reducing the energy consumption by 99.46%. Gordon [3] was designed by the University of California, San Diego. Unlike FAWN, which focuses on high performance and low energy consumption, Gordon's main purpose is to design the FTL to match the performance of processors and memory chips and take advantage of the concurrency feature among flash chips. The design includes dynamic address mapping, large physical page synthesis, and concurrency and pipeline mechanisms. Gordon integrates 256 GB flash memory and 2.5 GB dynamic random-access memory (DRAM) on a single board. It has been used at the San Diego Supercomputing Center in the USA for data-intensive computing fields such as astrophysics and genome sequencing.

From SSDs and flash cards to flash arrays and flash cluster systems, this evolution represents a horizontal scaling of building up flash storage systems. However, the FTL of flash devices affects the performance of flash hardware features, which in turn affects the construction of flash storage systems. This has resulted in the emergence of some new flash structures, such as open-channel SSD (OC SSD) and zone namespace SSD (ZNS SSD), which are vertical scaling approaches for building up flash storage systems.

14.1.1 OC SSD

In its early days, flash memory was mainly used as bare flash memory in embedded systems. Embedded flash implements FTL-related functions in the file system, including address mapping, garbage collection, and wear leveling.

At the early stages of SSD development, FTL was used to convert the read, write, and erase interfaces of flash memory to the traditional read and write interfaces to ensure compatibility with disk interfaces [4], as shown in Fig. 14.1a. In-device FTL. SSDs provide the same read and write interfaces as traditional disks and use the FTL to convert read and write requests into flash commands. This ensures good compatibility for deploying traditional file systems on SSDs, making it easy for SSDs to replace traditional disks.

Numerous studies have been conducted to optimize the file systems tailored to SSDs in this usage model. F2FS [5] is one of these implementations by Samsung. F2FS is an excellent example of improving the traditional log file system according to the characteristics of flash memory. Each segment of the file system log has the same size as the underlying flash memory block, and the data is grouped as hot or cold in order to improve the efficiency of SSD garbage collection. Furthermore,

14.1 Flash Storage System

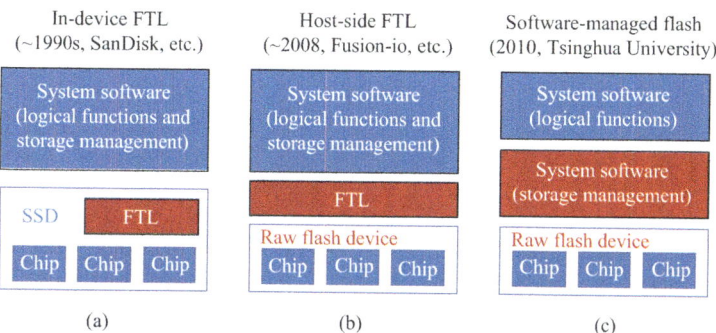

Fig. 14.1 SSD architecture evolution [4]. (**a**) In-device FTL. (**b**) Host-side FTL. (**c**) Software-managed flash

address mapping tables are used to avoid snowballing when updating metadata all the way to the root, reducing the number of metadata writes and improving the lifespan of SSDs.

However, FTL hides internal details, making it difficult for software to fully use hardware advantages. The FTL inside the SSD requires the SSD to have an embedded processor with excellent performance and a large in-device cache capacity. Later, Fusion-io proposed a design architecture for host-side FTL, as shown in Fig. 14.1b, as a way to fully utilize the host's processor and memory resources. Princeton University and Fusion-io implemented a new flash file system, DFS [6], based on host-side FTL, which moves the block allocation operation of the file system down to the FTL, thus avoiding the redundant management overhead of file system space management and flash device FTL management. Although the design of the host-side FTL effectively utilizes host-side resources, simple read and write interfaces are still used between the file system and FTL, limiting the use of semantic information.

Additionally, as shown in Fig. 14.1c, the storage team at Tsinghua University proposed the software-managed flash SSD architecture [7]. This architecture removes the FTL from the SSD and allows the host-side software to directly manage the flash media, which can further remove redundant management and fully utilize the internal parallelism of flash memory [8]. This architecture also gives birth to OC SSD. Based on this architecture, a series of key technologies and methods are proposed such as a reconstructable metadata management method for flash file systems [9], a key-value store acceleration method based on the open-channel architecture [10], a metadata design technology for distributed file systems [11], and an efficient transaction processing method supported by flash hardware [12, 13], as well as the Tsinghua-Solid Storage System (TH-SSS) by Tsinghua University. OC SSDs leave the FTL functions to the host by exposing the internal details of the device to the host, which is responsible for data placement and management. However, OC SSDs have some problems. First, the ecosystem of OC SSDs is not mature, requiring specific modifications to the storage stack and relatively large changes on the software side. Second, the internal management of SSDs differs among vendors, leading to compatibility issues [14]. Therefore, ZNS SSD and other forms were later proposed.

14.1.2 ZNS SSD

The zoned namespace (ZNS) SSD is an improvement of the OC SSD. The main idea is to expose the internal details of the flash memory to the host. However, with the introduction of zone-based storage, the ZNS SSD provides a unified storage interface and achieves NVMe-based standardization [15]. Supported by these improvements, the ZNS SSD provides a more simplified, unified, and standardized storage solution that reduces modifications to host software and drivers and improves the convenience of software development and system integration.

The ZNS SSD moves most of the functions in the FTL, including address translation and mapping, to the host for processing. This design makes the ZNS SSD more simplified and transparent, allowing the host operating system or storage management software to directly manage the storage device. In conventional SSDs, the FTL uses fine-grained mapping, which requires a large mapping table for address translation. In the ZNS SSD, mapping can be at block granularity, greatly reducing the overhead required for the mapping table. Since some of the mapping tables need to be stored in the DRAM of the SSD device, the ZNS SSD with block-granularity mapping also has significantly lower DRAM requirements. With the data management function moved to the host, internal garbage collection of the ZNS device is not needed. There is also no write amplification, which eliminates the need for reserved space, increasing the effective storage capacity.

The ZNS SSD is based on zone-based storage, an established concept that is more commonly used in the SMR HDD, where the logical addresses are divided into equal-sized zones with write constraints that are different from those regularly used. The ZNS SSD introduces some of its own unique concepts on the basis of zone-based storage. As illustrated in Fig. 14.2, the storage space of the ZNS SSD is

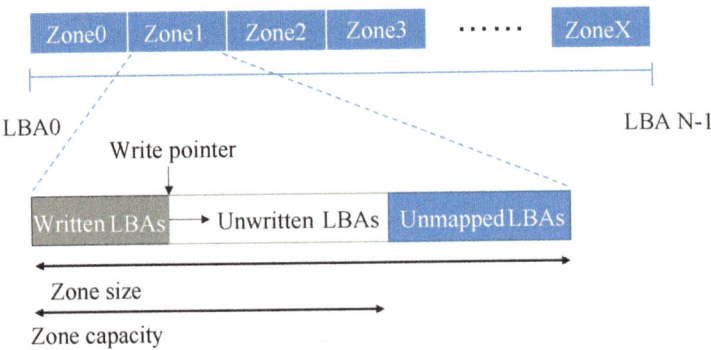

Fig. 14.2 Zone allocation of ZNS SSDs

divided into multiple zones that must be written into sequentially. Each zone has a write pointer to track the location of the next write, and a zone cannot be overwritten. The logical block addresses (LBAs) prior to the write pointer can be reused only after a zone has been reset. With zone capacity, a zone is divided into writable and non-writable parts. Zone capacity is less than or equal to the size of the corresponding zone. Zone capacity was introduced to allow zones to be aligned with the erase blocks of hardware media.

Each zone of the ZNS SSD has a state and can transition between states. The state of a zone determines the operation restrictions and resource allocation of the zone. There are six main states [16].

Empty: The zone is empty, meaning there is no data. Writes to a zone in this state are not allowed.
Full: The space of the zone is used up.
Open: Data can be written to, and the host can allocate resources to the zone.
Closed: The zone is still active but has its resources reclaimed by the host.
Read only: The zone can only be read.
Offline: The zone cannot be read or written, indicating that the zone is damaged and can no longer be used.

By defining and managing these zone states, ZNS SSDs achieve more efficient write and balanced erasures, as well as simplified data management and maintenance.

Since the resources of the host and device are limited, the zones that use resources must also be limited, including the number of open zones and the number of active zones that can be operated simultaneously, in ZNS SSDs. Open zones are those where write operations can be performed. The limit on open zones is determined by the host's internal resources (write buffers) and hardware media resources (number of channels). Active zones include all zones that are in the open and closed states where data can be stored. The zones to be written to are determined by transitioning between the open and closed states.

Zone-based storage devices do not support plug-and-play as traditional devices do. The use of ZNS SSD devices must adhere to the constraints and rules of zone-based storage. In order to integrate zone-based storage devices into a system, support is required from a specific software stack. Currently, Linux has supported zone-based storage through different integration methods, forming a zone-based storage software ecosystem. The first method is through a file system. File systems that comply with a zoned block device (ZBD) interface already support the sequential write constraint of zone-based devices and therefore can adapt to ZNS SSDs. One of the typical examples is F2FS. For general file systems, a device mapper is needed to handle the sequential write constraint of devices. The second method is to access directly through the original block interface, where applications can directly access data from the zone-based storage device. This method requires a device mapper to convert random writes to sequential writes. The third method uses a file access interface. In this case, a small file system plug-in is designed for a specific application that meets its characteristics. For example, RocksDB can access ZNS SSDs through the ZenFS file system.

ZNS products have promoted research regarding ZNS. Researchers have studied the characteristics of ZNS and proposed redundant array mechanisms, file systems, key-value store systems adapted to ZNS, and so on.

ZNS device characteristics [16]: ZNS must activate a zone before reading data from or writing data to it. The size of the internal buffer of ZNS has a limit on the number of activated zones, which may vary between vendors. Some vendors support as many as 4096 activated zones. However, ZNSs with higher upper limits of activated zones cannot more reliably guarantee performance isolation between zones, and tests have shown that some zones share bandwidth, which can lead to performance interference between their applications. To address this issue, researches have accurately determined the performance conflict zone groups of ZNSs, based on which a request remapping mechanism has been designed to balance requests between different conflict groups.

Key-value store system: Researchers at Western Digital Corporation comprehensively described the characteristics of ZNS devices [15] and designed a dedicated file system ZenFS for the key-value database RocksDB to run on ZNS. Based on their finding that ZNS SSDs have more stable performance and consume less memory within the device than traditional SSDs, they designed ZenFS, which includes a dedicated data zone and log zone, with the latter for storing the modification logs of super blocks and the pre-write logs of RocksDB. ZenFS is also designed with a zone allocation mechanism to write as many segmented files as possible that have similar life cycles in RocksDB to the same zone, making it easy to uniformly erase zones at some time in the future.

ZNS device–oriented file systems: Research has been conducted to design the file system ZNS+ [17] for ZNS devices. Considering the issue that the existing flash file system Flash-Friendly File System (F2FS) cannot run directly on ZNS, researchers modified the design of interspersed writes inside F2FS for ZNS+ and proposed an interspersed write scheme that adapts to the internal data organization of ZNS. Researchers also introduced new replication primitives on ZNS devices for ZNS+, which supports the direct execution of data replication inside ZNS based on the command from the host, avoiding the waste of ZNS bandwidth by data migration. By supplementing the primitives of ZNS, ZNS+ provides new ideas for future research. Garbage collection by the host without reading data into memory for consolidation has become a future research direction for ZNS-oriented storage software.

ZNS-oriented redundant array mechanisms: Redundant array mechanisms were originally designed to work on block devices. In order to fully utilize the high performance of sequential writes to disks and SSDs, researchers have proposed LogRAID [18, 19] to append data to arrays in the form of additional writes. Specifically, LogRAID converts data written to arrays into sequential writes, and to do this, it needs to maintain mapping between in-place writes and out-of-place writes. However, LogRAID may need to read data from the check area on the device and write again to the original location to complete the modification.

Unfortunately, it is difficult for existing ZNS interfaces to support this process. This led researchers at Carnegie Mellon University to propose RAIZN [20], a ZNS-oriented redundant array mechanism, to provide fault-tolerant zone abstraction across multiple ZNS devices. Specifically, RAIZN addresses the problem of partial checksum updates by reserving a separate zone on each ZNS device for storing temporary partial checksum updates and other metadata. When data is completely filled in the strip where the parity block resides, RAIZN will write the parity information of the separate zone back to the original address to avoid additional overhead while maintaining the mapping table at the redundant array layer.

ZNS-oriented memory swapping technology ZNSwap: To relieve memory pressure, a system will store some pages in memory to storage devices. With the development of storage technology, memory swapping technology applies not only to such full-memory pressure cases, but also to memory expansion cases for optimal system performance. However, in traditional memory swapping, since the SSD is opaque to the upper-layer storage software stack, invalid pages may be migrated during garbage collection, especially when device utilization is high. This seriously affects the swap-in and swap-out efficiency of the swap area. In contrast, ZNS devices are transparent to the host, do not have complex FTLs, and eliminate garbage collection, giving the host a higher degree of control over the device. ZNSwap [21] achieves a collaborative design of the operating system (OS) swapping technology and ZNS SSD to enable customized data management and placement through fine-grained management of the space in the swap area. Meanwhile, the host-side garbage collection operation is designed to replace TRIM commands, avoiding the replication of invalid data during swapping and eliminating the performance overhead caused by TRIM commands.

The ZNS SSD has been standardized into NVMe 2.0 by introducing the characteristics of ZNS devices, such as zone capacity, maximum active zone limit, and zone add-on functions [22, 23], on top of the original zone-based device specification. There is still much room for exploration regarding ZNS SSDs, such as zone-level tighter isolation to avoid interference between applications or users, which will be conducive to better data and privacy protection. Another example is size flexibility at the zone level to adapt zone size to application demand and capacity allocated at fine granularity, as opposed to the current practice of uniform sizing, to realize better performance. Besides, a complete ecosystem of ZNS SSDs is key to their widespread adoption. Working out how most applications can use ZNS SSDs with minimal modification to the software stack will be essential if mainstream storage devices are going to smoothly transition from traditional SSDs to ZNS SSDs in the future.

The future of flash memory architecture remains an open question. However, software-managed flash is still at the core of this issue of architectural exploration, determining how efficiently and collaboratively software and hardware can manage flash memory to leverage flash efficiency.

14.2 Near-Data Computing

Processor and storage are important components of modern computer systems. In recent years, processors' performance has been growing rapidly, but storage access speed has increased slowly. This imbalance has created what is called a "memory wall." In a computer system based on the von Neumann architecture, the processing unit and the storage unit are separated from each other, and applications need to frequently move data between the processor and storage. While being time-consuming, this type of massive data transmission consumes significantly more energy than computation, becoming the main contributor to system energy consumption. Similarly, this is called a "power consumption wall." The emergence of near-data computing technology provides us with an opportunity to resolve both of these problems [24]. This technology integrates the processor and storage to reduce the distance data has to travel. It is expected to both improve computer system performance and reduce energy consumption.

14.2.1 Near-Storage Computing

In modern computer architectures, the processor, memory, and storage media are separated and function independently of each other. When a computer processes data, the data is transmitted to the processor through an off-chip bus. However, the off-chip bus is long in distance and has a limited bandwidth, making data transmission a performance bottleneck for the processor. Near-storage computing places some logical computing units near the storage so that operations can be performed close to the data, reducing costly data movement overhead. Near-storage computing can be classified into near-disk computing and near-memory computing according to the storage layer.

14.2.1.1 Near-Disk Computing

Near-disk computing offloads data-sensitive operations to computing units close to the storage media to reduce load on the computer system's CPU and memory and improve application responsiveness. As SSDs are widely used, SSD-based near-disk computing solutions have attracted much attention from researchers. For example, CSSD [25] implemented programmable computing SSDs to accelerate graph deep learning algorithms using a collaborative hardware and software approach. Under this approach, a computing-capable field-programmable gate array (FPGA) is placed next to the SSD to preprocess data in the graph neural network algorithm. A software stack was also designed for corresponding hardware to compile user-written code into code that can run on the FPGA. Alibaba's AliFlash V5 SSD uses a near-disk computing architecture to accelerate offloading in relational database scenarios, which keeps latency low while increasing bandwidth.

14.2.1.2 Near-Memory Computing

Memory is often closer to the processor than storage media like SSDs, so the data movement overhead between the two is significantly lower. However, in some bandwidth-sensitive application scenarios, memory cannot meet the demand, and near-memory computing is still needed to improve performance. Dynamic random-access memory (DRAM) is currently the most widely used memory unit. Common near-memory computing solutions based on DRAM use a 2.5D/3D stacking technology or high bandwidth memory (HBM) encapsulation technology to integrate additional computing units inside the DRAM. For example, in the academic community, Max-PIM [26] supports complete bit-by-bit Boolean logic calculation by integrating exclusive nor (XNOR) circuits inside the DRAM. Based on this, it can concurrently search for both maximum and minimum values in the DRAM. This is useful in certain application fields such as big data sorting and graph computing. In SpaceA [27], 3D stacking is used to integrate the compute logic into DRAM, and a near-data computing architecture is used for sparse matrix-vector multiplication from both the perspectives of hardware design and data mapping. In TRiM [28], as the DRAM datapath has a hierarchical tree structure, researchers have added computing units to DDR4/5 at each level to augment the DRAM datapath, which can be used to optimize personalized recommendation systems. In the industry, Samsung announced in 2021 a programmable solution that integrates floating-point computation arrays, pipeline decoding control units, and local register file units in high bandwidth memory (HBM), which greatly improves computing performance and reduces power consumption [29].

14.2.2 Coupled Storage and Compute

In addition to near-storage computing realization by placing computing units near the storage, certain storage media can perform both computing and storage with the support of existing peripheral circuits. The computing architecture based on this kind of storage media is called the coupled storage-compute architecture. Coupled storage-compute technologies can be divided into three levels (Fig. 14.3). As the computing latency of these coupled devices increases, so too will their storage capacity, while the reads and writes will slow down.

SRAM-based coupled storage-compute technology: SRAM stores data by switching the state of internal transistors. When powered on, the data remains unchanged. SRAM is often used as a cache in computer storage systems because of its high access speed. SRAM-based coupled storage-compute is characterized by data computing directly in the cache, and the technology is highly reliable and scalable. At present, this technology is mainly implemented through analog computations in the voltage domain and time domain. In the voltage domain, this technology generally uses a digital-to-analog converter (DAC) first to convert digital numbers into voltages, then implements the computation through charge

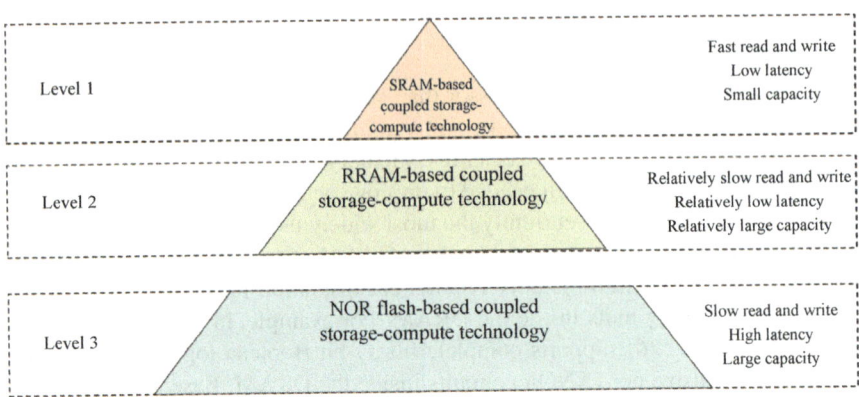

Fig. 14.3 Classification of coupled storage-compute technologies

sharing, and finally uses an analog-to-digital converter (ADC) to convert the analog quantities of the calculation result back into digital numbers. Using SRAM with six transistors and computing units, researchers were able to complete multi-bit data multiplication by inputting specific voltages [29]. In the time domain, this technology typically uses linear path delays or pulse bandwidths to represent multiple digits. In one study, a coupled storage-compute device using pulse-width modulation and eight-transistor sandwich-SRAMs was able to operate deep neural network (DNN) models [30]. The Chinese enterprise PIMCHIP has already developed a commercial SRAM-based integrated storage-compute unit, the PIMCHIP-S200. Jiutian Ruixin also launched the ADA20X, an SRAM-based coupled sense-storage-compute chip that can be widely used for vision applications.

RRAM-Based Coupled Storage-Compute Technology: Resistive random-access memory (RRAM) is a type of non-volatile memory that stores data by changing the resistance of cells. It can form a crossbar array structure by connecting RRAM cells with a series of word lines and bit lines. When an external voltage is applied to each word line, the current is transferred and converged to the bit lines according to Kirchhoff's law. By detecting the current at the end of each bit line, the total analog current in the corresponding column can be obtained. By utilizing this bit-line current summing property in conjunction with some peripheral circuits (such as DAC and ADC), RRAM crossbar arrays can perform local vector-matrix multiplication in the form of analog quantities. In recent years, RRAM-based coupled DNN storage-compute designs have received widespread attention from both academia and industry. For example, in PRIME [31], an RRAM-based microarchitecture and a set of circuits were designed for a DNN model to support operations at different layers of the model. This was used to show how model parameters of different scales could be mapped into the computing units of the architecture.

In ISAAC [32], a pipeline architecture was designed to realize parallel execution between different layers of the DNN model, and a new data encoding technique was defined to reduce the overhead of the ADC. In the industrial world, the semiconductor manufacturer TSMC is actively promoting the use of a RRAM-based coupled storage-compute architecture, and another manufacturer, Xinyuan Semiconductor, has also invested significant amounts into developing integrated storage-compute chips with RRAM as their core. Both of these applications could serve as the basic support for artificial intelligence (AI) applications.

NOR flash–based coupled storage-compute technology: NOR flash is a traditional type of non-volatile memory. Its basic storage unit is a floating-gate transistor, which realizes data storage through the introduction of electric charge. The NOR flash manufacturing process is very mature and cost-effective, meaning the potential application space for NOR flash–based coupled storage-compute technologies is quite broad. This has made it quite attractive to researchers. For example, a low-power two-terminal floating-gate transistor with multiple resistance levels can be used for analog neuromorphic computing [33]. A low-power analog circuit with digital input/output interfaces and configurable precision could also be used with an optimized sensing circuitry, DAC, and ADC to support energy-efficient vector-matrix multiplication [34]. Many companies are also pursuing the development and application of NOR flash–based coupled storage-compute technologies. The U.S. company Mythic, for example, launched its analog AI chip M1108AMP, which can be used in video analysis, visual detection, and other such fields. Zhicun (Witmen) from China also released their own NOR flash–based intelligent voice chip, the WTM1001.

14.3 Persistent Memory

Persistent memory boasts high integration, low static power consumption, no off-power data loss, and DRAM-like performance, presenting vast opportunities for storage systems. However, persistent memory hardware is very different from traditional storage media such as disks and flash memories in many ways. Building persistent memory systems is still not feasible due to three primary reasons. First, the software stack overhead is high. Persistent memory reduces the access latency of persistent data reads and writes from milliseconds to nanoseconds. However, traditional storage architectures are designed for storage media and require high software stack overhead to work with persistent memory, making it difficult to take advantage of the performance advantages of persistent memory. Second, the consistency overhead is high. Persistent memory provides data persistence at the main memory level, but the processor's on-chip cache system is still volatile. This means that persistent data will enter an inconsistent intermediate state upon a system failure. Traditional consistency technologies tend to introduce excessively high persistence latency, seriously reducing the performance of the persistent memory system. Third, the space utilization is low. Persistent memory is significantly pricier than traditional storage media. The space management mechanisms of traditional main memory easily cause main memory fragments. This significantly reduces the space utilization of persistent memory and increases system costs. Research institutions and enterprises are actively trying to tackle these issues and have already been able to reconstruct various storage systems based on persistent memory, such as file systems, key-value store systems, and distributed storage systems [35–37].

14.3.1 File Systems

A file system is the most basic module of an operating system (OS), organizing the device storage space into indexable file directory trees with files to facilitate user access. Organizing non-volatile memory into a file system is an important technical approach to ensure compatibility with existing applications. A simple way of doing this is to directly use an existing storage media file system to manage any non-volatile memory space. This solution can quickly improve performance, but it also has high software overhead, making it difficult to fully utilize the hardware advantages of persistent memory. The specific reasons are as follows: On the one hand, the overhead introduced by the unified abstraction of the OS obscures the high-performance feature of persistent memory. When the OS uniformly abstracts the file system, it shields the differences of different media in order to provide a unified interface. However, traditional storage media have significantly higher latency and lower bandwidth than persistent memory. This means the abstraction of traditional file systems cannot fully utilize the performance advantages provided by persistent memory. On the other hand, the byte-addressing feature of persistent memory cannot be fully utilized. Traditional file systems are designed based on storage media devices, which are accessed by the block, while persistent memory is accessed by the byte. This means direct access would cause serious data write amplification problems, as well as introduce consistency management issues. Researchers have so far tried to address these issues primarily through consistency guarantees, cache removal, and user-mode file systems. These solutions have significantly improved file system performance.

14.3.1.1 Consistency Guarantee Mechanisms

In 2009, Microsoft Research proposed BPFS [38], a file system based on byte-addressable persistent memory to accommodate access granularity mismatch between persistent memory (which accesses data by the byte) and existing file systems (which access data by the block). BPFS uses a tree structure as the basic data structure of the file system. In order to reduce the extra overhead caused by cascading updates in the system tree structure, BPFS exploits the byte-addressable feature of persistent memory and adopts a technique called short-circuit shadow paging to atomically update data. BPFS also decouples sequentiality and persistence, reducing the overhead caused by cache flushing.

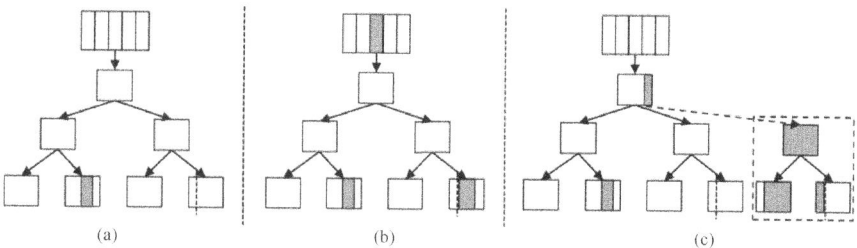

Fig. 14.4 Different update policies used by PMFS [39]. (**a**) In-place atomic updates. (**b**) In-place appending. (**c**) Partial copy-on-write

In 2014, Intel proposed their PMFS [39], in which the metadata update policy is redesigned based on the 8-byte data atomic update feature of persistent memory, as shown in Fig. 14.4. For small data updates, PMFS uses in-place atomic updates and fine-grained log appending mechanisms. For large data updates, PMFS uses a hybrid method of undo logs and copy-on-write (CoW) to ensure data consistency.

The University of California, San Diego, has also proposed a solution called NOVA [40]. This solution uses logs to organize metadata. For metadata modifications, NOVA uses appending modifications and atomically updating pointers to ensure the consistency of metadata. For data modifications, NOVA uses the CoW mechanism. For complex operations such as *rename* that involve modifications to multiple log structures, NOVA uses a logging mechanism to ensure crash consistency for such operations.

14.3.1.2 Cache Removal

Persistent memory and DRAM both deliver similar access performance. This means that the DRAM cache designed for storage media access is no longer efficient. The redundant data replication for the DRAM cache would also drag down the performance of persistent memory. Related research has addressed this problem in two ways: by removing the page cache and by removing the metadata cache.

Traditional file systems such as Ext4 and BtrFS have a direct access mode compatible with persistent memory, which allows users to remove the page cache by directly accessing data in persistent memory. File systems such as PMFS, NOVA, and BPFS use memory mapping to remove the page cache. Texas A&M University, for example, proposed the Storage Class Memory File System (SCMFS) [41] solution which optimizes data layouts by using page table mapping to make files in the file system have continuous address spaces, thereby improving program access performance.

In contrast, ByVFS removes the metadata cache and operates on metadata directly on the physical file system, taking advantage of persistent memory and improving performance.

14.3.1.3 User-Mode File Systems

Removing the cache can alleviate the performance limitations of virtual file systems (VFSs) on persistent memory to some extent, but VFS still brings a lot of unnecessary overheads, such as complex software execution logic and coarse-grained lock management. Aerie [42] is the first user-mode-based persistent memory file system, bypassing the VFS to fully utilize the performance of persistent memory. Strata [43] is also a user-mode file system, working on mixed media and allowing users to manage a variety of different storage devices at the same time, such as persistent memory, SSD, and HDD. Tsinghua University proposed its own kind, KucoFS [44], which can access file data in the persistent memory file system in user mode while still offloading complex metadata processing logic to the kernel.

14.3.2 Key-Value Store Systems

Key-value store systems provide operations, such as query, insertion, update, and deletion for key-value pairs, and are widely used in web search, e-commerce, and social networks where their scalability and real-time performance can go and play maximally. Persistent memory provides strong hardware support for key-value store systems to perform excellently and has a large capacity and low latency. At present, building key-value store systems based on persistent memory mainly focuses on index structure and space management.

14.3.2.1 Persistent Index Structure

The index structure is important in a key-value store system. It assists the key-value store system in finding key-specific data items. Index structures generally fall under two categories. One is hash tables, which are characterized by their scalability and low query overhead and support only single-point queries. The other index structures are tree indexes (such as the B+ tree), which are characterized by orderly organization of key-value pairs and support efficient range searches but have slow query speeds and high maintenance overheads. Currently, building data structures based on persistent memory needs to take into account the following considerations:

- Read-Write asymmetry.
  Persistent memory has significant asymmetry in read-write latency and bandwidth and is troubled with durability issues. A standard B+ tree has frequent sorting, balancing, and other operations that introduce significant write overheads, which, in turn, lead to serious performance and wear issues.
- Consistency optimization mechanisms.
  Sudden power failures or system crashes may make persistent memory data structures inconsistent, which can result in the loss of query functionality or even

data loss. Therefore, efficient consistency guarantee mechanisms must be provided for index structures of persistent memory.

In line with the index structure characteristics, researchers have already designed more fine-grained consistency update policies through system-provided persistence primitives. For example, the CDDS tree [45] is a consistent B+ tree designed by HP Labs for non-volatile main memory. It assigns a version number range to each data item. When performing an update operation, it generates a new version for each updated data item, which is then inserted into an appropriate location in the tree node. For a deletion operation, it can easily finish the task by setting the version number, and the entire process does not affect old data items. The CDDS tree reclaims these old data items at the appropriate time, thus ensuring that it can find the correct version of the data in case of a system error. However, consistency update mechanisms based on version numbers can cause exaggerated write amplification. Therefore, many subsequent works have further tried to reduce the consistency overhead of B+ trees from different aspects, such as by introducing an indirect query layer and by allowing intermediate inconsistent states.

14.3.2.2 Space Management

Space management of non-volatile main memory mainly deals with allocation and release operations. Main memory allocation/release operations may enter an inconsistent, invalid state due to system errors. This can result in a number of problems such as main memory leaks or wild pointer access once the system restarts. In addition, just as data in non-volatile main memory is retained after a system is shut down, fragments in persistent memory are retained. The lack of an effective mechanism to deal with fragments in its main memory will leave the fragments accumulating to seriously reduce the space utilization of the non-volatile main memory.

To reduce main memory fragments, Intel's Persistent Memory Development Kit (PMDK) uses different allocation policies for main memory allocation operations of different block sizes. For blocks smaller than 256 KB, it employs a split adaptation policy, which uses 35 allocation classes of different sizes to cut each 256 KB superblock into multiple smaller main memory blocks, all in multiples of 8 bytes, to thereby satisfy main memory allocation operations in certain ranges. Although this fine-grained split adaptation policy reduces the main memory fragments caused by allocation operations of blocks smaller than 256 KB to some extent, allocation operations of blocks larger than 256 KB are still prone to main memory fragmentation.

To eliminate finer-grained main memory fragments, Tsinghua University's LSNVMM [46] organizes the entire non-volatile main memory into a log structure. For all allocation operations, it adds new data directly to the end of the log rather than splitting the main memory superblock into fixed-size main memory blocks, thus eliminating most internal fragments. It eliminates external fragments too by migrating valid data, reclaiming unused main memory space, and organizing it into larger contiguous regions. Moreover, its defragmentation does not interrupt the normal operation, which reduces the impact on the performance of the whole system [47].

14.3.3 Distributed Storage Systems

The hardware performance of network and storage devices in distributed scenarios has been greatly improved with the development of persistent memory and high-speed network technologies (such as RDMA). However, directly integrating persistent memory and RDMA into an existing distributed storage system cannot leverage the performance of both. This is due to the fact that existing distributed software stacks and distributed protocols are designed for traditional network and storage devices. Existing research redesigns software stacks and distributed protocols to fully utilize the performance of persistent memory and RDMA in distributed storage.

14.3.3.1 Software Stacks

The latency overhead of traditional software stacks is insignificant for storage media accesses, but not for persistent memory and RDMA. This is where Tsinghua University's Octopus [48] (Fig. 14.5) comes in. It directly accesses a distributed persistent shared memory pool through RDMA to reduce redundant data replication and fully utilize the read/write bandwidth of the hardware. It also uses client-initiated data I/O operations to reduce the CPU and network load of the server and self-identified remote procedure call (RPC) protocol to achieve low-latency metadata access.

Orion [49] uses RDMA-capable networks to extend the NOVA standalone file system to distributed scenarios. Orion improves disaster recovery capabilities by maintaining multiple copies of metadata and data and reduces the processing pressure on servers by using one-sided RDMA primitives when reading logs from remote devices.

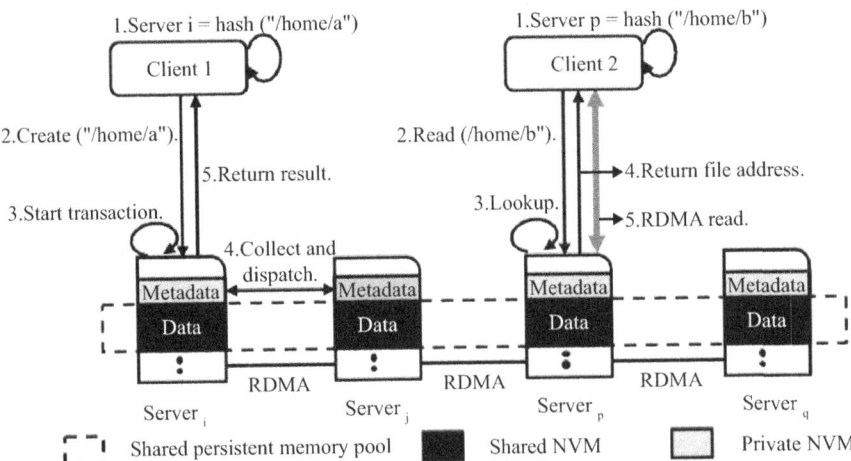

Fig. 14.5 Octopus architecture [48]

14.3.3.2 Distributed Protocols

Distributed protocols are a set of communication protocols used to achieve open, standards-based interoperability between a large number of computing systems. They mainly include copy protocols, cache coherence protocols, and transaction protocols.

- Copy protocols.

  Mojim [50] implements data fault tolerance for persistent memory systems based on RDMA. The data on the primary node is transferred to the persistent memory in the mirror node through RDMA, which reduces the CPU flushing overhead of the primary node. At the same time, the mirror node asynchronously backs up the data to multiple backup nodes in the background to provide higher reliability.
- Cache coherence protocols.

  Hotpot [51] provides a simple programming interface for applications based on the abstraction of distributed persistent shared memory, enabling single-node applications to fully utilize distributed storage resources. Hotpot uses local caches for data access acceleration and two additional distributed cache coherence commit protocols based on RDMA. The first supports concurrent write operations on the same cached page by different nodes through multi-stage commits. The second ensures that there is only one writer for a page at the same time through a centralized locking service.
- Transaction protocols.

  Microsoft's FaRM [52, 53] redesigned a distributed transaction protocol based on optimistic concurrency control for RDMA. This protocol focuses on three things. First, it adopts unreplicated coordinators to eliminate the coordinators' state replication overhead and simplify system recovery. Second, it merges replicas and transactions into a single layer, where the coordinator communicates directly with all the primary and backup replicas, in favor of lower system software. Lastly, it pushes data into the backup replicas through one-sided RDMA write primitives to reduce latency.

14.4 In-Network Storage

Programmable network devices support software-defined in-network packet processing, offering new insights into storage system design. They are typified by programmable switches and smart network interface cards (NICs) [54].

A programmable switch distinctively features a tailored programmable chip for in-network processing. Based on a reconfigurable match table architecture, the chip has multiple high-speed hardware pipelines, providing users with three programmable components: parser, register array, and match-action table. The parser specifies the protocol format of network packets, while the register array is a piece of high-speed static random-access memory (SRAM), which generally has only 10–20 MB of space for data storage. The match-action table specifies the modification and routing behavior applied to network packets when they flow through the switch with their header element meeting a specific condition (e.g., the UDP port number is 11) and then reads and writes the register array accordingly. Currently, programmable switches support line-rate packet forwarding, and the aggregated bandwidth reaches above 10 Tbit/s.

A smart NIC is an NIC chip attached with a piece of programmable hardware, which can be either an Arm CPU, a network processing unit (NPU), or a field-programmable gate array (FPGA), and is used to process network packets received by or sent from the NIC. The Arm CPU is the weakest in terms of processing power, but it is the easiest to program, in stark contrast to the FPGA.

With programmable switches and smart NICs, researchers are able to construct high-performance in-network storage systems to support data coordination, data scheduling, data caching, and other core tasks on network paths [54, 55].

14.4.1 In-Network Data Coordination

The programmable switches act as communication hubs of data coordination between the servers in distributed storage systems, such as typical Concordia [56] and SwitchTx [57].

Tsinghua University proposed Concordia, a distributed shared memory system that improves cache coherence based on programmable switches. In a distributed shared memory system, the coherence between the local caches that are used to reduce remote data access across servers is achieved at the costs of both expensive distributed coordination between servers and extra round trips, as well as CPU overhead for data sharing. Figure 14.6 illustrates the architecture based on which

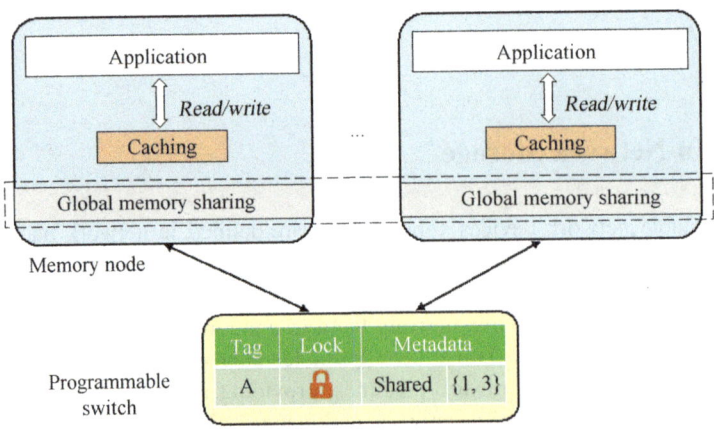

Fig. 14.6 Architecture of Concordia [56]

Concordia solves this classic issue. It uses a programmable switch to record the metadata of cache blocks, including their status and the list of servers holding these cache blocks. Upon receiving a request for cache coherence, the switch routes the request to the destination server set according to the metadata. The programmable switch uses efficient read/write locks to serialize concurrent and conflicting requests. Given the limited memory capacity on the programmable switch, an ownership transfer mechanism is introduced to Concordia, allowing the switch to only maintain the coherence of active cache blocks while leaving that of inactive ones to the servers.

SwitchTx is a distributed transaction processing system proposed by Tsinghua University to improve in-network coordination. Distributed transaction processing systems partition data across servers and use distributed concurrency control and submission protocols to provide transaction semantics to applications. These protocols incur huge overhead, such as network communications and CPU queuing, on the mission-critical paths of committing transactions, leading to huge latency and high conflict rates, which severely drag down system performance. SwitchTx features a scalable in-network coordination mechanism that abstracts a coordination task into "gather-and-scatter" operations and offloads them to programmable switches in a cluster. This helps shorten the communication length and reduce the CPU queuing overhead for transaction committing. Flow control based on transaction semantics also helps with the redesign of the admission control for transactions. As such, SwitchTx reduces the network overhead for distributed transaction processing, boosts system throughput, and reduces transaction latency.

14.4.2 In-Network Data Scheduling

In some cases, smart NICs help researchers schedule data requests. AlNiCo [58] is a typical example of this.

AlNiCo is a transaction scheduling system proposed by Tsinghua University. Two major trends in transaction processing systems have occurred in recent years. First, network bandwidth has significantly increased, and a single-node system can carry a large number of network requests. Second, modern servers are increasingly equipped with multi-core CPUs, causing resource contention among multi-core transactions. So it is crucial to schedule transaction requests to the most appropriate CPU cores for maximum parallel transaction processing on multi-core servers. However, scheduling is a double-edged sword as it creates extra computing overhead while improving transaction processing. Being a CPU-based method, it cannot ensure low latency for transaction processing. Smart NICs provide new opportunities for transaction scheduling, as they are on the critical path of request processing and can accelerate computing. AlNiCo uses FPGA-based NICs to intelligently schedule transaction requests to appropriate CPU cores to reduce transaction processing conflicts, as illustrated in Fig. 14.7. Specifically, it abstracts transaction requests, CPU core states, and workload characteristics into vectors that are suitable for FPGA processing and uses FPGA to accelerate scheduling decision-making. It

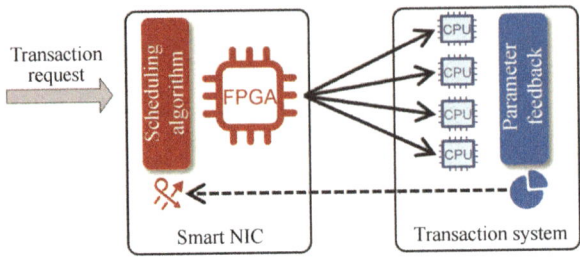

Fig. 14.7 AlNiCo for transaction scheduling [58]

Fig. 14.8 Architecture of the NetCache [59]

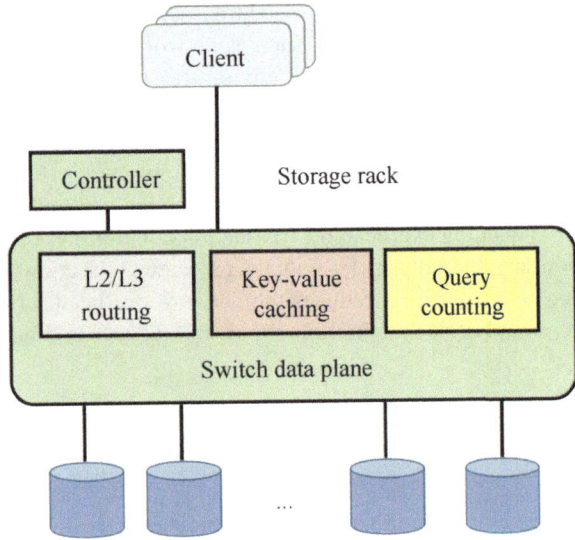

supports various concurrency control protocols, minimizing the latency for request scheduling and reducing transaction conflicts to boost system throughput.

14.4.3 In-Network Caching

By caching data of storage systems into their memory, programmable switches and smart NICs also reduce network round trips to increase throughput. NetCache [59] is a typical implementation for this.

NetCache is a load-balancing scheme of a distributed key-value store system proposed by Johns Hopkins University based on programmable switches. After a distributed key-value store system scatters data across servers, some data is more frequently accessed, which is termed workload skews, causing some servers to be overwhelmed with query requests, while others are not. Figure 14.8 shows the architecture of NetCache caching hotspot key-value pairs in distributed key-value store systems. It adopts programmable switches to quickly detect hotspot key-value pairs and save them to register arrays. Upon receiving a user read request, the switch returns the key-value pair if the request is a cache hit or otherwise routes the request to the destination server. Upon receiving a user write request, the switch labels the key-value pair as invalid and routes the request to the destination server if it is a cache hit. By unleashing the huge throughput of programmable switches, NetCache efficiently processes hot read requests, helping balance server loads.

14.5 Intelligent Storage

In recent years, AI has blown our minds in many cases. Deep learning, in particular, has advanced impressively. In 2017, DeepMind's AlphaGo, an AI system based on deep learning, defeated Ke Jie, the world's then-highest-ranked Go player, in the first game of a three-session match. In November 2022, OpenAI shook the world with ChatGPT, an AI chatbot that is stiff competition to humans in Q&A and text generation. Even in the data storage field, AI has unfolded in both ways: accelerating the development of storage systems (AI for storage) and leveraging high-performance storage to progress further (storage for AI).

14.5.1 AI for Storage

Fueled by AI, storage systems have become far more adaptive and perform on several fronts, mainly including learned index, automatic parameter tuning, and heuristic algorithm optimization.

14.5.1.1 Learned Index

Being a core component of a storage system, indexes map keys to the positions of data records. Considering that common index structures (such as B+ trees) are not space- and memory-efficient, researchers at the Massachusetts Institute of Technology (MIT) proposed the learned index in 2018 [60] to replace original index nodes with a simple model. The core principle is that the model learns the cumulative distribution function of keys and then predicts the positions of keys. In this way, a key is mapped via computation, rather than searching in an index node. Figure 14.9 is an example of learned indexes. To look up the data with a key of 510, the $H(x)$ function is called to compute and obtain its position subscript 1. Compared with other index structures, learned indexes require less space overhead while reducing several memory accesses to give extremely low access latency. The early learned indexes do not allow data to be inserted or deleted, and their re-training costs are high. For this, researchers have introduced various methods, and one of the most

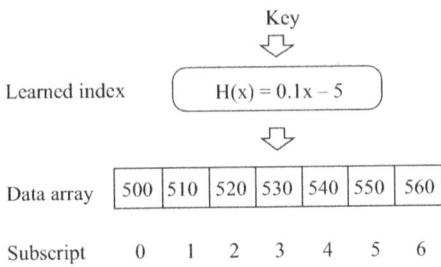

Fig. 14.9 Example of learned indexes

common ones is temporary buffers [61]. Based on this method, newly inserted data items are written into a temporary buffer, and the data in the buffer are periodically re-trained and combined with the original indexes.

Much research has been implemented to shift storage systems from index modules to learned indexes. Bourbon [62] and XStore [63] are two typical examples in this regard.

Bourbon applies the idea of learned indexes to log-structured merge (LSM) key-value data store systems. As described in Chap. 5, an LSM system is organized into multiple SSTables on storage devices, whereby each SSTable contains a certain number of ordered data items. The LSM system adds an index block to an SSTable to record indexes, realizing simple I/O operations. Similarly, Bourbon replaces the index blocks of SSTables with learned indexes to accelerate data lookups. It comes with several guidelines for boosting the efficiency of using learned indexes. For example, it prefers to learn the underlying SSTables, leveraging their long lifecycles to avoid frequent model invalidation and rebuilding.

XStore was engineered to work in ordered key-value store systems based on remote direct memory access (RDMA), where the servers use tree structures to maintain data and the clients use one-sided RDMA primitives to look up data directly. Considering the multiple network round trips required, these key-value store systems build index caches based on B+ trees or other structures to cache the mapping of keys to the remote addresses of data items. Such indexing excessively consumes client memory and gives rise to several memory accesses, with the latter being a major factor of huge latency, particularly when the RDMA featuring a very low network round trip time is used. By leveraging learned indexes, XStore builds high-performance index caches to achieve a trade-off between performance and space utilization. Compared with earlier methods, XStore reduces memory usage by 99% yet at a performance cost of only 20% [63].

14.5.1.2 Automatic Parameter Tuning

A storage system uses a massive number of configurable parameters to adapt the performance to different workloads and scenarios. For example, a Ceph distributed file system often has more than 1500 parameters. A manual parameter tuning task will take experienced users a long time to finish, and the results cannot be smoothly

14.5 Intelligent Storage

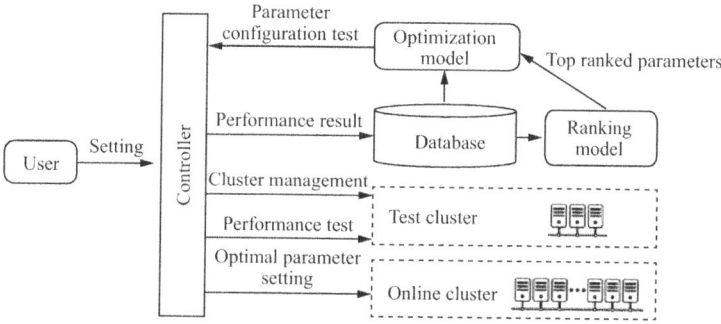

Fig. 14.10 Architecture of Sapphire [64]

ported to other hardware platforms. This has promoted many researchers to leverage machine learning for tuning, and one such recommendation is Sapphire [64] by researchers at Tsinghua University.

Sapphire uses an automatic approach to recommending optimal parameter configuration based on machine learning for distributed storage systems. Figure 14.10 shows the architecture of Sapphire, which mainly consists of a controller and machine learning model. The controller accepts settings from users, such as cluster setups and maximum iteration times, and manages distributed storage clusters by executing commands to make parameter configurations take effect. Also, the controller tests storage system performance through test tools and logs the results to databases. The machine learning model comprises a ranking part and an optimization part, with the former processing the test results to generate a parameter ranking list based on their impact on system performance. Based on the parameter ranking, the optimization part generates a search domain based on the top-ranked parameters and uses the search domain to locate the optimal parameter settings. For this, Sapphire uses a simulation-based approach to learning a small number of test clusters and building optimization models, based on which it recommends the most suitable parameter configurations for large online clusters.

14.5.1.3 Heuristic Algorithm Optimization

Heuristic algorithms, such as cache replacement and hot-cold data separation algorithms, improve storage system efficiencies but do not perform well in all scenarios. They are common in that their core is prediction, for example, of whether data is cold or hot, meaning all of them represent good use cases of AI technology. Here, we focus on two examples: Llama [65] proposed by Google and LinnOS [66] proposed by the University of Chicago.

Llama is a memory management system for C++ programs. In existing server applications, C++ objects are often short-lived and require very frequent memory allocation. This means the use of huge pages for memory access optimization will lead to severe memory fragmentation. Llama samples memory allocation (including contexts and some user-layer data) during a cold start, predicts the lifetime of objects, and uses system runtime information to train models. Based on the prediction results, it organizes data heaps to minimize fragmentation. It also temporarily caches the prediction results in the hash table to shorten inference time [65].

LinnOS was put forward to improve the predictability of SSD storage performance. SSD has various internal background operations, such as garbage collection and read repair. While necessary, these operations pose a huge threat to the access latency of read and write requests. To resolve this problem, LinnOS leverages the power of neural networks to predict SSD access latency. If an I/O is likely to have high latency, LinnOS cancels and redirects it to other SSDs to avoid latency spikes. Based on binary classification and with current and historical I/O queue lengths and earlier I/O latency records as the input, its predictions are more accurate and involve fewer parameters and computation workloads of neural networks [66]. With predictions still being offline, LinnOS does not adapt well to changing workloads.

14.5.2 Storage for AI

Soaring training datasets and model parameters have posed challenges to storage systems [67]. The slow increase in the storage space of accelerators, like graphics processing units (GPUs) and tensor processing units (TPUs), is falling behind the increasing demand of machine learning missions for storage space. At present, accelerators are outperforming storage systems in terms of read/write operations. In this context, many dedicated storage systems have been proposed to efficiently cover different machine learning procedures, including data loading, data preprocessing, and model training.

14.5.2.1 Data Loading

During training, datasets are loaded from local storage media or remote storage systems, where they are usually kept, to the accelerator's memory in batches. Prefetching and caching are typical methods to accelerate data loading. NoPFS [68] and SHADE [69] are two prime examples of these methods.

NoPFS is a data loading framework proposed by ETH Zürich for machine learning. When it comes to distributed training, frequent and random access to small data samples leads to congestion in shared file systems, and this is detrimental to efficient dataset loading. Modified access patterns, double buffering, and many other solutions are introduced to boost data loading efficiency. These solutions have significant limitations, such as damage to dataset randomization or extra hardware overhead. The key idea of NoPFS is to use the pseudo-randomness of a pseudo-random number generator to generate near-optimal prefetching and caching policies. To this end, it adopts a given random seed to predict when a given dataset sample will be accessed and performs access pattern analysis and performance modeling based on the prediction. In this way, it generates near-optimal sequences and paths to prefetch samples and generates caching policies accordingly.

SHADE is a data caching system proposed by Virginia Polytechnic Institute and State University and other institutions for machine learning tasks, which are not cache-friendly in terms of access mode: Random sampling is detrimental to data locality. SHADE introduces an importance-based approach for sampling datasets. It uses important data samples multiple times in a single training task to improve data locality. To this end, it scores the importance of training data and caches important data samples to improve the cache hit ratio.

14.5.2.2 Data Preprocessing

Preprocessing in a way that pumps more data to models than is ingested is critical to ensuring minimal impact on model training and maximum accelerator hardware utilization. tf.data [70] is a typical data preprocessing system.

tf.data is a data preprocessing system proposed by Google for machine learning. Its architecture is shown in Fig. 14.11. A CPU preprocesses data for machine learning at a rate far slower than the accelerator (such as a GPU and TPU) that ingests the data for training, meaning that preprocessing is a performance bottleneck. To solve this problem, tf.data uses a special API and runtime, based on which a centralized task dispatcher assigns preprocessing tasks to multiple nodes for execution, and training nodes then directly fetch preprocessed data from the preprocessing nodes. In addition, it enables preprocessing results to be cached on local storage devices to avoid repeated processing from causing extra overhead. Considering preprocessing differences between models and datasets, the runtime automatically tunes the degree of parallelism and memory buffer size of preprocessing to maximize performance.

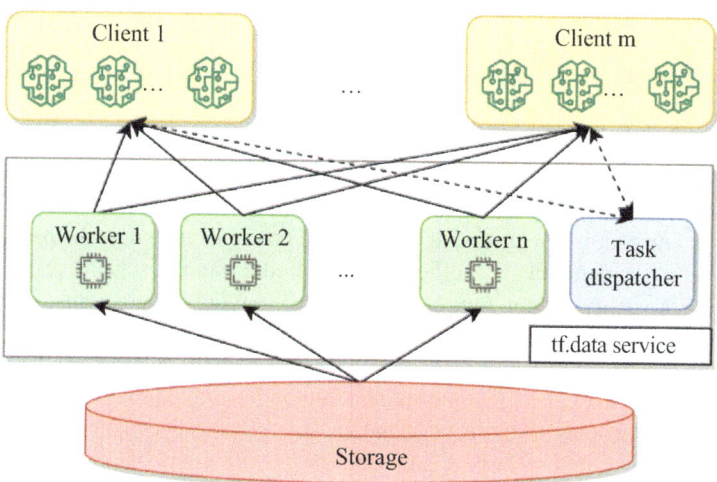

Fig. 14.11 Architecture of tf.data [70]

14.5.2.3 Model Training

Model parameters are increasing in scale far more rapidly than accelerators' storage space. This has encouraged researchers to propose innovative methods to increase accelerator storage capacity, such as distributed model data storage between accelerators and heterogeneous storage. In this section, we will look into some typical examples: FLECHE [71] and PetPS [72] for sparse accelerators and ZeRO [73] and Mobius [74] for dense accelerators.

FLECHE is an efficient GPU embedding table caching solution proposed by Tsinghua University for model recommendation based on deep learning. Such models have multiple embedding tables with an enormous number of sparse parameters, causing irregular and sparse accesses to DRAM to become a major source of performance impediment. It is common that existing GPUs cache hot parameters to reduce DRAM access. Different from the existing caching schemes, which are not efficient in terms of space utilization and GPU kernel maintenance overhead, FLECHE uses Hoffman encoding to re-encode feature IDs and manages all embedding tables in a unified way, in order to capture global hotspots and improve the hit ratios. Additionally, it supports kernel merging based on self-identification to reduce the overhead of kernel maintenance.

PetPS is a parameter server system proposed by Tsinghua University based on persistent memory. In order to provide real-time access to the trillions of parameters in industrial-grade sparse large models, the existing schemes distribute these sparse models to DRAM across multiple parameter servers. With the explosive increase in model parameters, these practices suffer problems such as high storage costs and slow recovery from crashes. PetPS turns to cost-effective persistent memory to store the parameters of sparse large models. Considering the high read latency of persistent memory, it introduces special hash indexes to minimize the reads of persistent memory through prefetching and offloads parameter serialization tasks to NICs to boost CPU efficiency.

Zero Redundancy Optimizer, or ZeRO, is a novel solution proposed by Microsoft for training large models. The key idea behind ZeRO is to split a model's parameters onto multiple GPUs, rather than keeping them all on a single GPU as data parallelism (DP) does, to enhance training performance despite a continuous surge in the number of parameters. The parameters stored on each GPU are a partition. When a partition is required for training, the GPU broadcasts its parameters to all GPUs. After the training is finished, it gathers and aggregates the gradients to update the parameters.

Mobius is a communication-efficient large model training scheme proposed by Tsinghua University for commodity GPUs. It helps avoid frequent cluster communications, which are necessary for the above ZeRO solution, from affecting training performance with commodity GPU servers whose communication link bandwidth is much smaller than GPU memory bandwidth and shared among multiple GPUs. Mobius gives full play to the heterogeneous storage resources on these servers to meet the storage needs of large model training and uses pipeline training to minimize communication overhead. It also models computing and communication based on hybrid linear planning to obtain optimal model partitions. Figure 14.12 illustrates the architecture of Mobius.

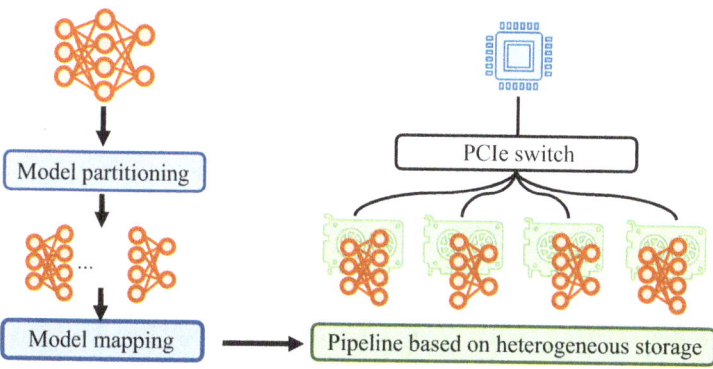

Fig. 14.12 Architecture of Mobius [74]

14.6 Edge Storage

Edge storage is a new paradigm of storage intended for the data generated by edge nodes on the Internet of things (IoT) and 5G networks, which are growing in terms of adoption across various industries. It scatters data onto nearby edge nodes for storage to dramatically shorten the physical distances from data generators to computing and storage devices, which in turn enables services to access the data at a high speed and low latency. However, the constraints at edge nodes on power supply, space, computing power, and communications pose severe challenges for real-time edge storage and processing.

Extensive design and research efforts have been made toward edge storage. The typical examples from the industry are the storage appliance TStor for edge nodes by Tencent, the Smart SSD that features computing support on storage devices by Samsung, the hyper-converged infrastructure for edge data storage FusionCube by Huawei, and the OpenYurt software platform for edge storage management by Alibaba. The world's renowned academic institutions, including Tsinghua University; the University of California, Los Angeles (UCLA); and Columbia University, also have already dug deep into a bunch of topics, such as devices [75], software and protocols [76–78], and data organization and search [79].

14.6.1 Edge Storage Devices

Edge storage devices are physical media for data storage. As is true with many other devices, their number of inherent hardware resources is a crucial performance factor. In addition to only limited storage and computation resources, the long physical paths from data sensing units to storage and computation units further restrict edge storage devices from providing real-time data storage and access. One research attempt to overcome this problem is converged sensing, storage, and computation, which means condensing sensing interfaces and storage and computation units into

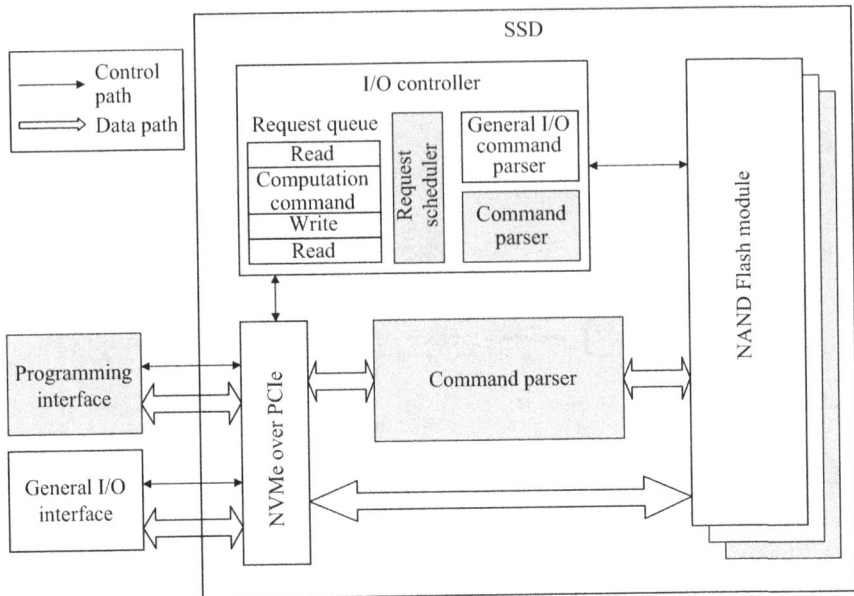

Fig. 14.13 Architecture of TH-iSSD [80]

one device, so as to shorten the physical data paths in return for lower data storage and access latency.

To achieve this, Tsinghua University proposed TH-iSSD [80]. Figure 14.13 illustrates its architecture of converged sensing, storage, and computation. It uses one hardware controller to implement the control logic of data sensing units, storage units, and computation accelerators, minimizing the data movement overhead. It is highly reconfigurable so that its sensing units and computation accelerators can be replaced to adapt to certain deployment cases with varying power supplies and application logic. It introduces priority-aware parallel I/O scheduling, with its scheduler dynamically reordering I/O requests in a fine-grained way, to overcome the issues of unbalanced read/write performance and erase-before-write restrictions. Therefore, the storage device's internal bandwidth can be fully utilized to deliver maximum performance. TH-iSSD provides file abstraction, based on which the storage units are not managed as raw block devices without a file system, and extensive host code modifications are not necessary, allowing users to focus on computation logic without worrying about data placement.

14.6.2 Edge Storage I/O Stack

Edge storage I/O stacks handle the I/O requests of data in edge nodes. An edge storage I/O stack is composed of hierarchical software and hardware components, including user space, file system, page cache, generic block layer, device driver, and

14.6 Edge Storage

block device. It mainly provides data exchanges between storage devices (such as hard disks and solid-state disks) and applications.

Edge nodes are usually heterogeneous in their hardware configurations and operating systems. This makes it incredibly challenging to efficiently abstract and manage their I/O stacks in such a way that the I/O stacks match the drivers and interfaces of different applications that run on different devices.

λ-IO [81] is a typical attempt by researchers from Tsinghua University to design unified storage I/O stacks for heterogeneous devices. Using an architecture illustrated in Fig. 14.14, it proposes a trio of interface, runtime, and scheduling designs to efficiently manage computation and storage resources. First, it extends both host and device I/O stacks to provide applications with extended programming interfaces. Alongside original I/O operations, the applications can submit λ requests over the extended interfaces to customize, load, and call computational logic when reading and writing files. This allows developers to maintain their familiar programming style to access and process file data, without having to focus on how computation tasks are executed and scheduled. Second, by enhancing normal I/O stacks with the Extended Berkeley Packet Filter (eBPF), it allows the runtime to cross the host-and-device boundary and support pointer access and a dynamic-length loop while introducing additional information to support dynamic verification. Third, it utilizes a dynamic request scheduling mechanism, in which the execution time of kernel and storage device requests are modeled to enable requests to be quickly scheduled to the faster side, improving efficiency.

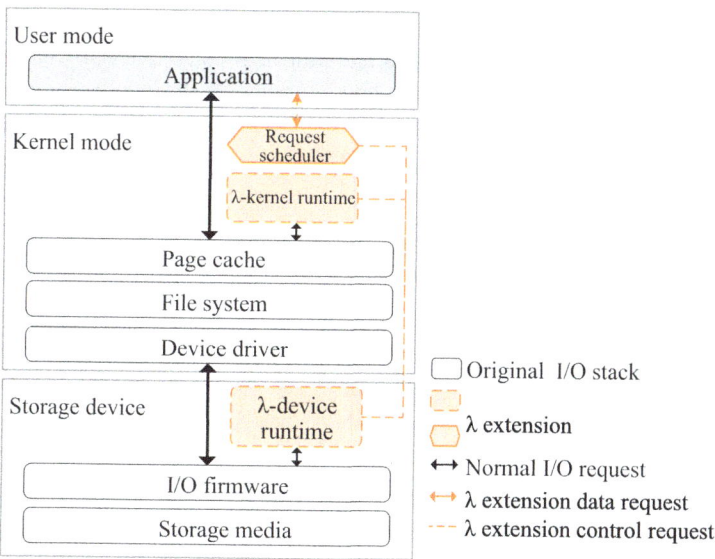

Fig. 14.14 Architecture of λ-IO [81]

14.6.3 Edge Data Organization and Search

Edge data organization and search store data on edge devices or edge servers based on a certain architecture. They search for specific data in a bank of data using a certain method. As a common storage system that places data as key-value pairs and searches for data by key, the distributed key-value store boasts low storage and access latency, lending itself to edge data storage. The existing key-value stores normally adopt randomized data placement policies, such as consistent hashing and hash slot sharding, without accounting for the time-varying patterns of data storage clients and their key-value access requests, which results in an average long request latency in the systems where such policies are running.

This has attracted plenty of research interest and effort regarding how to optimize these data placement policies, and one of these representative examples is Portkey [82], a distributed key-value store method that dynamically adapts data placement to time-varying client mobility and data access patterns. It offers new insights into distributed key-value stores in edge nodes with highly dynamic environments, by identifying the time-varying mobility and latency patterns of edge applications, formulating the data placement as an online optimization problem, and using a greedy algorithm to implement fast data placement that is approximate to the optimal decision. Portkey mainly involves two processes: data collection and data placement decision-making. For data collection, Portkey adopts a series of lightweight techniques to create succinct latency sketches. Specifically, it probes the end-to-end latency from a client to the server and, in subsequent time windows, uses locality-aware reprofiling to start latency information recollection when a client moves. Regarding data placement decision-making, Portkey operates on keys separately through an adaptive solver and uses a greedy assignment to deal with the host storage constraints. The assignment prioritizes key-value pairs that most affect overall data storage performance, which means it balances storage requirements with access frequency. For example, keys that are frequently accessed but only by a single client can skip the adaptive solver. This greedy heuristic algorithm prioritizes fast data placement over delayed optimal ones.

Figure 14.15 illustrates the workflow of Portkey. Portkey is integrated as a software module into an existing data storage system. The client data store tracks each key's access requests, intelligently monitors the end-to-end client-server latencies, and uploads the access and latency information to the adaptive placement engine at an application-defined window size. Once all client information is received, the engine computes a global network distance matrix. Binding the matrix with the client's access key sets, the placement solver performs fast near-optimal global key-value placements and delivers migration instructions to the appropriate data storage servers.

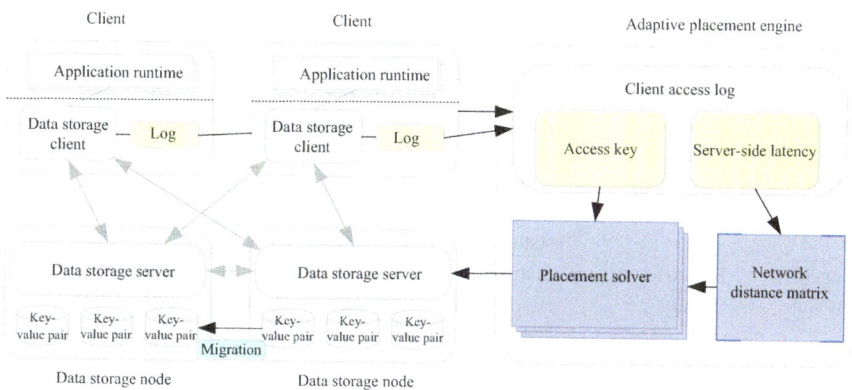

Fig. 14.15 Workflow of Portkey [82]

14.7 Blockchain Storage

Blockchain is an emerging decentralized, immutable, and traceable ledger technology that involves cryptography, peer-to-peer (P2P) networks, consensus algorithms, and smart contracts. Recent years have seen blockchain technology develop rapidly, with its applications expanding from earlier digital currency and financial services to healthcare, education, government services, supply chains, copyright protection, and Internet of things (IoT) security, showing tremendous benefits. With a large majority of blockchain services leveraging distributed storage approaches to ensure secure trusted evidence and query traceability, the problems experienced with blockchain-oriented storage systems in terms of low efficiency, huge overhead, and slow query hamper blockchain from further advancement, becoming a major issue for implementing blockchain technology.

14.7.1 Blockchain Storage System Overview

Mostly, blockchain may use a file system, key-value database, relational database, or other systems for data storage, depending on its design, purpose, and data access frequency. Table 14.1 provides a summary of common blockchain data storage systems and their characteristics.

File systems mainly target block data, storing it primarily as binary codes in files and searching for it by file number. Although write performance is moderate, read performance is poor. For example, bitcoin [83] uses a file system to store its block data.

Key-value databases use simple key-value methods to store data, and the typical implementations include LevelDB, RocksDB, and CouchDB. LevelDB and

Table 14.1 Typical blockchain storage systems

Storage system	Relational semantics	Storage content	Read/write balance	Typical blockchain cases
File system	Weakest	Blockchain data	Write intensive	Bitcoin
LevelDB	Weak	Block, index, status data	Write intensive	Bitcoin, Ethereum
RocksDB	Weak	Block, index, status data	Relatively balanced write/read	FISCO-BCOS
CouchDB	Relatively strong	Status data	Read intensive	Hyperledger fabric
MySQL	Strong	Block, index, status data	Balanced write/read	FISCO-BCOS

RocksDB are typically used for storing block, index, and status data on a blockchain. LevelDB stores data in sequential batches in non-volatile memory through log-structured merge (LSM) trees, and its moderate read performance makes it ideal for the earlier stages of a blockchain system when intensive write is necessary. For example, the blockchain application Ethereum [84] adopts LevelDB to implement its smart contract functionality. RocksDB can be considered an enhanced LevelDB and thereby offers an overall higher performance in that flexible read and write adjustments are supported. One of the common blockchain services that adopt the enhanced RocksDB is FISCO-BCOS [85]. CouchDB boasts a fairly good read performance and complex query functionality, becoming the status data storage solution for Hyperledger Fabric [86].

MySQL is a common relational database. Although providing balanced read and write performance, its overall performance is weaker compared with other storage systems. FISCO-BCOS uses MySQL as its storage engine to handle queries in complex scenarios.

14.7.2 Blockchain Storage System Optimization

Blockchain onboards business users mainly with source tracing and evidence storage, both of which involve the storage and query of data on the blockchain. Being fully decentralized, blockchain turns to the full backup of node ledgers for data consistency. While keeping data consistent, up-to-date, and shareable across a network, this multi-node redundancy mechanism drives up the data volumes across the nodes. Besides, the blockchain structure uses chain hashing to keep data tamper-proof yet at the price of low tracing efficiency. Till now, the following major solutions have been proposed for optimization:

14.7.2.1 On-Blockchain Content Pruning

This optimization targets the data that was placed into a blockchain ledger at a remote past time but has been rarely accessed and is of little value. It removes such data and logs the removal operation as a transaction record on the blockchain to downsize the large volumes of data across storage nodes. For example, developers have designed a block data pruning policy to address the storage capacity issues experienced by bitcoin wallets. This policy constructs a full set of unspent transaction outputs (UTXOs) and then discards historical transaction data, which is to remove the old data, to save local storage space.

14.7.2.2 Blockchain Sharding

Sharding alleviates processing and storage pressures by scattering data onto different servers and tailoring each server to process only the local portion of the data. Inspired by this principle, developers divide a multi-node blockchain into small groups, and the nodes within each group form a smaller blockchain, namely, a blockchain shard. A policy is introduced to distribute the transactions of the blockchain system into the shards for processing. This brings multifold advantages. First, a single node does not have to keep or process all transaction data. Second, the overall performance is not limited to the capabilities of individual nodes. And third, the blockchain system is able to fulfill multiple transactions in parallel.

14.7.2.3 Blockchain Storage Structure Optimization

In a blockchain system, the storage structure has a direct impact on the data storage performance. UStore [87] is a distributed data storage system built around a Git-like (Git is a popular source code versioning system) data structure and synthesizes the advantages of many distributed systems and databases. It provides improved performance and richer query functionality than ordinary key-value store systems. Later, based on UStore, the storage engine ForkBase [88] was designed to support multi-version data, with each version of data content being uniquely identified. In addition, a novel index structure called a pattern-oriented split tree was purposed in ForkBase to efficiently identify and eliminate data deduplication and thereby minimize data redundancy.

14.7.2.4 Off-Blockchain Storage Support

Auxiliary storage optimizes blockchain by working with off-blockchain storage methods. The core principle behind this optimization is that on-blockchain storage focuses only on the minor portion of key data and transfers the major chunks of data to off-blockchain storage systems. The indexes of transferred data are saved on the blockchain to ease queries. In a data access process, the blockchain system first queries the index of the target data and then fetches it from the off-blockchain storage systems. The auxiliary storage systems often have fairly high capacity, and this way makes full use of the capacity resources of the auxiliary storage systems to supplement blockchain systems with extra space.

14.8 Disaggregated Data Center Architectures

Data is increasing exponentially around the globe, putting data centers under unprecedented pressure for storage and management. Built on server architecture, existing data centers are suffering a shortfall in resource utilization, scalability, and performance as they are under increasing strains to meet service needs. In recent years, disaggregated data center architectures have come to the attention of academic and industry professionals. In these architectures, hardware resources are disaggregated into different pools, such as processor pools, memory pools, and storage pools, which are all interconnected through high-speed networks. This enables administrators to scale the hardware resources of one or more pools for applications to share on demand. This being said, the disaggregated architectures have sharply different memory access patterns, storage tiers, fault tolerance models, and software overhead, bringing new challenges to building system software that is friendly to disaggregated architectures [89, 90].

14.8.1 Background

In existing data centers, server nodes are interconnected using networks to facilitate dynamic scale-up with servers as the minimum unit. Though conducive to storing the swelling volumes of data, this approach does not differentiate the hardware resource requirements of big data services and has led to issues regarding resource utilization, scalability, and performance [89, 90].

1. Data storage and server replacement periods are not aligned.

   The massive amount of data generated by artificial intelligence (AI), big data, and other services is stored for a period of between 8 and 10 years, depending on its lifecycle. In existing data centers, servers are replaced following processor upgrades at an interval of 3–5 years. This stark difference in time means system resources may be wasted, because a CPU upgrade in the server will come with data migration mandated from replaced storage space to a new one.

2. Memory utilization is imbalanced spatially and temporally.

   Memory shares a large chunk (up to 50%) of a server's cost but has extremely low utilization. Average memory utilization is only 45% for Google clusters and is also below 65% for Alibaba clusters. Two factors are credited to this situation. A server's memory is mostly configured based on its peak demands, which may be equated to inadequate usage during an off-peak hour. Within a data center, memory usage also varies greatly between different servers at one time, indicating an overall memory waste.

3. Cloud-native applications require elastic computing and storages.

   Being cloud-native means that applications such as databases and serverless deployments require resources to be elastically allocated. This means scaling storage resources based on the changing amount of data and computing resources

based on the density of requests. For example, in a cloud-native database, data is stored in a back-end object system, and the virtual machines (VMs) executing transactions are dynamically added or removed in response to SQL requests' changing traffic volumes. In serverless applications, containers are created independently on the basis of per-function requests.
4. Data center taxes are high.

Cloud data centers are flexibly provisioned in the form of virtualized physical resources to tenants who need varying data encryption and compression services. This virtualization for network and storage eats up a considerable portion of a server's CPU resources, namely, "data center taxes." For example, such infrastructure software running in Google's data centers incurs as much as 30% of data center taxes. The impact is threefold. First, the taxed CPU resources cannot be sold to tenants, shrinking the profit margins of cloud service providers. Second, infrastructure software will encroach into the cache and other resources of foreground tasks, causing them to underperform. Third, continued high-performance virtualization is impossible due to the fact that general-purpose CPU performance improves much slower than I/O peripherals. Special processors are key to repealing data center taxes by offloading infrastructure tasks. However, given the CPU's central role in a server while also treating a special processor as a peripheral, this approach will be unable to ensure efficient access to server resources.

14.8.2 Architecture Features and Key Technologies

Disaggregated data center architectures are preferred for overhauling the low resource utilization and poor inflexibility of existing data centers. Figure 14.16 [89] illustrates the architecture of a disaggregated data center. Based on the architecture, hardware resources are disaggregated into pools, which are interconnected over high-speed networks. The memory pool mainly consists of dynamic random-access memory (DRAM) resources and is tasked with the low-latency caching of data (e.g., that of process space) for various applications. The storage pool contains low-speed HDD and high-speed SSD resources (NAND or Optane). Noteworthily, persistent memory supports byte addressing and data persistence and therefore can increase the capacity of both the memory and storage pools [89–92]. The compute resource pool heterogeneously consists of CPU, GPU, and FPGA pools as well as other resources. The interconnecting networks can be RDMA, NVMe-oF, and CXL. Remote direct memory access (RDMA) enables compute resource pools to directly access memory pools and persistent memory pools. The NVMe over Fabrics (NVMe-oF) enables compute resource pools to directly access HDD and SSD pools. The Compute Express Link (CXL) supports mutual access between all resources, but features weaker scalability, and is therefore more suitable for small data centers.

In an architecture with disaggregated hardware resources, data flows on network paths and is exchanged between hardware resources. To avoid network access upon

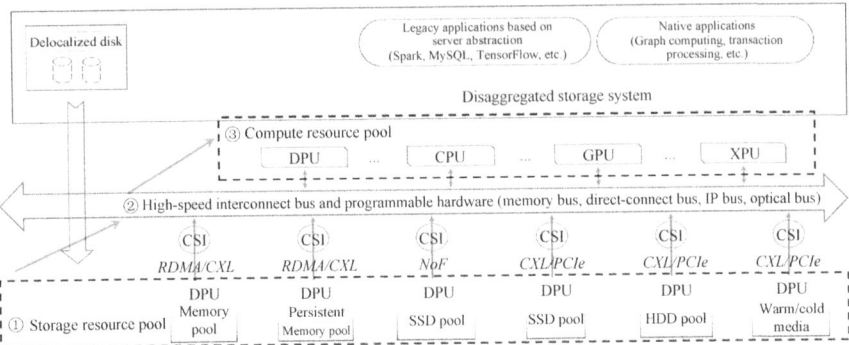

Fig. 14.16 Architecture of a disaggregated data center [89]

each CPU read and write, the compute resource pools are configured with a handful of local memory for caching, while the memory and storage pools (e.g., Arm CPU core of smart NICs) are configured with a tiny part of compute resources for management tasks, such as space allocation and garbage collection. Given the high importance of networks, path acceleration is necessary through smart NICs, programmable switches, and other programmable devices. The smart NICs function as the control plane of resource pools, performing tasks such as resource initialization and exception processing and also offloading the infrastructure such as virtualization, to reduce CPU resource consumption. The programmable switches support line-rate processing and are positioned in the center, lending themselves to accelerate hardware coordination and minimize software overhead.

The most distinct difference separating new architecture from legacy architectures is its deeper storage-computing decoupling, which is no longer simply limited to CPUs from storage, but also achievable on storage and computing hardware. It builds all hardware into independent resource pools, such as compute resources, memory, and HDD/SSD pools, making it possible for hardware resources to be truly expanded and flexibly shared without impacting one another. The finer-grained division breaks the current processing logic centered around general-purpose CPUs, leaving behind the tasks they are not good at, such as data processing and aggregation, to special accelerators and DPUs. In this way, the optimal combination of hardware resources is realized globally to produce an optimal energy efficiency ratio. The disaggregated data center architecture is superior to traditional data centers in many aspects.

1. High resource utilization.

    As hardware resources are disaggregated into pools to facilitate resource sharing, applications reuse memory and storage resources over time. This changes peak-hour resource planning from the level of a single server to an entire data center, while enabling different hardware to be separately scaled. This means a certain resource like memory can be increased without adding new CPUs and other resources as needed in a server-based architecture, minimizing resource waste.

2. High flexibility.

   By analyzing the changing resources required by individual application executions, a disaggregated data center architecture instantly allocates the resources that have become idle to the most desired applications. This helps rapidly adapt to changing loads, such as burst traffic, to ensure the ultimate scalability of applications. Facing a failure of Moore's law, data centers are turning to heterogeneous accelerators such as traffic processing unit (TPU) and FPGA, to boost hardware capacity. Given the limited and fixed slots on server motherboards, the existing server architecture is struggling to fulfill the growing demand for novel compute power. Without relying on server-based architecture, the disaggregated data center architecture enables new devices to be increased only by constructing a resource pool and connecting it to networks.

3. Lower data center taxes.

   The disaggregated data center scraps the CPU-centered architecture, enabling compute resources to equally access network, memory, and storage resources. This means that infrastructure such as virtualization can be easily offloaded to an FPGA device and a smart NIC to free up CPU resources, lowering data center taxes.

The disaggregated systems perform unified storage resource management and provisioning to the processors in the compute resource pools based on the following key technologies:

1. Interface abstracting.

   The memory resource pool depends on interface abstraction to expose memory space to remote applications, runtime, operating systems, and processor hardware. This makes the performance and compatibility trade-off the most critical point of consideration during design. The incumbent memory interface abstraction is mainly based on operating system memory exchange, automatic non-uniform memory access (AutoNUMA), a dedicated user library, and the JAVA runtime. They differ from the perspective of interconnection technologies, compatibility, minimum page management granularity, and performance. Table 14.2 provides a detailed comparison.

Table 14.2 Differences in interface abstraction

Mechanism	Interconnection	Compatibility	Minimum management granularity	Performance	Overhead
Memory exchange	RDMA	High	Memory page (4 KB)	Low	Processing page fault interrupts
AutoNUMA	Memory semantic bus (CXL)	High	Memory page (4 KB)	Medium	Hot page scanning
Dedicated user library	RDMA	Low	Any	High	None
JAVA runtime	RDMA	High	Any	Medium	Java runtime

2. Data switching.

The full use of computing nodes' local memory is crucial to reducing network accesses, indicating the significance of memory data switching to ensure that hot data is placed into the local memory and other data in remote memory. To achieve this, the precise information collection of hot page memory, efficient data migration, and responsive data prefetching are necessary. In typical cases, applications access the disaggregated memory resources as instructed by the CPU, without direct intervention from system software. This means that there may be few opportunities to collect memory access statistics. The common hot memory tracing mechanisms include software instrumentation, page table tagging, and CPU hardware counting. Data migration cannot start unless the cold and hot page distribution of an application is accurately obtained, and it aims to migrate the frequently accessed pages to the local memory and others to the remote memory, depending on the page's storage locations. Current research focuses on migration timing, route selection, and front- and back-end coordination. Data prefetching predicts possible data access operations based on current access characteristics to prefetch the target pages to the local memory. With disaggregated memory, page prefetching means transferring data across networks, and its accuracy is paramount because any prefetching errors would be understandably very costly.

3. Disaggregated memory management.

Memory cannot be shared across compute nodes without efficient data management and correct concurrent reads and writes. This involves designing concurrent indexing, distributed transaction protocols, and data partitioning policies. With limited compute resources, memory pools usually use one-sided RDMA primitives to implement indexing and protocols for memory disaggregation. This means that designing RDMA-friendly data structures and coordinating concurrent operations will be the major challenge for disaggregated memory management.

4. Disaggregated file and object management.

Storage resource management needs richer semantics for upper-layer applications than memory resource management does in terms of object interfacing and directory tree structure-based file interfacing. Storage reliability and persistence must also be considered alongside other indicators. Given that storage pools have limited compute resources, it is critical to support lightweight and efficient object and file management.

5. Smart hardware offloading.

In disaggregated data centers, disaggregated storage systems can use programmable switches, smart NICs, and other programmable network hardware to offload data management tasks and distributed protocols for the purpose of reducing software overhead and improving performance. Memory nodes have limited compute power, having to rely on one-sided RDMA to access remote memory resources. With RDMA semantics limited to just read/write and atomic instructions, round trips are unavoidable in complex use cases, reducing overall system performance. A considerable amount of research has been devoted to leveraging the smart NICs, programmable switches, and DPUs that form the

14.8 Disaggregated Data Center Architectures

storage pools to extend RDMA semantics and storage protocols under the disaggregated architecture.

14.8.3 Future Trends

The research on disaggregated memory architecture has demonstrated several major new trends.

1. Process fault tolerance of disaggregated memory.

    Unlike an existing data center where the memory space of a process is physically located on the local server, the memory space of a process in a disaggregated data center may cross multiple memory nodes. Understandably, this inevitably enlarges the process's failure domain, which means that a memory crash on any of these residing nodes will cause the process to lose data and then become inoperative. This demonstrates how it is crucial and challenging to ensure process fault tolerance in disaggregated data centers. Current replication mechanisms entail multiplied memory usage, and this goes against the initial idea of boosting resource utilization through disaggregation. Until now, there has been just limited research concerning process fault tolerance based on erasure coding, which is an aggressive solution to perform fault tolerance on all data in a memory pool without considering the recoverability of some sort of data, such as those having a checkpoint or snapshot. Therefore, a selection mechanism must be introduced for this mechanism to achieve reliability with minimal overhead. In addition, the erasure coding can be offloaded to programmable devices, such as programmable switches. Furthermore, efforts must include how to quickly migrate processes from a faulty compute node to a normal one.

2. System design in heterogeneous networks.

    In disaggregated data centers, the networks interconnecting different resource pools are certainly heterogeneous. At the rack layer, memory sharing between servers within a rack is based on CXL, while these cross-rack cases at the cluster layer will be based on RDMA. Heterogeneous networks differ in performance and interfacing functionality—in this case, while CXL features low latency and supports synchronous operations and native Load/Store instructions, RDMA has high latency, along with asynchronous operations and reads/writes in peripheral I/O mode. In this sense, it is necessary to consider disaggregating the memory and storage resources at the network level to adapt to the heterogeneous topology. Given that the storage resources are heterogeneous, too (as they may include a mixture of persistent memory, high-speed SSDs, and low-speed disks), the coordination between heterogeneous storage resources and their heterogeneous interconnecting networks will also be of significant importance. In heterogeneous networks, the programming models that bring applications to bear need to be considered as well: unified management by the operating system or direct exposure of network attributes to upper-layer applications.

3. System design for heterogeneous compute power.

   Research into disaggregated data center architectures has so far mainly focused on memory and storage resource management, with little attention being paid to compute resources. With the rise of cloud computing and AI, it is common that data centers have heterogeneous compute resources, such as CPUs, GPUs, FPGAs, and AI accelerators. In a disaggregated architecture, they will form heterogeneous compute resource pools as well. Therefore, it is worth trying to find ways to maximize their performance. The main challenges are twofold. First, the compute power of all heterogeneous compute resource pools must be appropriately encapsulated to ensure that it can be dynamically allocated to individual computing tasks. Second, to ensure efficient data exchange between heterogeneous compute resource pools and memory and storage pools, the remote memory and storage resources must be properly abstracted to ensure that these heterogeneous CPUs, GPUs, FPGAs, and AI accelerators can locate, retrieve, and read/write data at a high speed.

14.9 High-Density New Storage

With surging data volumes comes an explosive increase in the demand for data storage. This puts the spotlight on the "capacity wall" issues raised by current storage media like disks and tapes. There are two reasons for this: The storage density of the media grows way slower than data volumes, and the storage media lifespan is far shorter than the period during which data is expected to be stored.

Common storage media now posts an approximately annual 20% capability increase, far behind the rate at which data grows. The Rethink Data report by Seagate in 2020 based on IDC analysis found that enterprise data would grow annually by 42.2% on average from 2020 to 2022, more than twice the growth rate of storage density. Relying on magnetic signals (such as HDDs and tapes) or electrical signals (such as SSDs) to keep data, these storage media generally have a lifespan of between 5 and 10 years, which is much shorter than that of data in most cases. Therefore, higher density and longer storage periods are of categorical urgency. Fortunately, many new technologies have been proposed, including shingled magnetic recording, optical storage, and DNA storage.

14.9.1 Shingled Magnetic Recording

Realizing that the current HDD storage density is stretching the physical limit of about 1 TB/in^2, leading disk vendors like Western Digital and Seagate to shift their research to a new HDD technology called shingled magnetic recording (SMR) [93]. SMR utilizes existing magnetic heads and platter media and introduces minor ameliorations in the processing techniques to increase storage density, producing a

14.9 High-Density New Storage

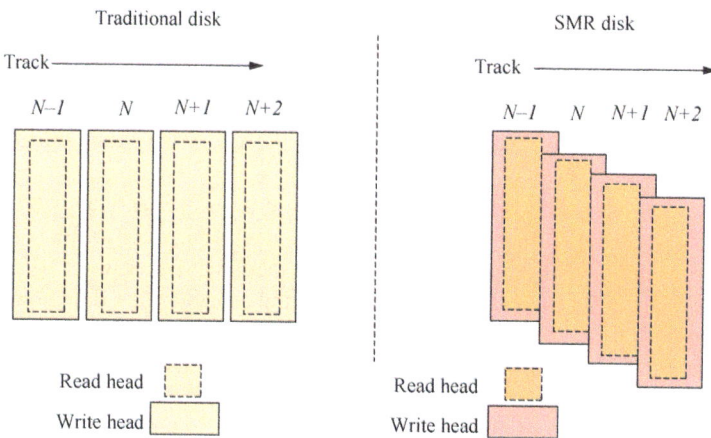

Fig. 14.17 Comparison between SMR and HDD disks

10–25% capacity gain per unit area. Specifically, its magnetic tracks are partially arranged in a shingled formation such that there are more magnetic tracks in a given platter area, as illustrated in Fig. 14.17.

SMR write heads overlap with adjacent magnetic tracks, and this change in structure means random writes and update-in-place writes are not supported. The technologies that resolve this issue mainly include the following:

- Redirect on write (ROW): All writes are aggregated and executed in a similar way to that used in log-structured file systems. This helps resolve the issue that SMR disks do not support random or update-in-place writes.
- Garbage collection: ROW leads to space fragmentation, disabling data from being rewritten to SMR disks, which reduces space utilization. Garbage collection migrates a valid data block in a write unit to another write unit, overcoming this problem.
- Hot and cold data separation: Garbage collection incurs repeated data migrations and consequently increases write operations, leading to write amplification. Cold and hot data separation distributes cold and hot data to different write units based on identified lifecycles, reducing write amplification.

14.9.2 High-Density Optical Storage

The storage capabilities of magnetic devices, high-density HDD disks or tapes alike, decrease over time, having an adverse impact on data correctness. Due to this, data in HDD storage systems needs to be migrated every 3–5 years. Likewise, data migration is needed every 10 years in tape-based storage systems. Optical storage is a popular research field concerning the long-term storage of massive data.

Blu-ray storage has been commonly implemented by big Internet firms to keep cold data. However, research has demonstrated its physical limit which prevents one disk from having more than 40 layers that provide a total capacity of up to 1 TB. This has compelled players in the industry to explore holographic optical storage, super-resolution optical storage, and glass storage with the aim of boosting optical storage density.

While traditional optical disks store data as a series of pits and lands on the surface of the medium (of which the sizes of the pits and lands directly affect capacity), the holographic optical storage expands from the medium surface to its 3D space for data storage, to achieve a higher storage density and capacity. Based on holographic optical storage, a single disk now provides a capacity of 2.5 TB. New breakthroughs in holographic materials can increase capacity to 4 TB or more.

Traditional optical disks, CD, DVD, and Blu-ray ones alike, all realize data writing and reading based on 1/0 modulation, differing only in the wavelengths of laser beams used. Although with a shorter wavelength available, the discernible laser spot sizes at the recording point are subject to the limits of diffraction, making it difficult to boost the capacity density. Super-resolution optical storage [94] uses dual-beam recording to break through the single-beam diffraction limit, reducing the minimum size of recording points to allow for a higher capacity density. Based on super-resolution optical storage, the capacity of a single disk can theoretically reach 1 PB.

Glass storage [95] fully exploits multiple properties of light including polarization, wavelength, and intensity along with the 3D space of the medium to store data in glass. Using these multiple properties of light to denote data at a single recording point, it increases capacity density. To date, it allows a single disk to provide a capacity of 360 TB.

14.9.3 DNA Storage

DNA storage [96] uses synthetic deoxyribonucleic acid (DNA) as a medium, encoding and storing binary data into DNA bases: adenine (A), cytosine (C), guanine (G), and thymine (T). It has the advantages of high storage density, 1 g of DNA being able to keep 2 PB of data, and a long storage time that lasts many lifetimes.

There are six steps to DNA storage, as illustrated in Fig. 14.18, with the first three concerning data writing and the other ones for data reading. Encoding is the first step, aiming to map binary data to DNA sequences, which is followed by step 2 of synthesizing DNA strands based on the DNA sequences. In step 3, DNA strands are cloned into a vector, completing data storage. To read data, XOR tests are performed on base pairs in step 4 to extract DNA molecules. In step 5, the DNA molecules are sequenced and combined to form DNA sequences. The final step is to decode the information from the DNA sequences into binary data.

At the Flash Memory Summit in November 2020, Microsoft, Western Digital, Twist Bioscience, and DNA sequencer Illumina established a DNA storage alliance with the aim of developing a business ecosystem for DNA storage. DNA storage is

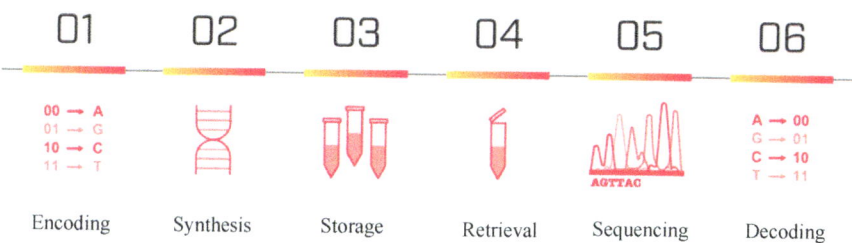

Fig. 14.18 DNA storage procedures

still in its infancy and faces several key problems, such as overwrite, random write, and high reliability of DNA storage of data. Although these issues can be easily resolved in other systems, it will still take a while of extensive research to find solutions to DNA storage.

14.10 Summary

Data storage systems are a critical form of information technology infrastructure for the digital economy. Many new high-performing technologies are emerging to deliver massive data storage with high reliability and high scalability. This chapter discussed the overall development trends of storage technology in new storage modes, non-volatile storage systems, and application storage optimization. We presented representative frontier academic research, including in-storage computing, in-network storage, disaggregated storage, edge storage, and high-density storage for new storage modes, flash and persistent memory storage in the domain of non-volatile storage systems, and intelligent and also blockchain storage for application storage optimization. These cutting-edge technologies will unlock new opportunities for the information technology industry and are of great significance to the high-quality growth of the digital economy.

14.11 Practice Questions

1. **What challenges face flash storage systems, and what are the solutions?**
    Answer: The challenges are twofold. First, the software performance of flash storage systems is inefficient. High-performance flash SSDs provide low latency and high concurrency, which increases the proportion of software CPU overhead. Second, SSDs of flash storage systems mask internal details, making it difficult to exploit hardware characteristics. For challenge one, the storage protocol and software stack can be redesigned to focus on the performance and collaboration of CPU and I/O operations. For challenge two, we can change the FTL in SSDs and design storage systems that are aware of hardware characteristics.

2. **In what ways are open-channel SSD and ZNS SSD similar to and different from each other?**

    Answer:

    Differences: Open-channel SSDs push all functions (address mapping, wear leveling, garbage collection, etc.) to the host, which requires host-side software support. ZNS SSDs push only some functions to the host for processing, such as address translation and mapping. With the introduction of zone-based storage, unified storage interfaces are implemented, and a more simplified and standardized solution is provided.

    Similarities: Both expose the internal details of the flash memory to the host, making it easier to improve the software efficiency of the flash memory storage system.

3. **What problems in computer systems does near-data computing resolve?**

    Answer: The near-data computing paradigm integrates storage and computing units to reduce the physical distance during data calculation. This improves data processing performance, reduces energy consumption, and overcomes the memory and power consumption barriers in computer systems.

4. **What are the advantages of near-disk and near-memory computing, and their application scenarios?**

    Answer: Near-disk computing is applicable to data that is stored in storage media and needs to be flushed to disks after one calculation. It supports data compression, erasure coding, and key-value database compaction. Near-memory computing is used for processing data that is stored in memory and requires high bandwidth. It is applicable to scenarios such as in-memory databases and artificial intelligence (AI).

5. **What are the advantages and disadvantages of persistent memory compared to traditional storage media such as DRAM and flash memory?**

    Answer:

    Advantages: Persistent memory is non-volatile, meaning it will not lose data after a power failure, something not supported by DRAM. Further, its manufacturing capacity is larger than that of DRAM and at a lower price. Compared with flash memory, persistent memory has faster read/write speeds and higher throughput.

    Disadvantages: Persistent memory is slower than DRAM and more expensive per unit capacity than flash memory.

6. **What problems may we have when using persistent memory to build file systems and key-value store systems? What are the solutions?**

    Answer: There are three problems: First, the software stack overhead is high. Persistent memory reduces the latency of persistent data reads and writes from milliseconds to nanoseconds. However, traditional storage architectures are designed for storage media, so it is difficult to take advantage of the performance of new storage devices. Second, the consistency overhead is high. Persistent memory provides data persistence at the main memory level, but the processor's on-chip cache system is still volatile. This means that persistent data will enter an inconsistent intermediate state upon a system failure.

Traditional consistency technologies tend to introduce excessive high persistence latency, seriously reducing the performance of the persistent memory system. Third, the space utilization is low. The persistent memory is significantly pricier than traditional storage media. The space management mechanisms of traditional main memory easily cause main memory fragments. This significantly reduces the space utilization of persistent memory and increases system costs.

Solution: Consistency guarantee, cache removal, and user-mode file systems can help optimize file systems based on persistent memory. Index structures and space management can also optimize key-value store systems.

7. **What are the theoretical grounds for caching data in network devices to improve application performance?**

    Answer: In-network data caching shortens network transmission paths and thereby reduces the latency and network bandwidth consumption for data transmission. Caching a small amount of hot data in network devices also reduces and balances server loads. Theoretically, for a storage system with a cluster scale of N, caching $O(N \log N)$ data items suffices for load balancing among N servers.

8. **Why are programmable network devices unable to process complex operations?**

    Answer: Programmable network devices, like smart switches or routers, are tailored for high-speed data packet transmission. They have limited processor and memory resources that are dedicated only to fast data forwarding. This means they lack the computing power needed for complex operations, like in-depth data packet analysis, encryption and decryption, and advanced protocol processing.

9. **To which scenarios are learned indexes applicable? What are their drawbacks?**

    Answer: Learned indexes help reduce memory consumption and increase query speed. They can be widely used for read-intensive tasks but are not suited for write-intensive workloads.

10. **Large language models lead to drastic increases in parameters. What will this mean for storage?**

    Answer: Large language models comprise massive model weights and parameters, which need large systems and high-performance devices for storage. Compression, sparsity, and quantization technologies can help reduce the storage space of large language models.

11. **The disaggregated memory architecture features multi-tier memory. Compared with existing heterogeneous external memory, what challenges will this new architecture have on memory management?**

    Answer: The multi-tier memory architecture typically includes different tiers of memory, like DRAM and NVRAM, with varying speeds and capacity. This means more complex management in terms of data migration as well as hierarchical policies and decision-making. A faulty or damaged memory will negatively affect the normal operations within the entire resource pool. Remote memory may also lead to higher access latency, causing processor pipelines to be blocked and reducing processor efficiency.

12. **The disaggregated memory architecture leads to higher memory access latency, which is lethal to memory-intensive applications. How can this be avoided?**

    Answer: (1) Efficient, accurate hot and cold data migration and replacement between multi-tier memory; (2) asynchronous memory access that allows for application execution and memory access in parallel

References

1. Shu J, Lu Y, Zhang J, et al. Research progress on non-volatile memory-based storage system. Sci Technol Rev. 2016;34(14):86–94.
2. Andersen DG, Franklin J, Kaminsky M, et al. FAWN: a fast array of wimpy nodes. In: Proceedings of the ACM SIGOPS 22nd symposium on operating systems principles (SOSP). Big Sky: ACM; 2009. p. 1–14.
3. Caufield AM, Grupp LM, Gordon SS. Using flash memory to build fast, power-efficient clusters for data-intensive applications. In: Proceedings of the 14th international conference on architectural support for programming languages and operating systems (ASPLOS). New York: ACM; 2009. p. 217–28.
4. Lu Y, Yang Z, Shu J. Revisiting the architecture and system of flash-based Storage. J Comput Res Dev. 2019;56(1):23–34.
5. Lee C, Sim D, Hwang J, et al. F2FS: a new file system for flash storage. In: Proceedings of the 13th USENIX conference on file and storage technologies (FAST). Santa Clara: USENIX; 2015. p. 273–86.
6. Josephson WK, Bongo LA, Flynn D, et al. DFS: a file system for virtualized flash storage. In: Proceedings of the 8th USENIX conference on file and storage technologies (FAST). Berkeley: USENIX; 2010. p. 85–99.
7. Lu Y, Shu J, Zheng W. Extending the lifetime of flash-based storage through reducing write amplification from file systems. In: Proceedings of the 11th USENIX conference on file and storage technologies (FAST). Berkeley: USENIX; 2013. p. 257–70.
8. Zhang J, Shu J, Lu Y. ParaFS: a log-structured file system to exploit the internal parallelism of flash devices. In: 2016 USENIX annual technical conference (USENIX ATC). Denver, CO: USENIX; 2016. p. 87–100.
9. Lu Y, Shu J, Wang W. ReconFS: a reconstructable file system on flash storage. In: The 12th USENIX conference on file and storage technologies (FAST), San Jose, CA; 2014. p. 75–88.
10. Zhang J, Lu Y, Shu J, et al. FlashKV: accelerating KV performance with open-channel SSDs. ACM Trans Embed Comput Syst. 2017;16(5):1–19.
11. Li S, Lu Y, Shu J, et al. LocoFS: a loosely-coupled metadata service for distributed file system. In: The international conference for high performance computing, networking, storage and analysis (SC). Denver: ACM; 2017. p. 1–12.
12. Lu Y, Shu J, Guo J, et al. LightTx: a lightweight transactional design in flash-based SSDs to support flexible transact. In: 31st IEEE international conference on computer design (ICCD). Asheville, NC: IEEE; 2013. p. 115–22.
13. Lu Y, Shu J, Guo J, et al. High-performance and lightweight transaction support in flash-based SSDs. IEEE Trans Comput. 2015;64(10):2819–32.
14. Bae H, Kim J, Kwon M, et al. What you can't forget: exploiting parallelism for zoned namespaces. In: Proceedings of the 14th ACM workshop on hot topics in storage and file systems. New York: ACM; 2022. p. 79–85.

15. Bjørling M, Aghayev A, Holmberg H, et al. ZNS: avoiding the block interface tax for flash-based. In: 2021 USENIX annual technical conference (USENIX ATC 21). Berkeley: USENIX; 2021. p. 689–703.
16. Zoned storage. NVMe zoned namespaces (2020-10-09) [2023-06-08].
17. Han K, Gwak H, Shin D, et al. ZNS+: advanced zoned namespace interface for supporting in-storage zone compaction. In: OSDI 21. Berkeley: USENIX; 2021. p. 147–62.
18. Kim J, Lim K, Jung Y, et al. Alleviating garbage collection interference through spatial separation in all flash arrays. In: USENIX annual technical conference. Berkeley: USENIX; 2019. p. 799–812.
19. Colgrove J, Davis JD, Hayes J, et al. Purity: building fast, highly-available enterprise flash storage from commodity components. In: Proceedings of the 2015 ACM SIGMOD international conference on management of data. New York: ACM; 2015. p. 1683–94.
20. Kim T, Jeon J, Arora N, et al. RAIZN: redundant array of independent zoned namespaces. In: Proceedings of the 28th ACM international conference on architectural support for programming languages and operating systems. New York: ACM; 2023. p. 660–73.
21. Bergman S, Cassel N, Bjørling M, et al. ZNSwap: un-block your swap. In: 2022 USENIX annual technical conference. Carlsbad: USENIX; 2022. p. 1–25.
22. Shin H, Oh M, Choi G, et al. Exploring performance characteristics of ZNS SSDs: observation and implication. In: 2020 9th non-volatile memory systems and applications symposium (NVMSA). Piscataway: IEEE; 2020. p. 1–5.
23. Nick T, Trivedi A. Understanding NVMe zoned namespace (ZNS) flash SSD storage devices (2022-01-03)[2023-06-08]. arXiv:2206.01547.
24. Mao H, Shu J, Li F, et al. Development of processing-in-memory. Sci Sin Inf. 2021;51(2):173–205.
25. Kwon M, Gouk D, Lee S, et al. Hardware/software co-programmable framework for computational SSDs to accelerate deep learning service on large-scale graphs. In: Proceedings of the USENIX conference on file and storage technologies (FAST). Berkeley: USENIX; 2022. p. 147–64.
26. Zhang F, Angizi S, Fan D. Max-PIM: fast and efficient max/min searching in DRAM. In: Proceedings of the design automation conference (DAC). Piscataway: IEEE; 2021. p. 211–6.
27. Xie X, Liang Z, Gu P, et al. SpaceA: sparse matrix vector multiplication on processing-in-memory accelerator. In: Proceedings of the international symposium on high performance computer architecture (HPCA). Piscataway: IEEE; 2021. p. 570–83.
28. Park J, Kim B, Yun S, et al. TRiM: enhancing processor-memory interfaces with scalable tensor reduction in memory. In: Proceedings of the international symposium on microarchitecture (MICRO). New York: ACM; 2021. p. 268–81.
29. Lee S, Kang S, Lee J, et al. Hardware architecture and software stack for PIM based on commercial DRAM technology: industrial product. In: Proceedings of the annual international symposium on computer architecture (ISCA). Piscataway: IEEE; 2021. p. 43–56.
30. Yang J, Kong Y, Wang Z, et al. 24.4 sandwich-RAM: an energy-efficient in-memory BWN architecture with pulse-width modulation. In: Proceedings of the international solid-state circuits conference (ISSCC). Piscataway: IEEE; 2019. p. 394–6.
31. Chi P, Li S, Xu C, et al. Prime: a novel processing-in-memory architecture for neural network computation in ReRAM-based main memory. ACM SIGARCH Comput Archit News. 2016;44(3):27–39.
32. Shafiee A, Nag A, Muralimanohar N, et al. ISAAC: a convolutional neural network accelerator with in-situ analog arithmetic in crossbars. ACM SIGARCH Comput Archit News. 2016;44(3):14–26.
33. Danial L, Pikhay E, Herbelin E, et al. Two-terminal floating-gate transistors with a low power memristive operation mode for analogue neuromorphic computing. Nat Electron. 2019;2(12):596–605.
34. Mahmoodi MR, Strukov D. An ultra-low energy internally analog, externally digital vector-matrix multiplier based on NOR flash memory technology. In: Proceedings of the design automation conference (DAC). Piscataway: IEEE; 2018. p. 1–6.

35. Shu J, Chen Y, Hu Q, et al. Development of system software on non-volatile main memory. Sci Sin Inf. 2021;51(6):869–99.
36. Chen Y. Research on key technologies for persistent memory storage system. Beijing: Tsinghua University; 2021.
37. Lu Y, Shu J. Persistent memory: from a system software perspective. Commun China Comput Federation. 2019;15(1):15–20.
38. Condit J, Nightingale EB, Frost C, et al. Better I/O through byte-addressable, persistent memory. In: Matthews J, editor. SOSP'09: proceedings of the 22nd symposium on operating systems principles. New York, NY: ACM; 2009. p. 133–46.
39. Dulloor SR, Kumar S, Keshavamurthy A, et al. System software for persistent memory. In: Bultermann D, Bos H, editors. EuroSys'14: proceedings of the 9th European conference on computer systems. New York, NY: ACM; 2014. p. 1–15.
40. Xu J, Swanson S. NOVA: a log-structured file system for hybrid volatile/non-volatile main memories. In: Brown A, Popovici F, editors. FAST'16: proceedings of the 14th USENIX conference on file and storage technologies. Berkeley, CA: USENIX Association; 2016. p. 323–38.
41. Wu X, Reddy ALN. SCMFS: a file system for storage class memory. In: Lathrop S, editor. SC'11: proceedings of 24th international conference for high performance computing, networking, storage and analysis. New York, NY: ACM; 2011. p. 1–23.
42. Volos H, Nalli S, Panneerselvam S, et al. Aerie: flexible file-system interfaces to storage-class memory. In: Bultermann D, Bos H, editors. EuroSys'14: proceedings of the 9th European conference on computer systems. New York, NY: ACM; 2014. p. 1–14.
43. Kwon Y, Fingler H, Hunt T, et al. Strata: a cross media file system. In: Chen H, Zhou L, editors. SOSP'17: proceedings of the 26th symposium on operating systems principles. New York, NY: ACM; 2017. p. 460–77.
44. Chen Y, Lu Y, Zhu B, et al. Scalable persistent memory file system with kernel-userspace collaboration. In: Proceedings of the USENIX conference on file and Storage technologies (FAST), vol. 21. Berkeley: USENIX; 2021. p. 81–95.
45. Venkataraman S, Tolia N, Ranganathan P, et al. Consistent and durable data structures for non-volatile byte-addressable memory. In: Proceedings of the 9th USENIX conference on file and storage technologies (FAST). Berkeley: USENIX; 2011.
46. Hu Q, Ren J, Badam A, et al. Log-structured non-volatile main memory. In: Proceedings of the USENIX annual technical conference (ATC). Berkeley: USENIX; 2017. p. 703–17.
47. Chen Y, Lu Y, Luo S, et al. Survey on RDMA-based distributed storage systems. J Comput Res Dev. 2019;56(2):227–39.
48. Lu Y, Shu J, Chen Y, et al. Octopus: an RDMA-enabled distributed persistent memory file system. In: Silva D, Ford B, editors. USENIX ATC'17: proceedings of the 23rd conference on USENIX annual technical conference. Berkeley, CA: USENIX; 2017. p. 773–85.
49. Yang J, Izraelevitz J, Swanson S. Orion: a distributed file system for non-volatile main memory and RDMA-capable networks. In: Merchant A, Weatherspoon H, editors. FAST'19: proceedings of the 17th USENIX conference on file and storage technologies. Berkeley, CA: USENIX; 2019. p. 221–34.
50. Zhang Y, Yang J, Memaripour A, et al. Mojim: a reliable and highly-available non-volatile memory system. In: Ozturk O, Ebcioglu K, editors. ASPLOS'15: proceedings of the 20th international conference on architectural support for programming languages and operating systems. New York, NY: ACM; 2015. p. 3–18.
51. Shan Y, Tsai SY, Zhang Y. Distributed shared persistent memory. In: Curino C, editor. SoCC'17: Proceedings of the 8th symposium on cloud computing. New York, NY: ACM; 2017. p. 323–37.
52. Dragojević A, Narayanan D, Hodson O, et al. FaRM: fast remote memory. In: Mahajan R, Stoica I, editors. NSDI'14: proceedings of the 11th USENIX conference on networked systems design and implementation. Berkeley, CA: USENIX; 2014. p. 401–14.
53. Dragojevića A, Narayanan D, Nightingale EB, et al. No compromises: distributed transactions with consistency, availability, and performance. In: Miller E, editor. SOSP'15: proceedings of the 25th symposium on operating systems principles. New York, NY: ACM; 2015. p. 54–70.

References

54. Wang Q, Li J, Shu J. Survey on in-network storage systems. J Comput Res Dev. 2023;60(11):2681–95.
55. Wang Q. Research on key technologies of network-storage co-design for distributed in-memory storage system. Beijing: Tsinghua University; 2023.
56. Wang Q, Lu Y, Xu E, et al. Concordia: distributed shared memory with {in-network} cache coherence. In: 19th USENIX conference on file and Storage technologies (FAST 21). Berkeley: USENIX; 2021. p. 277–92.
57. Li J, Lu Y, Zhang Y, et al. SwitchTx: scalable in-network coordination for distributed transaction processing. Proc VLDB Endow. 2022;15(11):2881–94.
58. Li J, Lu Y, Wang Q, et al. AlNiCo: SmartNIC-accelerated contention-aware request scheduling for transaction processing. In: 2022 USENIX annual technical conference (USENIX ATC 22). Berkeley: USENIX; 2022. p. 951–66.
59. Jin X, Li X, Zhang H, et al. Netcache: balancing key-value stores with fast in-network caching. In: Proceedings of the 26th symposium on operating systems principles. New York: ACM; 2017. p. 121–36.
60. Kraska T, Beutel A, Chi EH, et al. The case for learned index structures. In: Proceedings of the 2018 international conference on management of data. New York: ACM; 2018. p. 489–504.
61. Tang C, Wang Y, Dong Z, et al. XIndex: a scalable learned index for multicore data storage. In: Proceedings of the 25th ACM SIGPLAN symposium on principles and practice of parallel programming. New York: ACM; 2020. p. 308–20.
62. Dai Y, Xu Y, Ganesan A, et al. From Wisckey to Bourbon: a learned index for log-structured merge trees. In: Proceedings of the 14th USENIX conference on operating systems design and implementation. Berkeley: USENIX; 2020. p. 155–71.
63. Wei X, Chen R, Chen H. Fast RDMA-based ordered key-value store using remote learned cache. In: Proceedings of the 14th USENIX conference on operating systems design and implementation. Berkeley: USENIX; 2020. p. 117–35.
64. Lyu W, Lu Y, Shu J, et al. Sapphire: automatic configuration recommendation for distributed storage systems (2020-07-07) [2023-0609]. arXiv:2007.03220.
65. Maas M, Andersen DG, Isard M, et al. Learning-based memory allocation for C++ server workloads. In: Proceedings of the twenty-fifth international conference on architectural support for programming languages and operating systems. New York: ACM; 2020. p. 541–56.
66. Hao M, Toksoz L, Li N, et al. LinnOS: predictability on unpredictable flash storage with a light neural network. In: Proceedings of the 14th USENIX conference on operating systems design and implementation. Berkeley: USENIX; 2020. p. 173–90.
67. Feng Y, Wang Q, Xie M, et al. From BERT to ChatGPT: challenges and technical development of storage systems for large model training. J Comput Res Dev. 2024;61(4):809–23.
68. Nikoli D, Böhringer R, Ben-Nun T, et al. Clairvoyant prefetching for distributed machine learning I/O. In: Proceedings of the international conference for high performance computing, networking, storage and analysis. New York: ACM; 2021. p. 1–15.
69. Khan R, Yazdani A, Fu Y, et al. SHADE: enable fundamental cacheability for distributed deep learning training. In: 21st USENIX conference on file and storage technologies (FAST 23). New York: ACM; 2023. p. 135–51.
70. Murray D, Šimša J, Klimovic A, et al. tf.data: a machine learning data processing framework. Proc VLDB Endow. 2021;14(12):2945–58.
71. Xie M, Lu Y, Lin J, et al. Fleche: an efficient GPU embedding cache for personalized recommendations. In: Proceedings of the seventeenth European conference on computer systems. New York: ACM; 2022. p. 402–16.
72. Xie M, Lu Y, Wang Q, et al. PetPS: supporting huge embedding models with persistent memory. Proc VLDB Endow. 2023;16(5):1013–22.
73. Rajbhandari S, Rasley J, Ruwase O, et al. Zero: memory optimizations toward training trillion parameter models. In: SC20: international conference for high performance computing, networking, storage and analysis. Piscataway: IEEE; 2020. p. 1–16.
74. Feng Y, Xie M, Tian Z, et al. Mobius: fine tuning large-scale models on commodity GPU servers. In: Proceedings of the 28th ACM international conference on architectural support for programming languages and operating systems. New York: ACM; 2023. p. 489–501.

75. Ruan Z, He T, Cong J. INSIDER: designing in-storage computing system for emerging high-performance drive. In: Proceedings of the USENIX annual technical conference (ATC). Berkeley: USENIX; 2019. p. 379–94.
76. Qiao Y, Chen X, Zheng N, et al. Closing the B+-tree vs. LSM-tree write amplification gap on modern storage hardware with built-in transparent compression. In: Proceedings of the USENIX conference on file and storage technologies (FAST). Berkeley: USENIX; 2022. p. 69–82.
77. Nawab F, Agrawal D, El Abbadi A. Dpaxos: managing data closer to users for low-latency and mobile applications. In: Proceedings of the international conference on management of data (SIGMOD). New York: ACM; 2018. p. 1221–36.
78. Chen X, Song H, Jiang J, et al. Achieving low tail-latency and high scalability for serializable transactions in edge computing. In: Proceedings of the European conference on computer systems (EuroSys). New York: ACM; 2021. p. 210–27.
79. Gupta H, Ramachandran U. Fogstore: a geo-distributed key-value store guaranteeing low latency for strongly consistent access. In: Proceedings of the international conference on distributed and event-based systems (DEBS). New York: ACM; 2018. p. 148–59.
80. Shu J, Fang K, Chen Y, et al. TH-iSSD: design and implementation of a generic and reconfigurable near-data processing framework. ACM Trans Embedded Comput Syst. 2023;22(6):96:1–96:23.
81. Yang Z, Lu Y, Liao X, et al. λ-IO: a unified IO stack for computational storage. In: Proceedings of the USENIX conference on file and storage technologies (FAST). Berkeley: USENIX; 2023. p. 347–62.
82. Noor J, Srivastava M, Netravali R. Portkey: adaptive key-value placement over dynamic edge networks. In: Proceedings of the ACM symposium on cloud computing (SOCC). New York: ACM; 2021. p. 197–213.
83. Nakamoto S. Bitcoin: a peer-to-peer electronic cash system (2008-10-31) [2023-06-09].
84. Buterin V. A next-generation smart contract and decentralized application platform. white paper, vol 3(37); 2014. p. 2-1.
85. FISCO. Fisco-Bcos (2020-01-09) [2023-06-09].
86. Hyperledger Foundation. Hyperledger Fabric Project (2017-08-05) [2023-06-09].
87. Dinh A, Wang J, Wang S, et al. UStore: a distributed storage with rich semantics (2017-02-09) [2023-06-09]. arXiv:1702.02799.
88. Wang S, Dinh A, Lin Q, et al. Forkbase: an efficient storage engine for blockchain and forkable applications (2018-02-14) [2023-06-09]. arXiv:1802.04949, 2018.
89. Shu J, Chen Y, Wang Q, et al. Progress on the storage systems for disaggregated data centers. Sci Sin Inf. 2023;53(8):1503–28.
90. Shu J. Technology prospects of new storage-compute separation architecture. Commun China Comput Federation. 2022;18(11):53–60.
91. Yang Z, Wang Q, Liao X, et al. TeRM: extending RDMA-attached memory with SSD. In: Proceedings of USENIX conference on file and storage technologies (FAST); 2024. p. 1–16.
92. Wang Q, Lu Y, Shu J. Building write-optimized tree indexes on disaggregated memory. SIGMOD Rec. 2023;52(1):45–52.
93. Suresh A, Gibson G, Ganger G. Shingled magnetic recording for big data applications. Carnegie Mellon University. Parallel data lab technical report CMU-PD L-12-105; 2012.
94. Jiang M, Zhang M, Li X, et al. Research progress of super-resolution optical data storage. Opto Electron Eng. 2019;46(3):180649.
95. Anderson P, Black R, Cerkauskaite A, et al. Glass: a new media for a new era. In: 10th USENIX workshop on hot topics in storage and file systems (HotStorage 18); 2018.
96. Li B, Song NY, Ou L, et al. Can we store the whole world's data in DNA storage. In: 12th USENIX workshop on hot topics in storage and file systems (HotStorage 20); 2020.

Made in the USA
Monee, IL
03 May 2026